R   rned on or before

Sensory Science Theory and Applications in Foods

ift Basic Symposium Series

Edited by
INSTITUTE OF FOOD TECHNOLOGISTS
221 N. LaSalle St.
Chicago, Illinois

Sensory Science Theory and Applications in Foods

Harry T. Lawless
Cornell University
Ithaca, New York

Barbara P. Klein
University of Illinois at Urbana-Champaign
Urbana, Illinois

Marcel Dekker, Inc.   New York • Basel • Hong Kong

ISBN 0-8247-8537-1

This book is printed on acid-free paper.

MARCEL DEKKER, INC.
270 Madison Avenue, New York, New York 10016

Current printing (last digit):
10 9 8 7 6 5 4 3 2 1

666.07
SEN

PRINTED IN THE UNITED STATES OF AMERICA

Preface

The Fourteenth Annual Basic Symposium "Advances in Sensory Science" sponsored by the Institute of Food Technologists and the International Union of Food Science and Technology was held in Anaheim, California, on June 15–16, 1990. The symposium speakers were drawn from a variety of disciplines, from industry and academia, and share a common interest in sensory perception and its measurement. Current research in the sensory sciences, both theoretical and applied, is summarized and critically discussed in the fifteen chapters in this volume of proceedings by experts active in their respective disciplines. The book is a unique addition to both the sensory and food science literature, in that it brings together presentations from the subject matter areas that form the basis of sensory analysis in the food and consumer products industries. We hope that it will challenge the reader to think about different aspects of sensory science and its applications.

The first chapter sets the stage for bridging the gap that exists between psychologists and physiologists concerned with understanding sensory perception and those who seek to apply the theories. The authors of Chapters 2 to 4 address the physiological mechanisms of taste, odor, and chemical irritant perception and how we measure these sensations. Also, the temporal aspects of sensation are discussed in Chapters 2 and 3.

A unique approach to sensory perception is presented in Chapter 5, where the laws of psychophysics and sensory physiology are explained by a unifying mathematical theory. In Chapter 6, another concept of unification of the senses is discussed in light of the theory of synesthetic metaphor and the use of cross-modal language. The possible applications of neural networks to sensory systems are described in Chapter 7. Neural networks are a means of modeling the process by which humans distinguish between stimuli and ultimately make informational decisions.

In sensory evaluation, descriptive analysis is frequently used. For a panel of judges to be used as an analytical tool, they must be trained to agree on the sensations they perceive and the language used to describe the sample. As discussed in Chapter 8, panel training is a matter of concept formation and alignment, a form of calibration of the judges with respect to sensory qualities and their quantitation. Such efforts at sensory training are part of the development of expertise. Learning, training, or sensory calibration changes people's abilities to taste. Comparisons of how learning affects language in one realm, that of wine-tasting expertise, are discussed in Chapter 9, together with some theoretical frameworks for thinking about interactions of language and perception.

Individual differences in perception identified in sensory studies can be viewed as "nuisance variables" or can provide clues to product differences that might otherwise be ignored, a topic that is examined in Chapter 10. Furthermore, understanding sources of individual differences can reduce error variance, thus improving the sensitivity of sensory test methods.

In Chapters 11 to 13, the methods used for language development and descriptive analysis of complex foods or beverages are critically compared. Language development is a critical element of sensory evaluation, one that has received far too little attention and experimental study. Chapter 11 provides an overview of the profiling techniques currently being used and discusses methods for developing carefully thought out hybrids. The ways in which appropriate terms for evaluation of products are selected are described in Chapter 12. The use of statistical procedures, specifically Procrustes analysis, to compensate for differences in individuals' use of terms and scales is examined in Chapter 13.

Qualitative research methods are growing in popularity as a tool for sensory researchers. The qualitative methods find particular utility in terminology development and in exploratory studies, since consumer responses and purchasing behavior can be affected by factors other than the product itself. As discussed in Chapter 14, qualitative research is a means of obtaining information not readily available from trained panel evaluations.

Finally, statistical analysis of sensory data is a critical component of quantitative sensory methods. Most techniques of statistical inference are aimed at indicating sensory differences. However, sensory claims made for products in the media require substantiation by establishing sensory equivalence. This goal is common to many sensory tests, but must be approached differently from a statistical perspective, as explored in Chapter 15.

This book, then, provides a multidisciplinary approach to our understanding of how humans perceive the objects in their environment and how they use this information for description and selection of food and other consumer products. Ultimately, how we apply sensory science and techniques is dependent on our ability to evaluate their usefulness and interpret the results.

The symposium was approved and supported by the IFT Basic Symposia Committee, which included Drs. Barbara P. Klein (Chair), Merle D. Pierson (Past-Chair), Cavit Akin, Peter Bluestein, Allen V. Clark, Nancy E. Fogg-Johnson, Navam Hettiarachchy, Patricia A. Kendall, Ronald Richter, Les Smoot, and Louise Wicker. John Klis (Director of Publications) and Anna May Schenck (Associate Scientific Editor) and other IFT staff members provided support and coordination for the smooth operation of the symposium and publication of the proceedings.

Our final thanks go to the speakers and the participants in the symposium on advances in sensory science. Their professionalism and enthusiasm made it a pleasure to co-chair the symposium and co-edit this volume.

Harry T. Lawless
Barbara P. Klein

Contributors

Edgar Chambers IV, Ph.D.   Department of Foods & Nutrition, Kansas State University, Manhattan, Kansas

Margery A. Einstein   SensTek, Inc., Mercer Island, Washington

Maximo Gacula, Jr., Ph.D.   Biostatistics and Consumer Testing Department, The Dial Corporation Technical Center, Scottsdale, Arizona

Barry G. Green, Ph.D.   Monell Chemical Senses Center, Philadelphia, Pennsylvania

Bruce P. Halpern, M.Sc., Ph.D.   Department of Psychology and Section of Neurobiology and Behavior, Cornell University, Ithaca, New York

Barbara P. Klein, Ph.D.   Division of Foods and Nutrition, University of Illinois at Urbana-Champaign, Urbana, Illinois

Harry T. Lawless, Ph.D.   Department of Food Science, New York State College of Agriculture and Life Sciences, Cornell University, Ithaca, New York

Lawrence E. Marks   Department of Psychology, John B. Pierce Laboratory and Yale University, New Haven, Connecticut

Kenneth H. Norwich, M.D., Ph.D.   Institute of Biomedical Engineering, University of Toronto, Toronto, Ontario, Canada

Robert J. O'Connell, Ph.D.   Neurobiology Group, The Worcester Foundation for Experimental Biology, Shrewsbury, Massachusetts

Michael O'Mahony, Ph.D.   Department of Food Science and Technology, University of California, Davis, California

Dolores C. Oreskovich, Ph.D.   Division of Foods and Nutrition, University of Illinois at Urbana-Champaign, Urbana, Illinois

John R. Piggott, M.Sc., Ph.D.   Department of Bioscience and Biotechnology, University of Strathclyde, Glasgow, Scotland

Elizabeth A. Smith   Kansas State University, Manhattan, Kansas

Peter M. Smith   Research Information Services, S. C. Johnson & Son, Inc., Racine, Wisconsin

Gregg E. A. Solomon   Department of Psychology, Harvard University, Cambridge, Massachusetts

David A. Stevens, Ph.D.   Frances L. Hiatt School of Psychology, Clark University, Worcester, Massachusetts

John W. Sutherland, Ph.D.   Department of Mechanical and Industrial Engineering, University of Illinois at Urbana-Champaign, Urbana, Illinois

L. Greg Walter   Research Information Services, S. C. Johnson & Son, Inc., Racine, Wisconsin

Contents

Sensory Science Theory and Applications in Foods

1

Bridging the Gap Between Sensory Science and Product Evaluation

Harry T. Lawless

New York State College of Agriculture and Life Sciences
Cornell University
Ithaca, New York

INTRODUCTION

It is possible to view the scientific literature on human perception and sensory function as largely unconnected to the applied technology of sensory evaluation. Several factors contribute to this unfortunate situation. Academic psychologists concerned with theoretically meaningful phenomena are often unaware of the potential applications of their research in product evaluation. Higher and higher degrees of specialization and narrowing of research focus on highly specific effects and one or two methodological paradigms all contribute to forces that support the ivory-tower isolation. Academic researchers are understandably prone to stick to areas that they know best, that are productive in terms of publications, and that are fundable. Academic scientists often have the luxury of pursuing an area in great depth—they tend to mine particular veins of ore until they are played out. Industrial practitioners can also suffer from narrowness of focus, usually due to their understandable orientation toward substantive products, rather than perceptual processes. Under the very real business pressures of the need to develop new products under time constraints, evaluation practitioners often have little time for reviewing literature in which the connections to applied practice are not immediately apparent. Similarly, they have little time to devote to the application of new and yet unproven methodologies when operating in a practical, results-oriented product development support service.

Part of the problem arises from what Brinberg and McGrath (1985) termed the "conflicting desiderata" of laboratory control vs. real-life applicability. In many cases it is difficult or impossible to maintain scientific control in order to study relationships between variables without confounding effects, while establishing the worldly applicability of those effects. Scientists in the academic, theoretical milieu often decide to resolve this conflict in the direction of laboratory control, while applied practitioners understandably seek external validity. However, even though effects discovered in the laboratory may be obscured by other overriding factors in uncontrolled real-world research, this does not mean that they are inoperative. Conditions or circumstances may arise in which small effects become major contributors to product perception.

The distinction between controlled research and generalizable research may be somewhat less applicable to sensory evaluation in the sense that both modes of research lie within the mission of industrial scientists. Sensory evaluation practitioners are concerned both with analytical specification of perceived product differences as well as predicting consumer acceptance in the real world. These two objectives lead to two "styles" of performing sensory evaluation, a dichotomy that has been coined as Type I and Type II sensory evaluation by O'Mahony, in order to emphasize the differences in goals, orientation, and methodological approach. In analytical (Type I) sensory evaluation, laboratory control is paramount, in order to maximize the sensitivity of the measuring instrument. In hedonic-predictive sensory evaluation (Type II), the ability to generalize to the consuming public is of greater concern. Different types of panelists, serving procedures, sensory methods, and statistical models for analysis may be chosen by the sensory practitioner, depending upon whether the objective of the test falls in the Type I or Type II mode. While basic research findings may have limited direct application to Type II research, they may be more directly applicable to Type I analyses, which after all are conducted under similar conditions of controlled (albeit sometimes artificial) laboratory environments.

A further contributing factor to a lack of application of sensory principles to product evaluation concerns the traditional mission of sensory evaluation departments as testing services. Sensory departments that act merely as testing functionaries tend to accept products from research and development clients without participating fully in the design, formulation, and optimization strategies for those products. A common symptom in this way of conducting business is a focus on testing *products,* rather than examining *variables*. A related symptom is a tendency to perform series of paired tests, rather than multi-product designs with systematic variation along one or more continua reflecting ingredient or process variables. This stands in stark contrast to psychophysical approaches to understanding the range and effects of physical variables on sensory perception. Understanding dose-response functions cannot be achieved in two-product studies. Examples of the psychophysical style of research concerning effects of physical variables on human responses can be found in Chapters 2 and 3. The research of these scientists illustrates important effects of the variables of time and temperature on oral chemosensory perception of taste and irritation. Time and temperature have obvious relevance to food stimuli. Foods are served at different temperatures. Sensations change over time as food is introduced, masticated, and swallowed. On the other hand, some decisions about taste intensity, quality, and palatability can be made very rapidly by the ingesting panelist (see Chapter 2).

An additional difference between the style of research in the applied and academic worlds is in the approach to statistical inference. In the academic world, research is largely exploratory. Tests are designed to reject the null hypothesis, indicating that an effect was present, i.e., that two treatments were unlikely to be equivalent. There is a high risk in rejecting a true null, and academic researchers protect against this outcome by setting conservative values for alpha risk. Finding an effect where none exists may launch a whole series of needless experiments or send cohorts of hopeful graduate students down blind alleys of unprofitable thesis research. Conversely, failing to reject the null is viewed as an ambiguous condition—something might still have happened. Perhaps the treatment does produce an effect, but perhaps we didn't set up a wide enough range of treatments or control some interacting variables (See Chapter 10) and thus the effect was obscured. In industry, on the other hand, failing to reject the null may be taken as evidence for equivalence. Equivalence is an important conclusion in ingredient substitutions and in packaging or process changes when there is a target or standard food to be matched sensorially. Equivalence is an issue at the heart of quality control and also plays an important part in claim substantiation. Several approaches to the demonstration of product equivalence are discussed in Chapter 15. A related issue concerns a lack of attention to effect size in basic research. Finding any difference at all can be a promising indication of new principles or laws of perceptual functioning for the basic researcher. However, industry often makes a distinction between statistical significance and practical significance. Even very reliable effects may be of little consequence. Degree-of-difference is an important consideration, not the mere presence or absence of any difference at all. As discussed below in the section on individual differences in bitter taste perception, some published findings in the basic science literature are quite small and have heretofore been of little practical consequence.

One goal of this book is to find potential connections between basic research and practice and to point to evolving applications of sensory methods that apply basic principles. This chapter is divided into five sections to explore relationships between basic research, primarily in the chemical senses, and sensory evaluation practice. The first section discusses the dependence of food acceptance on sensory properties, and emphasizes the general principle that understanding sensory function should aid in the optimization of the perceived attributes of a product. The second section illustrates two spectacular failures to connect well-understood psychophysical phenomena to food perception. These phenomena are taste adaptation and individual differences in taste perception of bitter substances. A final topic in this section concerns the somewhat confused literature on how (and if) the senses of taste and smell interact and will examine differences between psychophysical findings and those published in the applied food literature. The third section of the paper explores complexity of food systems as one reason for the difficulty in connecting basic research to practice. A potential solution to the complexity problem is to interpolate assessments at intervening stages of perceptual processing in order to build more powerful explanatory models. The fourth section discusses some fortuitous and some deliberate applications of basic principles to sensory work. The fifth and final section is concerned with future challenges in connecting sensory practice to basic principles, including two specific opportunities: better understanding of terminology development, and better understanding of individual differences and their physiological basis.

THE UNSPOKEN MODEL

A common view of product acceptability is one that predicates the hedonic reaction to a product on its sensory properties. At first glance, this seems almost too obvious to bother stating. However, several factors conspire to make this explicit relationship rarely heard in spite of how obvious or trivial it may seem to sensory evaluation practitioners. One factor that contributes to the ease with which food scientists can sometimes ignore this relationship is that there are also many nonsensory aspects to food quality (Lawless, in press). Quality is closely synonymous with value, and it is often a goal of product marketers and product developers to maximize perceived quality. However, a number of factors contribute to the perceived quality of a product in addition to its sensory properties. Microbiological integrity, other safety factors, and nutritional content are all important. Brand image, price, and competitive positioning all affect a consumer's perception of quality. A bit closer to sensory properties is the consumer's perception of product *consistency*. A major quality factor for many people is the sense of predictability or reliability of the product, which translates into value in return for money spent. Even more closely related to a sensory basis for food quality is the negative focus than defines quality as a lack of perceivable defects. This type of analysis has been the foundation of historical commodity judging, for example, in the area of dairy products (Bodyfelt et al., 1988). Finally, products are purchased for and consumed during specific situations. Thus a product is not only acceptable or preferred in some general sense, but is appropriate for a purpose (Schutz, 1988). We may find lobster highly acceptable, but few people consume it on a daily basis. Such factors point out that real-life product preferences and food choices are more complex than simple results from a blind-labeled acceptance test would suggest. However, in the long run, a product must have some constellation of sensory properties that consumers find appealing on a blind basis, or the all-important factor of repeat purchase has a dim outlook at best. An interesting question in current marketing-oriented food industry concerns whether products are receiving sufficient testing to ensure some probability of success. For example, were the over 200 oat bran products launched in 1989 all optimized for flavor and texture appeal? Does lack of testing and sensory optimization contribute to the 80–90% failure rate of new food product introductions (Shapiro, 1990)?

These concerns aside, what mathematical relationships can we use to model the connection between acceptance and perceived sensory properties? Table 1.1 shows a simple polynomial expression by which food acceptance is predicted by a weighted function of the perceived intensities of various sensory attributes. Regression techniques can be used to fit such models to empirical data (Moskowitz, 1983). Weighting coefficients are positive for desirable attributes, i.e., when the consumer feels that the more of this characteristic the product has, the better the product. Negative coefficients are for product defects in which the more intense the sensory attribute, the less acceptable the product. Higher order terms may be added to describe continua with optima, in which you reach some most desireable level, after which you have too much of a good thing. This is achieved with a small negative coefficient for a squared term, as is often applied

TABLE 1.1 An Acceptance Model

Acceptance $= k_1(S_1) + k_2(S_2) + \ldots K_n(S_n)$ (simple linear model)
$\qquad\qquad\qquad + k_{12}(S_1 * S_2) \ldots$ (with interaction terms)
$\qquad\qquad\qquad + k_{1m}(S_1)^a{}_m \ldots$ (higher order polynomial)
where $S_n =$ perceived intensity of attribute n
Intensity, S_n can also be predicted by equations such as
$$S_n = K_n I^b{}_n$$
where b is the characteristic of the power function exponent and I is the stimulus intensity in physical units (e.g., molarity)

to sweetness optimization (Lawless, 1977). Interaction terms may take into account the simultaneous balancing of two or more attributes, as in the importance of sweet-to-sour ratios in fruit products and wines. Perceived intensity, in turn, can be modeled on the basis of ingredient concentrations, by relationships such as Steven's power law or alternative functions. It is important for the product formulator to realize that these dose-response functions may be curvilinear and not assume simple linear functions. One cannot assume, for example, that sweetness will necessarily be doubled by doubling the amount of sugar in a product. There is usually a law of diminishing returns. Understanding such relationships, as well as the weighting coefficients for sensory contributions to overall acceptance, can help the product formulator in engineering acceptable products. The formulator who understands sensory thresholds (minimum perceivable levels), psychophysical functions (dose-response curves), and sensory interactions will be in a superior position in terms of understanding the probable perceptual impact of changing variables. Interactions between ingredients or between sensations such as in masking or synergy can be of great importance. It is difficult, for example, to increase the sweetness of saccharin beyond a certain level due to the increasing inhibitory effects of its bitter component. Many such interactions are documented in the psychological or sensory literature (Lawless, 1986), and being aware of them would seem to have obvious rewards for the sensory practitioner and product developer.

What are the potential limitations of an acceptance model like that in Table 1.1? Linear regression has proven to be a powerful tool. Furthermore, modeling on the basis of simple polynomials is so common in science and engineering that it seems a natural starting point. Many acceptability relationships, however, may be highly nonlinear, discontinuous, or otherwise impossible to model with simple equations. Some sensory qualities may have simple binary influences on acceptability, for example, especially with sensory defects. Any stray hairs at all lying on a pat of butter render it inappropriate for consumption. Other input-output relationships may be so complex as to defy attempts at modeling with equations. Other approaches may be brought to bear, and the future holds promise of an increasing array of modeling tools that will be available to the sensory scientist and practitioner. For example, the recent elaboration of neural network or connectionist models in psychology and artificial intelligence may find application in acceptance modeling. Although many sensory evaluation professionals would be satisfied with knowing that they helped to optimize the simple blind-judgments of product acceptability, neural network modeling could be applied to the *simultaneous* optimization of a constellation of desired characteristics, including not only overall acceptance, but appropriateness in certain situations, appeal to certain market segments, and marketing considerations such as value perceptions. Virtually any vector of complex consumer-perceived attributes can be modeled with a neural net as a function of sensory attributes or other physical factors, as long as they are measurable (see Chapter 7).

The relationship of acceptance to the intensity of various sensory characteristics emphasizes the value of understanding and quantifying sensory attributes of products in determining their eventual success. Furthermore, those who understand basic principles of human sensory function, human perception, and human judgment will not only design better, more informationally powerful tests, but be better able to assist their product development counterparts in the engineering of products that meet specific sensory objectives. In spite of this importance, the degree of application of psychophysical principles to food science is disappointing. Doubtless, some of this lack of connection lies in attitudes concerning the appropriate activities for each area of research, as well as how scientists are rewarded in academic vs. industrial settings (Lawless and Klein, 1989). A commonly held attitude is that it is the business of commercial food industries to perform research on products and the business of academic laboratories to break ground on new theories and technologies. An increasingly competitive government funding picture suggests that applied questions will not engender grant support unless they can be connected to clinical abnormalities such as eating disorders or hot topics (e.g., value-added processing) concerning specific commodities. The potential for industrial support of basic research is also grim. Concerns about the eroding of potential competitive advantages and the priority of patents over scientific publications are common considerations that protect profits but inhibit the progress of public science. Thus a gap between the basic science of sensory function and practical questions of food acceptance remains unfilled.

THE ROCKY ROAD—CONNECTING BASIC RESEARCH TO PRACTICE

Making connections between well-established sensory research findings and industrial practice is not often easy. A classic example is the experimentally robust psychophysical phenomenon of sensory adaptation. Under conditions of constant stimulation, taste and smell sensations tend to abate, a phenomenon by which the brain protects itself from information overload (O'Mahony, 1986). The brain has a useful tendency to ignore stimuli which are unchanging and are therefore of low informational value. Adaptation has been amply documented for the sense of taste under conditions in which solutions are flowed over controlled ares of the tongue and in which tactile and thermal sensations are minimized (McBurney, 1966). However, examination of the practical implications of this phenomenon on testing methods has shown only small effects that are often clouded by other contributing factors. In a series of experiments with simple salt solutions, for which gustatory adaptation is well understood, performance in discrimination tests is dependent upon a number of additional effects such as contributions of salivary sodium, stimulus learning, differences in sub and supra-adapting taste sensitivity, as well as sensory adaptation (O'Mahony and Odbert, 1985; O'Mahony and Goldstein, 1987; Vie and O'Mahony, 1989). Perhaps more importantly, if stimulation of the tongue is intermittent rather than constant, taste adaptation is effectively negated and sensation intensity is maintained. Under some conditions, not only is intensity maintained, but enhancement rather than declining sensation is the rule (Meiselman and Halpern, 1973). When intermittent stimulation occurs in actual eating and drinking, adaptation may have a minor role at best. In the psychological and physiological literature, taste adaptation has been widely studied, due to its physiological correlations with phasic nerve responses and its contribution to the understanding of phenomena such as water tastes (Bartoshuk, 1968). However, its relevance to food perception, especially in eating situations, remains to be determined.

Does PTC Status Affect Food Perception?

Other taste phenomena have suffered similar fates of poor generalization or poor predictive validation in food perception. One of the most profoundly influential discoveries in flavor chemistry was the observation in the 1930s by A. L. Fox that a compound, then called para-ethoxy phenylthiocarbamide, was intensely bitter to some people and nearly tasteless to others (Fox, 1932). This observation gave rise to a host of studies using related compounds. Early studies of this effect used a simpler molecule, phenylthiourea (phenylthiocarbamide or PTC), and later studies used propylthiouracil (PROP) due to its lower toxicity and lack of sulfury odor. A number of important findings appeared quickly in the literature. First, thresholds for these substances followed a bimodal distribution, with about ⅓ of a Caucasian population being thus classified as an insensitive "nontasting" group, among people with otherwise normal taste sensitivity. Families of compounds with the functional group $N - C = S$ showed the effect (Harris and Kalmus, 1949). Genetic studies revealed a simple Mendelian pattern of inheritance, with nontasting status requiring two recessive alleles (Blakeslee, 1932). Anthropological surveys became fashionable, showing, for example, that racial groups of African or Oriental origin were nearly all sensitive tasters of PTC (Cohen and Ogdon, 1949).

While the mechanism underlying this phenomenon at a physiological level has never been elucidated, much interest and a fair amount of speculation was stimulated among physiologists and physiologically oriented psychophysical researchers. One appealing idea was that nontasters lacked a gustatory receptor type. Other theories were proposed from time to time, including one that saliva was a necessary factor in determining the presence of the effect (Cohen and Ogdon, 1949). The saliva hypothesis was rejected on the basis of experiments in which PTC was flowed over the tongue, and all saliva was rinsed away by extended preadapting rinses with water. Under such conditions, PTC tasters and nontasters do not change their patterns of responsiveness (e.g., Lawless, 1979a). A somewhat better substantiated notion was that at least two types of bitter taste response mechanisms are present in humans. The presence of two independent mechanisms is consistent with two observations. First, responses to PTC, as assessed in threshold studies, were uncorrelated or very weakly correlated with responsiveness to many other bitter compounds such as quinine, whose salts were considered prototypical classical bitter exemplars (Fischer and Griffin, 1963). Second, adaptation of the tongue to PTC left quinine responsiveness largely unaffected, and conversely, quinine adaptation had no effect on responsiveness to PTC (McBurney et al., 1972; Lawless, 1979a).

In recent psychophysical literature, a number of unsuspected correlations with other compounds have been observed. Some observations suggested a taster–nontaster dimorphism for creatine and creatinine, two components of meat flavor, but that dimorphism, as well as its correlation with PTC status, has been largely discredited (Lawless, 1979a). Tasters and nontasters of PTC were shown to differ in their responsiveness to the bitterness of caffeine (Hall et al., 1975) and tasters and nontasters of PROP (a related compound, propylthiouracil) to the bitterness of saccharin (Bartoshuk, 1979). Surprisingly, PROP taster status was also related to *sweet* responsiveness to sucrose, saccharin, and neohesperidin dihydrochalcone (Gent and Bartoshuk, 1983). These effects are often small, and require careful psychophysical procedures to be observed. Furthermore, it is often problematic to rule out differences in overall sensitivity that might contribute to an apparent taster–nontaster difference. However, given such correlations, it seems reasonable to ask whether PTC status might influence food likes and dislikes. For example, could one's sensitivity to PTC be correlated with reactions to the bitterness of caffeine in coffee or bitter side tastes of saccharin in reduced-calorie sodas? Furthermore, since bitter tastes will tend to have a suppressive or inhibitory effect on other taste qualities present, tasters and nontasters should not only differ in their bitter responses to complex food tastes, but in the expression of other tastes in food (Lawless, 1979b).

The potential practical implications were apparent to A. L. Fox at the beginning of PTC research. In 1954, at a meeting of the American Chemical Society, he presented a paper, "Why Johnny Likes Spinach and Mary Doesn't," which presented some data from a survey of food likes and dislikes. Manuscript versions of this talk are in limited circulation. His frequency breakdowns were cross-referenced to both PTC status and responses to sodium benzoate. Sodium benzoate is a complex-tasting salt, which evokes different responses from different people. Unfortunately, benzoate classification was somewhat unreliable (Hoover, 1956), perhaps because the original method did not control for response biases among people as to the taste qualities that were elicited. Other studies that have examined food likes and dislikes among PTC tasters and nontasters have found little or no effect. The only significant finding that appears in the literature is a general relationship between PROP status and the number of foods one dislikes (Fischer et al., 1961).

One provocative study by Greene (1974) examined PTC status in a population exposed to naturally occurring goitrogens. This is reasonable since PTC-related compounds occur in some plants, and in underdeveloped mountainous districts in which iodine supplementation is unavailable, some advantage might accrue to people sensitive to bitter compounds if their taste sensations led to the avoidance of goitrogenous plants. Greene assessed PTC status and visuo-motor maturation in two Ecuadorian Andean communities in which goiter and cretinism were endemic. One community had received iodine supplementation and the other had not. In the community that had not received iodine supplementation, there was a small but statistically significant correlation between taste sensitivity and neurological maturation. In the iodine-supplemented community, no such relationship was evident. This supports the notion that bitter-sensitive individuals might be at an advantage in avoiding food crops in which naturally occurring goitrogens are prevalent. Other studies have noted slightly increased proportions of nontasters among patients with nodular goiter compared with the normal population (Harris et al., 1949; Azevedo et al., 1965), although the causal determinance of this trend remain unclear.

In summary, it is surprising that such a robust phenomenon, which has been so heavily studied in laboratory psychophysics, genetics, and anthropology should tell us so little about food flavor perception and/or food likes and dislikes. Two explanations come readily to mind. The first is the uncomfortable notion that taste perception simply has a weak influence on food habits. This idea is unpalatable, especially when the influence of unpleasant flavors on food rejection is considered. Bitter taste defects in milk, for example, are sufficient to cause immediate rejection on the part of both dairy inspectors and consumers (Bodyfelt et al., 1988). A second explanation is that individual food habits are determined by so many diverse factors that a relationship to individual taste sensitivities is easily obscured. Nonetheless, the gap between laboratory sensory science and practical sensory implications is striking in this case.

Do Taste and Smell Interact?

Another example of an apparent discrepancy between the food literature and chemosensory psychophysics is the issue of how taste and smell interact. It is a common belief among food scientists, as well as consumers, that taste and smell are somehow related or are closely linked in the perception of flavor. Some of this assumed relationship derives from the generic use of the word "taste" to mean all aspects of food flavor. However, if we restrict the use of the word taste to mean sensations from nonvolatile substances perceived in the oral cavity, and add the further restriction that gustation proper does not involve tactile or irritating chemical sensations, we are left with a technical definition that includes such sensations as sweet, sour, salty, and bitter. Regardless of whether these four sensation qualities are taste primaries or merely distinct taste sensations among many other possibilities, we can ask whether these taste sensations interact with aromas and volatile flavors that are sensed by the olfactory apparatus. This taste/small distinction is thus made on anatomical grounds. A number of experiments have addressed this question in the psychophysical laboratory.

Murphy et al. (1977) examined perceived odor intensity, perceived taste intensity, and overall perceived intensity of mixtures of sodium saccharin with the volatile flavor compound ethyl butyrate. A second experiment (Murphy and Cain, 1980) examined the same ratings for sucrose–citral and NaCl–citral mixtures. The pattern of results was consistent in the two studies. Intensity ratings showed about 90% additivity. That is, when framed as a simple question about the ways in which gustatory and olfactory stimulation combine to produce overall impressions of flavor strength, there is little evidence for interactions among the two modalities. There was one notable exception to this rule, however. The volatile compounds ethyl butyrate and citral contributed to judgments of "taste" magnitude, a reliable illusion in both studies. When a flavorous solution is placed into the mouth, untrained subjects have a hard time distinguishing the volatile sensations as odor and misattribute them to taste. This illusion is eliminated by pinching the nostrils shut during tasting, which prohibits the retronasal passage of volatile materials and effectively cuts off the volatile flavor impressions. Aside from this mislabeling, the psychophysical evidence points to more independence of taste and smell than interaction, in contrast to popular belief.

However, a different result was seen in real products. Von Sydow et al. (1974) examined ratings for taste and odor attributes in fruit juices that varied in added sucrose. Ratings for pleasant odor attributes increased and those for unpleasant odor attributes decreased, which von Sydow interpreted as evidence for a psychological effect as opposed to a physical interaction. For example, attentional mechanisms could influence this shift. Sucrose also suppressed "harsh" tastes such as bitterness, sourness, and astringency. Such unpleasant tastes may have drawn attention away from volatile characteristics in juices of low sweetness and harsher character. When the juices were more "in balance," panelists' attention may not have been so captured by the harsh tastes, causing a higher probability for recognition of volatile character and thus higher average ratings. A somewhat similar effect was reported by Perng and McDaniel (1989), in which blackberry juice flavor was rated by a trained panel, at varying levels of sucrose and acidity. Sucrose enhanced flavor ratings, and juices with high acid level showed lower ratings.

The pattern is potentially complicated by the ways in which interactions may depend upon the particular flavorants and tastants which are combined. Wiseman and McDaniel (1989) reported some enhancement of fruitiness of orange and strawberry solutions by aspartame, as compared to little or no effect for sucrose, and a somewhat greater enhancement for orange than strawberry. Frank and Byram (1988) found sweetness to be enhanced by strawberry odor, but not by peanut butter odor. Later studies with greater numbers of tastants showed general suppression of sodium chloride saltiness by volatile flavors, but more complex interactions with other tastants (Frank et al., 1990). Further research is needed to see whether these empirical findings can be generalized into any coherent rules, or whether tastant–flavorant interactions will remain a matter for case-by-case study with little or no systematic pattern. One potentially profitable avenue for research is the degree of cultural experience panelists have with particular combinations, i.e., a potential influence of learned expectancies.

The instructions that are given to subjects in these studies may have profound effects, as in many other sensory methods. Lawless and Schlegel (1984) studied citral–sucrose mixtures using both direct scaling and "indirect" Thurstonian scale values derived from triangle test performance. The direct scaling, simple ratings of perceived sweetness and lemon character, showed the pattern of independence noted by previous psychophysical workers. However, in the triangle tests, a pronounced interaction was seen in the case of a pair of mixtures, which was so highly discriminable as to yield a larger-than-predicted scale value. Furthermore, another pair which was barely discriminable according to the triangle tests received significantly different sweetness ratings. This result indicates that when subjects' attentions are directed to specific attributes, as in ratings, paired comparisons or forced-choice procedures, they may find products to be much more discriminable than when areas of difference are unspecified, as in traditional discrimination procedures such as the triangle test or duo-trio test (Ennis, 1990).

The ratings that subjects are instructed to make will also influence apparent taste-volatile interactions. Frank et al. (1989) reported that strawberry odor could enhance the sweetness of sucrose–strawberry solutions, an effect reminisicent of the enhancement reported by Wiseman and McDaniel (1989) and also the mislabeling of volatile sensations as taste intensity estimates originally observed by Murphy et al. (1977). Further study of this effect revealed that when subjects were instructed to make total intensity ratings and then partition them into their components, no significant enhancement of sweetness by strawberry odor was seen (Frank et al., 1990). That is, when subjects were able to psychologically "unload" their strawberry impressions on a flavor rating scale, the interaction with sweetness was attenuated. Restriction of ratings to only sweetness in a control experiment restored the enhancement effect. This finding has broad implications for the ways in which sensory evaluation, particularly descriptive analysis in which multiple attributes of complex foods are rated, should be conducted. It also suggests some caution in substantiating claims for various synergies or enhancement effects in which ratings are restricted to too few attributes. Respondents may choose to "dump" some of their impressions into the most suitable category or the only allowable response if the attribute they perceive is otherwise unavailable on the ballot. Alleged enhancements such as the effect of ethyl maltol on sweetness should be viewed with caution unless the response biases inherent in dumping responses or in mislabeling smells as tastes can be ruled out. Ethyl maltol is especially problematic in this regard since it has an odor sometimes characterized as "sweet" (Civille and Lawless, 1986).

THE COMPLEXITY ISSUE

One property that discourages the systematic application of psychophysical methods to food systems is the degree of physical and sensory complexity of foods. Even the simplest product will vary in a number of dimensions, each capable of interacting with other sensory attributes or other physical ingredients at various levels. Thus a single attribute or ingredient can produce a variety of sensory impressions, with varying impact on overall consumer acceptance of the product. Consider even a simple system like an engineered fruit-flavored beverage. Such a product will have at least the following major components: a sweetener, an acidulant, odor, and volatile flavor. The percevied intensities of these components may interact (or not) through such mechanisms as taste mixture suppression, which would act to decrease the impact of the tastants. These four components have a potential of six two-way interactions, four three-way interactions, and one four-way interaction. As if this were not complex enough, there may also be emergent properties or Gestalt-like effects, in which the pattern of sensory inputs forms a complex recognizable whole that is unpredictable from consideration of the components singly. Foods are by their nature multivariate; most psychophysical approaches to understanding sensory function have dealth with single continua producing simple sensory effects, e.g., those that can be rated on a single attribute scale. It is tempting to resort to simple empiricism with food products and take the attitude that fundamental variables and all their possible interactions are too complex to understand. This contributes to the reduction of sensory evaluation to a routine of testing products, rather than being an approach to systematic investigation of variables. This is unfortunate, since a psychophysical approach toward variables would yield information that can sometimes be generalized to other situations, products and problems.

Model Building

How can the complexity of foods and food perceptions be dealt with? A variety of multivariate techniques are available for building empirical relationships between responses of interest (e.g., acceptance) and other variables such as perceived attributes or ingredient concentrations. One problem is that any two variables can be examined for correlation. Correlation programs available on canned computer software packages have no inherent intelligence. There is nothing that prohibits a well-meaning investigator from examining the relationship between two variables, regardless of whether such a relationship has any logical or causal validity. This sometimes leads to research which yields only small correlations among variables. Classical examples of this sort can be found in the attempts to predict eating behavior on the basis of measured attitudes or personality characteristics. The causal chain is too long and the connections too weak to yield strong predictive relationships. One potential solution is to shorten the causal chain, by examining predictor variables that are more closely linked to the variable of interest; to understand a complex phenomenon piecemeal by building understanding through the assembly of a set of more direct contributing interrelationships.

A first approach to model building is often to look for simple additive relationships among variables. For example, suppose we are interested in how two ingredient variables contribute to the overall acceptance of a product. We might wish to optimize the product by finding that combination of ingredients that maximizes its sensory appeal. It is tempting to examine the overall acceptability of the product as a simple additive function of the ingredients, through multiple regression, or to seek the optimum empirically through response surface techniques. Factorial designs are often employed to examine each level of one variable at many levels of another. But does it make sense to examine acceptance as a function of ingredients, while ignoring the sensations that are produced by those ingredients, as well as their potential interactions? An alternative to jumping from physical ingredients to acceptance is to interpolate the sensory perceptions between those two sets of variables. Then we build two sets of relationships: first, psychophysical relationships specifying how sensations arise from ingredients and how sensations interact, then combination rules that define how component *sensations* (not physical ingredients) contribute to overall acceptance. A simple example follows.

In a psychophysical study of pleasantness ratings of model solutions of taste mixtures, quinine and sucrose levels were varied in a factorial design to examine the contributions of an innately unpleasant (bitter) sensation and an innately pleasant (sweet) sensation to overall hedonics (Lawless, 1977). Instead of attempting to predict overall acceptance on the basis of sucrose and quinine concentrations, prediction was made on the basis of observed sweetness and bitterness levels in the mixtures themselves. This was done since the marginal sweetness and bitterness levels, i.e., of unmixed sucrose and unmixed quinine, were poor predictors of what transpires in mixtures. Sweet and bitter sensory signals inhibit one another (Lawless, 1979b), and hence mixtures are less sweet and less bitter than their marginal (unmixed, equimolar) counterparts. This can create emergent properties in overall hedonic ratings that are difficult to predict from simple-minded factorial models based addition of marginal values for ingredient levels. One example is when a small amount of quinine is added to an intense level of sucrose (e.g., 1.0 M, Table 1.2). The quinine is unpleasant when tasted alone, and yet hedonic ratings go up when it is added to sucrose, an apparently nonsensical or at least nonadditive relationship. However, if two sensory factors are considered, this result makes sense. First, the quinine bitterness suppresses sucrose sweetness. Second, the relationship between sweetness and acceptance is nonlinear. It is a quadratic polynomial with a negative squared term that makes the function turn over at high levels, which is equivalent to saying that the sweetness function has an optimum, after which stimuli become too sweet and acceptance falls off. This relationship is shown in Fig. 1.1.

The paradox of the positive hedonic result in the case of adding a negative stimulus is explained by the fact that the suppressive effect of a small amount of quinine changes the sweetness level toward its optimum. That is, it shifts the sweetness level back up the function toward its peak. This is similar to the reaction of reduced-calorie soda drinkers when faced with a nondiet soft drink. Often, they find it too sweet and lacking in the bitter aftertaste of the intensive sweetener, to which they are accustomed. The point is that the brain, on deciding whether something is like, does no analyze chemical concentrations. Rather the

TABLE 1.2 Intensity and Pleasantness Ratings of Sucrose, Quinine, and Mixtures

Sweetness, 1.0 M sucrose	67.3
Hedonic rating	+20.2
Bitterness, 10^{-6} M quinine sulfate	3.5
Hedonic rating	− 1.1
Sweetness, mixture	58.0
Bitterness, mixture	1.0
Hedonic rating	+24.1

Data are normalized means from 150-mm line scale markings, with sweetness and bitterness ratings representing distance from the left-end zero point, and hedonic ratings representing distance from a central zero point (positive indicating pleasantness and negative indicating unpleasantness). See Lawless (1977) for details.

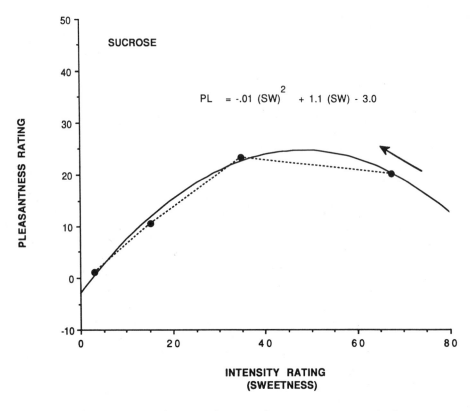

FIG. 1.1 Pleasantness as a function of perceived sweetness intensity. The four points are for 0.03, 0.10., 0.32, and 1.0 M sucrose in water. Adding small amounts of a bitter substance to 1.0 M sucrose causes sweetness to decrease, and therefore pleasantness to increase in the direction of the arrow. *(Replotted from Lawless, 1977.)*

conscious decision reflects upon sensations, which are a result of both receptor output and higher levels of processing. These sensations interact, sometimes in inhibitory ways. The practical implication of the quinine–sucrose mixtures illustration is that understanding is gained through logical interpolation of intervening steps. What appears to be paradoxical in terms of unmixed ingredients and their contributions is explained when sensory interactions are measured through the scaling of component sensations. In practical terms, better models can be built when descriptive specification (i.e., by a trained, calibrated panel) of a product is interpolated between consumer acceptance and formulation. Jumping directly from ingredient formulas to acceptance may permit one to find an empirical optimum, but it yields no causal principles to be used in the future.

Using Interpolated Sensory Responses—CHARM Analysis

Another potential application of an interpolated sensory response in understanding complex phenomena is in the area of flavor chemistry. Naturally occurring flavorous materials are complex mixtures of hundreds or even thousands of chemical compounds. Furthermore, since many plants share common biochemical machinery, it is not surprising that they produce similar profiles of flavor compounds. For example, a common set of terpene aroma materials can be found in many herbs and fruits, and yet the human observer's nose decodes this complex pattern and recognizes them as distinct. For example, we have little difficulty in discriminating oregano from lavender, in spite of their similar volatile profile. It is more difficult, however, to conclude from the list of ingredients and even their concentrations, what distinguishes oregano from lavender. One problem is that very potent aroma compounds may have strong sensory impact, even in minute amounts. Conversely, other compounds may be present in high concentrations but produce little sensory impact. In order to estimate the sensory impact of a given aroma compound, the sensitivities and responsiveness of the human observer must be worked into the causal model at some point.

Various procedures for sniffing the effluents from gas chromatographs are currently in use to separate the components of a complex natural material and assess their sensory impact. One such method is CHARM analysis (Acree et al., 1984). In this procedure, an observer is seated at the outlet port of a stream of cooled, humidified air into which the effluent of a gas chromatograph is mixed. The observer responds to the onset and offset of a smell in the airstream by pressing or releasing a computer mouse button. Various dilutions of the product are run through the chromatograph, until the observer smells no odors whatsoever. The number of dilutions at which an observer responds to a given odor is used to calculate a type of odor unit, called CHARM value, roughly speaking a reciprocal threshold. The assumption is that the greater the number of dilutions at which a compound is smelled, the greater its impact in the original product. Since responses are cross-referenced to a meaningful time base via a retention index, they can be uniquely identified on other GC runs when coupled with mass spectrometry. Figure 1.2 shows a flame ionization detector response to a GC run of a sample of orange juice, with the human (CHARM) response below (Marin et al., in press). The higher the CHARM peak, the more impactful the compound, since peak height is related to the number of dilutions at which the compound was smelled. Several aspects of this instrumental/sensory comparison are of interest. First, there are many peaks in the FID detector at which there is no human response, i.e., there are many compounds of low impact. Second, there is a zone of active human response after peak 1083 in which the FID trace is virtually flat. Thus there is little instrumental detection of a number of com-

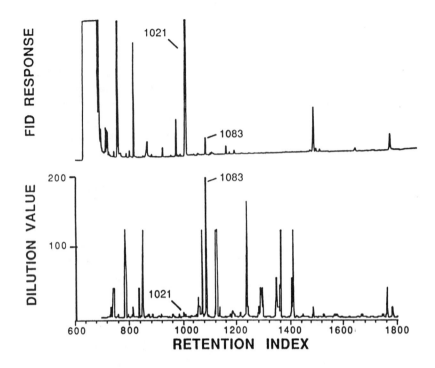

FIG. 1.2 Flame ionization detector response (upper trace) to components of orange juice. Peak 1021 is limonene and peak 1083 is linalool. CHARM response chromatogram (lower trace) to the same juice. Peak height in the CHARM chromatogram is proportional to the number of dilutions at which the observer noticed a smell at that retention time. *(From Marin et al. (in press), by permission.)*

pounds to which the human nose is exceptionally responsive. Also, note the rather weak CHARM response to peak 1021, which is limonene, often considered a prototypical contributor to orange flavor. In contrast, there is a robust response to the peak numbered 1083, which is the common terpene material, linalool. By itself, linalool is not considered to be very orange-like in character, yet the CHARM analysis suggests that it may have a potent contribution to orange aroma in juice.

Of course, other methods are available to cross-reference sensory impact to instrumentally measured flavor and aroma components (Buttery et al., 1989). Thresholds could be measured in other ways or even taken from the literature, although a great deal of caution is needed when different measurement procedures are used for threshold determination (ASTM), 1978). An important contribution of such approaches is the interpolation of a measure of sensory impact between knowledge of the physical ingredients and higher level human perceptions such as descriptive profiles or the identification of an aroma as prototypically "orange." In using CHARM values as a screen for entry into a neural network, for example (see Chapter 7), one could eliminate many nonimpactful compounds and build better relationships through the elimination of extraneous noise in the model and potentially spurious chance correlations.

PRINCIPLES AND APPLICATIONS: SOME PARALLELS

Although the direct application of perceptual principles to applied product evaluation has been sometimes disappointing, there are sometimes correspondences between the psychological literature and the ways in which practical techniques have developed. Since sensory evaluation professionals use human observers as their measuring instruments, it is reasonable that they would design methods that take into account the abilities, biases, and limitations of those observers and develop procedures that have some consistency with psychological research findings. In some other cases, the application of psychological principles has been direct, or a mixture of direct application and fortuitous correspondence.

The Nine-point Scale and Human Information Transmission

One example of a sensory method with perhaps fortuitous as well as deliberate relationships to psychological principles is the nine-point Quartermaster hedonic scale (Table 1.3). This instrument has proven to be a durable and useful method for the assessment of food likes and dislikes by consumers. It has a number of

TABLE 1.3 Quartermaster Nine-point Acceptance Scale

Verbal anchor	Traditional scale value
Like extremely	9
Like very much	8
Like moderately	7
Like slightly	6
Neither like nor dislike	5
Dislike slightly	4
Dislike moderately	3
Dislike very much	2
Dislike extremely	1

salient properties: it is balanced, bipolar, contains a neutral point, and has approximately equal psychological spacing between scale points, giving it roughly interval scale properties. It is simple to understand and easily used by untrained judges. What factors contribute to the utility of this instrument?

As a general scale type, category scales, with discrete responses that are assigned numerical values, have been widely used in both psychology and sensory evaluation. In terms of their discriminative power, i.e., the ability of the instrument to register perceived differences among stimuli or products, nine-point category scales for sensation intensity and quality have proven to be as good or better than other methods such as graphical line marking scales or magnitude estimation (Lawless and Malone, 1986a,b). Like the nine-point hedonic scales, integer scales for sensation intensity are easily grasped by untrained subjects and straightforward to use, requiring only the name of the attribute and verbal end-anchors to confirm the direction of increasing sensation intensity. But why nine points and not more or less? A reasonable hypothesis might be that greater numbers of scale points would lead to increased discriminative power. However, increasing the number of scale points tends to yield diminishing returns in terms of product discrimination.

One possible reason for the general appropriateness of nine-point category scales is the limited ability of humans to process information. In the 1950s, a number of engineers and psychologists began to examine human information transmission capacities in controlled tests. These tests examined the ability of subjects to identify stimuli as the number of alternatives increased. For simple stimuli which varied in only one perceptual dimension, subjects were usually able to identify only about five to seven alternatives, regardless of how widely spaced they might be on a perceptual continuum (discriminability not being a problem) (Miller, 1956). It is obvious that people are able to break through this limit as the dimensionality of the stimuli increases. For example, we are able to name hundreds, if not thousands, of faces, which are highly multidimensional stimuli. The same may be true for odors (Desor and Beauchamp, 1974). Furthermore, the training of observers on some continua may aid in increasing their information transmission to some degree. However, it is nonetheless striking and countintuitive that people should perform so poorly in such a simple task.

Returning to the question of how many category points are sufficient, having only nine alternatives would seem to provide ample room for subjects to make assignments of stimuli to unique categories based on intensity distinctions along simple continua. Most sensory rating scales are designed to reflect simple continua along dimensions of perceived stimulus strength. Little or no advantage is to be gained by the addition of more scale points when the perceptual and memory capacities of subjects do not lead to further distinctions being made. In reviewing this issue, Cox (1980) concluded that five to nine scale points (in other words, seven plus or minus two) are usually sufficient. It is fair to question whether the assignment of rating scale values is analogous to a task in which stimuli are named or identified. However, human biases such as Parducci's frequency effect (Parducci, 1965), in which subjects tend to use category scale points an equal number of times, suggest a correspondence in these two seemingly different tasks. Although the development of the nine-point hedonic scale preceded the analysis of information transmission, there is a comfortable fit with the capacities of human judges to make isolated assignments of responses to single stimulus presentations.

Some relationships of the nine-point scale to psychological principles were more deliberate. Jones and Thurstone (1955) used a variation of Thurstonian scaling to examine the value of different hedonic phrases (e.g., like, dislike, average, bad), the degree of overlap between similar terms, and the nature of the underlying distributions of responses from large numbers of people to those terms. Fifty-one descriptive phrases were rated by thousands of military personnel on simple numerical rating scales to indicate the degree to which they were hedonically positive or negative. Jones et al. (1955) developed the nine-point scale to use terms from those rating studies that were approximately equally spaced in psychological scale units (although the extreme terms, dislike extremely and like extremely, were psychological outliers) and that had low standard deviations, indicating good interindividual agreement on their magnitude. Terms were avoided that showed cumulative frequency distributions that were nonnormal, usually indicating mixed opinions. For example, the word "average" is neutral in connotation to many people, but somewhat negative to others. They also examined the test-retest reliability of various forms of the scale, including unbalanced scales and those lacking a neutral point. The balanced scales with neutral points were generally more reliable, although the differences in correlations were small.

In summary, at least part of the success and longevity of the nine-point hedonic scale has to do with its careful development, and consistency with the abilities and limitations of the human observer as an information processor.

Issues in Good Design and Experimental Control

A second example of a correspondence to perceptual principles can be seen in the reconstruction of tomato paste aroma from a simple mixture of only seven compounds. Using chromatographic procedures, Buttery and colleagues (1989) were able to identify many important compounds in tomato paste aroma. An important assumption in their methods is the idea that compounds that are present below their thresholds were unlikely to contribute substantially to the overall aroma character. Thus of all the various compounds present in the product, seven of those present in levels above their thresholds (as measured individually) were considered for inclusion in a synthetic tomato paste solution. A mixture of dimethyl sulfide, beta-damascenone, 3-methyl butanal, 3-methylbutyric acid, eugenol, methional, and 1-nitro-2-phenylethane was judged by an experienced pannel to be very similar to the aroma of good quality tomato pastes. Concentrations used in the synthetic mixture were guided by the concentrations of the components found in the natural product. While there are some hints that subthreshold compounds may synergize to produce noticeable flavor impact (Day et al., 1963), a useful rule of thumb is still that those compounds present above their sensory thresholds are likely to be important contributors to the overall aroma and flavor of a food. This rule of thumb actually illustrates two principles: that human responses need to be considered in predictive flavor synthesis and that thresholds are useful reference measures.

Principles of good sensory testing can be seen in a study of sensory-instrumental correlations with boar taint due to androstenone. In this study (Thomson and Pearson, 1977), two sensory assessments of boar odor were correlated with instrumental measures of androstenone content. In one assessment, a flexible panel of from three to five packing house personnel rated boar taint intensity on a 0–6 scale, and a consensus value was arrived at for each sample through discussion. In a second assessment, three lab workers, screened for androstenone sensitivity, rated the same samples on a 0–9 scale, under controlled atmosphere conditions, and data were averaged. The instrumental correlation improved from 0.27 (not significantly different from zero) in the first test to 0.47 ($p < 0.05$) in the second test. Several principles common to psychological experiments are evident in the second, more accurate procedure. Screening panelists for sensitivity, a good principle for olfactory assessments in general, suggests an awareness of the specific anosmia for this compound. Keeping the same three panelists constant would help eliminate between-panelist variance introduced by the inclusion of different people at different times as in the packing house assessments. Control of the testing room air is in keeping with a minimization of olfactory adaptation and/or masking, potentially a problem in a packing house atmosphere. Use of a slightly extended category scale takes better advantage of information transmission capacities, as well as helping to avoid the statistical problem of range restriction in regression. Finally, consensus procedures may give undue weight to the opinions of dominant panelists, a potential problem in all small-group interactions, and certainly a potential source of bias in this study, in which one of the packing house evaluators was an official inspector. Statistical averaging, common in most sensory tests, but USDA still avoided in some "expert" evaluation schemes, gives equal weight to all respondents.

Sensory professionals may note a familiar pattern in the first set of odor assessments. Each of the factors mentioned above is a seemingly small alteration in procedure which might be justified on the basis of convenience, time, or cost. Using a floating pool of available panelists, performing tests under convenient but somewhat odiferous (noisy) conditions, avoiding statistics, using a traditional or customary scaling procedure—each of these shortcuts probably changes the pattern of results in only small ways, individually, but the net result in the boar taint study was a potentially different conclusion about the strength of a sensory relationship. In order to perform sensitive sensory analyses, practitioners must guard against such seemingly insignificant degradations in procedural details, and strive to perform the best tests that resources allow.

Issues of experimental design and statistical analysis were cornerstones of the advances produced by the development of Quantitative Descriptive Analysis. This technique, introduced by Stone, Sidel, and colleagues (Stone et al., 1974; Stone and Sidel, 1985) in the early 1970s, offered potential advantages over previous descriptive analysis procedures that were based on consensus methods. While the technique itself has undergone changes and adaptations over the years in the hands of different laboratories and researchers, many of the original advances provided by this procedure remain. Considerations of experimental design were incorporated into the descriptive method, including factorial designs with replications, and repeated-measures or within-subjects analyses, and general use of analysis of variance, all very powerful statistical methods with established history in behavioral research (Rosenthal and Rosnow, 1984). The within-subjects design, in which every panelist judges every product and replicate, allows between-panelist variance to be partitioned, analogous to the added power of a dependent-samples or correlated t-test over an independent-samples t. Thus differences between products can be estimated without the contaminating variance from individual subject idiosyncracies such as tendencies to use different portions of rating scales. Replication was emphasized in this method, which facilitates assessments of judge and panel consistency. Factorial designs allow assessments of statistical interactions, which can also provide insights into panelist agreement concerning how terms are applied to the set of products at hand. Such diagnostic information can be valuable in the selection of attributes, in decisions about the need for further training and in the weight given to particular attributes in conclusions about product differences. The application of statistical and design principles, combined with a sensitivity toward the difficulties in using human observers as measuring instruments, led to the quantitative descriptive procedures being a step forward from previous profiling methods based on consensus.

FUTURE CHALLENGES AND OPPORTUNITIES

Sensory Terminology

Great strides have been made in the quantification of sensory phenomena, and many of the quantitative procedures used in sensory evaluation stand on firm ground. A bit shakier is the question of how the qualitative world of multi-dimensional products is to be carved up an analyzed. Functionally, this problem arises in situations when new descriptive techniques must be developed for a specific product or product category. What attributes are to be chosen that characterize the salient sensory attributes of the product? The descriptive system can be thought of as a type of model, in which words corresponding to per-ceptions provide a quantitative specification of subjective experience. The words are a set of dimensions forming a multidimensional space, and each product represents a point or vector in that set of dimensions. It is important to realize that the sensory specification is only a model, a metaphor of sorts (see Chapter 6) and that the sensory specification can never be equated with the sensations themselves. However, one can assess the utility of the model as well as the utility of individual terms on a set of objective criteria. Broadly speaking, a descriptive system will have several overriding goals—the specification of sensory proper-ties of a product, delineation of points of sensory difference between products, communication of subjective experiences among individuals interested in the products, and facilitation of decisions particularly for research or business directions about further actions to be taken with the products (Meilgaard et al., 1979).

These general concerns about whole descriptive systems are related to more specific criteria for the utility of specific attributes. Some of these criteria are defined by the classical measurement concerns about reliability (repeatability, precision) and validity (prediction, correlation with other meaningful measures). Considering reliability, a term should be used in the same way by an individual upon repeated occasions (intraindividual consistency), be used in the same way by different individuals on a panel (interindividual consistency), and be used by the panel consistently upon subsequent evaluation sessions or experimental replications (consistency of panel mean scores across replicates). Considering the validity issue, one can ask whether a term in a trained panel analysis relates in any way to the acceptability judgments or to consumption measures or perhaps even to purchase decisions of consumers. Another criterion is psychophysical validity in the sense of correlations with physical product attributes such as ingredients or instrumental assessments. Another criterion is how the term is learned by the panel, i.e., the facility with which they become calibrated or aligned as to their concept of what is meant by the term (see Chapter 8). Is a physical reference standard readily available, or a verbal definition referring to common experience that all can agree upon? How long does it take the panel to become trained as to the use of the attribute? Finally, does the term have communication value to people outside the technical community, or is it likely to be misunderstood or require unwieldy definition?

Lack of overlap among attributes or orthogonality is another criterion. Often, synonymous terms are deleted at the discretion of the panel leader or following discussion among the panelists. A desirable evaluation scheme is one in which scales are not redundant, i.e., are minimally correlated. Principal component analysis or factor analysis can be used to examine the pattern of intercorrelation and suggest ways in which attribute lists can be reduced. A desirable goal is to keep the evaluation ballot as short as possible, while still covering all the important attributes that differentiate the product category. Keeping the evaluation as streamlined as possible improves panelist motivation and ability to concentrate. In practice, a balance is usually struck between the need for parsimony and the need to provide a complete and detailed description of all the sensory nuances of a product (see Chapter 12). The nature of the product and the business objectives of the evaluation may influence the degree of sophistication of the terminology system. Lyon (1987) demonstrated the use of multiple criteria in term selection, including significance testing, frequency of panel usage, multivariate analyses, and professional judgment in the development of a ballot for attributes describing warmed-over flavor in chicken meat. This case study showed that terminology development can be a multi-stage process involving several criteria to evaluate the evaluation itself.

Closely related to the reliability issue, which is after all a question of error variance, is the value of the term in discriminating among products. Low error variance leads to high test sensitivity, all other factors such as panel size being equal. Testing for significant differences is perhaps the most common statistical analysis in sensory practice. Test sensitivity can be described as a question of discriminative power, which is a function of the products selected for the test, the error variance, and the sample size or number of panelists. Using probability levels as indicators of statistical discrimination is common, for example, in retaining attributes during terminology development that differentiate products at $p < 0.05$ or less. However, this criterion must be applied with care and a good measure of professional judgment since the p-values will depend not only on the standard deviations (i.e., the error) of the data, but the number of panelists and the particular samples chosen for the test. Table 1.4 shows various measures that correct for sample size that are used in the techniques of meta-analysis (Rosenthal, 1987). Meta-analysis is a popular procedure in behavioral research in which results from different studies can be compared or combined. It is useful in the estimation of effect sizes based upon a body of published literature in which different methods and sample sizes were employed. Many of these measures are related to the difference between sample means divided by standard deviations (not standard errors), or to conversion of t values to Pearson's r. These have a useful property in that they can be compared statistically. For example, if effect sizes are converted to Person's rs, the rs can then be compared after conversion to Fisher zs. This could allow a conclusion, for example, that one analysis method or comparison was (statistically) significantly more sensitive to product differences, having produced a significantly larger effect size. Another simple

TABLE 1.4   Effect Size Indicators Used in Meta-Analysis

Standardized Differences Between Means:	
Cohen's d	$(M_1 - M_2) / S_{pooled}$
	where S = standard deviation, N as divisor
Glass's delta	$(M_1 - M_2) / S_{control\ group}$
	where S = standard deviation, N-1 as divisor
Hedge's g	$(M_1 - M_2) / S_{pooled}$
	where S = standard deviation, N-1 as divisor
Transformations of t:	
Rosenthal's r	$[t^2 / (t^2 + df)]^{1/2}$

statistic usable with individual products is the coefficient of variation, which is the standard deviation divided by the mean. This takes into account the general trend for Weber's Law, that more intense stimuli tend to be more variable, and so adjusts for the overall intensity level.

However, even the meta-analytic measures still depend upon one fortuitous variable, namely the products chosen for the test. A particular set of samples may simply not differ with regard to the attribute in question. One danger is that later investigations may include products that do differ on that scale—this demands a degree of foresight, knowledge of the product category, and professional judgment. Another approach is to eliminate the mean differences from consideration and simply compare standard deviations. A general rule of thumb is that standard deviations for trained panels should be less than 20% of scale range, and preferably in the range of 10%, on the average. For example, standard deviations should fall in the range of about 0.75–1.5 scale points on a nine-point category scale, 15–30 mm on a 150-mm line scale, to demonstrate that a panel is using an attribute with acceptable consistency. Even this measure, however, is not without its pitfalls. Some attributes are simply "easier" to use or less noisy than others. Odor intensity is notoriously variable, as evidenced by the large difference thresholds for smell. Visual attributes, on the other hand, are usually easily discriminable by panelists. Furthermore, standard deviations are often lower when products fall at the ends of bounded scales, i.e., are very high or very low in that attribute, than when the product is intermediate in intensity (e.g., Lawless, 1989). Once again, fortuitous selection of products that are extreme on the scale might yield low variance and produce a false sense of security.

In summary, many of these criteria can be used to measure the utility of a term after it has been generated. What is lacking in the sensory community is some consensus on the procedures by which attributes should be generated in the first place. Several philosophies exist, including a search of the literature (Piggott and Canaway, 1981), group discussion among panelists with the range of products at hand (Stone and Sidel, 1985), recourse to expert knowledge of product attributes or defects (Bodyfelt et al., 1988), or even direction from consumers. The safest path may be one that uses a variety of sources for the initial suggestion of terms. Two procedures are currently in vogue. One common procedure is to have group discussions, either among the descriptive panelists themselves as an official part of the ballot development in a quantitative descriptive procedure or of other observers, similar to a traditional focused group discussion used in exploratory market research (Marlow, 1987) (see also Chapter 14). A second technique that has been recently employed is a variation of the repertory grid method used for studying personal construct systems in psychology. In this procedure, representative samples are presented in groups of three (triads), and subjects are asked to describe the ways in which any two are similar and may be distinguished from a third. The oddest sample is then eliminated and another substituted; triads continue to be changed in this fashion until the product set is exhausted. This procedure successfully elicits terms from individuals that may then be used for the construction of idiosyncratic sets of rating scales, as in free-choice profiling (McEwan and Thomson, 1988). This latter technique purports to avoid the need for panel standardization entirely by providing subjects with their own terminology. Another version of this technique involves having each subject sort the sample set into groups and describe the ways in which each subgroup differs (Steenkamp and Van Trijp, 1988). Further division of subgroups continues, generating more scales on the basis of reported rationales, until no further division seems logical to the subject, i.e., when the groups are homogeneous in all sensory respects. Following this individualized term generation, products are rated and the multivariate specifications can be subjected to Generalized Procrustes Analysis to arrive at a common configuration (Marshall and Kirby, 1988; McEwan et al., 1989). These methods are as yet in their adolescence in terms of practical development and application and their advantages relative to other procedures and potential pitfalls are not yet clear (see Chapter 13).

Another psychological literature that has been brought to bear on the question of sensory terminology concerns concept learning and categorization (see Chapter 8). Concept learning helps break down the term-learning and term-usage process into substages. Stages include exposure to good examples for the terms and the differentiation of good examples from poor examples and from nonexamples, i.e., the learning of prototypies and category boundaries. A practical issue is whether one or multiple reference standards are optimal for concept learning. By "optimal" we can apply the criteria of speed of learning, degree of retention, and the precision with which the term is used. Another question is whether sensory terms should be organized as hierarchies where general categories subsume more subtle or more specific differentiations (Lawless, 1988). Hierarchical schemes can help the researcher escape pseudo-problems such as the question of whether sweetness is a basic taste or whether there are multiple qualities of sweetness. Those two propositions can both be true within an hierarchical model. Also, different levels of the hierarchy may be applicable at different points in panel training. As expertise grows, finer differentiations become possible. Also, panelists can learn to use superordinate groupings (broader, more inclusive terms) to specify the relationships of previously unconnected items.

Individual Differences

Other major opportunities concern the understanding of how people differ (see Chapter 10), particularly in the sense of smell. Individual differences in olfactory function have received a great deal of recent experimental study. Aging and clinical pathologies both can cause general deficits in olfactory perception (Stevens and Cain, 1987). These are general deficits in the sense that they seem to affect a variety of compounds tested. In contrast are the specific anosmias (Amoore, 1977), in which a person of otherwise normal olfactory acuity belongs to a group with a measured threshold over 2 standard deviations above the "normal" population mean. Population distributions of thresholds for such odors resemble the classical bimodal distributions for PTC tasting and non-tasting. A variety of these have been documented, including anosmias to the malty odor of isobutyraldehyde, the camphoraceous odor of cineole, the fishy odor of trimethylamine, among others (Wysocki and Labows, 1984). Recently, Marin (1989) used CHARM analysis to explore the relationship of age and gender to olfactory responsiveness to several fruity esters and minty compounds. While the small number of subjects in each group makes generalization difficult from these data, there were major differences among subjects in their reactions to some of the minty compounds, an effect that could be related to the specific anosmia for carvone (Pelosi and Viti, 1978).

Some anosmias have been studied across families of related compounds to determine which chemicals may be most potent in producing differences between anosmic and normal groups (Amoore, 1977; Wysocki and Labows, 1984). Correlations between various documented anosmias are also being examined in order to suggest linkages or commonalities in physiological mechanisms (see Chapters 4 and 10). As noted by O'Connell (in Chapter 4), the physiological basis for specific anosmia remains unclear and could occur as a malfunction in various parts of the olfactory reception and transduction mechanisms. Simple absence of a receptor site is unlikely to be the whole story. This is shown in the literature on the most thoroughly studied anosmia, the deficit in sensitivity to androstenone. Wysocki, Beauchamp, and colleagues have demonstrated that this individual characteristic has a strong genetic component as measured in twin studies, that there is a developmental trend insofar as children are rarely anosmic but the frequency of the deficit increases after puberty, especially among males and that there is environmentally influenced flexibility—exposure to the compound induces increased sensitivity (Wysocki and Beauchamp, 1984; Dorries et al., 1989; Wysocki et al., 1989).

If stable differences in olfactory sensitivity are common, they have several implications for the perception of fragrances, aroma materials, and volatile flavors in food. Some people are anosmic to the malty aroma of isobutyraldehyde (Amoore et al., 1976), a common flavor material. For example, a malty defect in fluid milk can occur from infestations of *Streptococcus lactis* v. *maltigenes,* which produces isobutyraldehyde and related compounds (Bodyfelt et al., 1988). Amoore himself suggested that the anosmic defect might influence the perceptions of dairy judges in recognizing this defect. Another dairy defect arises from hydrolytic racidity, characterized by the production of short-chain free fatty acids. One of the first well-documented specific anosmias was that to isovaleric acid (Amoore et al., 1968). Of interest here is that other workers in dairy science have recently examined thresholds to a wide array of branched-chain free fatty acids, and noted that some subjects were almost completely unresponsive to some compounds (Brennand et al., 1989). Returning to the anosmia to androstenone, this compound is a strong contributor to boar taint in pork, which is of considerable commercial significance (Thompson and Pearson, 1977).

Specific anosmia could influence food flavor perception in a number of ways. In addition to the simple detection of an off-flavor such as a malty defect in milk or boar taint in pork, the presence of these compounds may also interact with other flavor materials that are present. An inhibitory interaction is common among tastes and among odors, usually referred to as mixture suppression in taste and odor counteraction in olfaction (Lawless, 1986). However, individuals who are unresponsive to an odor may show no inhibition of the other flavors, which should therefore appear stronger. As noted above, such a relationship was found for sucrose and PTC mixtures (Lawless, 1979b). Tasters of PTC who perceived the bitterness shows suppression of sucrose sweetness in mixtures, while nontasters perceived sucrose to be the same intensity regardless of whether PTC was present or not. Similarly, anosmics may overrepresent their perceptions of other flavor notes, since these other notes are not inhibited by the compounds to which they are unresponsive. Another possible interaction is a qualitative rather than intensive shift in aroma or flavor perception. These possibilities warrant further study and would help determine whether specific anosmia is a merely a physiological curiosity or a phenomenon of any practical importance.

If several experts are asked to describe the flavor of a wine, it is not uncommon to have drastically different volatile notes stressed by different individuals (Lawless, 1984). Certainly part of this lack of agreement arises in the lack of calibration among winetasters in their usage of flavor and aroma terminology (see Chapter 9). However, it is also possible that individual differences in the mosaic of olfactory sensitivities among people could also contribute. For those involved in sensory evaluation, the existence of anatomically based individual differences has implications for how people should be selected for panel training and how that training should proceed. In the case of dairy work, subjects who are anosmic to malty flavors or lacking in sensitivity to rancid defects may be screened from panels that are likely to come across those flavor notes. Similarly, it would be questionable to employ a person as a pork inspector if that individual was anosmic to androstenone. Since the occurrence of compounds to which subjects are anosmic cannot always be predicted, and since the type and number of specific anosmias in human olfaction is still largely unknown, a certain amount of individual differences in panelist perceptions may have to be tolerated in the day-to-day functioning of sensory panels. A common experience of people participating in panel training is that some attributes are difficult to learn or understand, while other terms are easy to grasp. Also, what is easy for one participant may be difficult for another and vice versa. Individual differences in olfactory anatomy and physiology may be at the root of such differences. This also suggests that attempts to calibrate a panel beyond a certain point may be futile or at least not cost-efficient. In other words, there may be physiological limits on the degree to which panel consensus can be enforced through training.

There is one finding that challenges the potential impact of specific anosmias, however. Stevens et al. (1988) measured olfactory thresholds over 20 occasions for three healthy young observers for three compounds. While mean thresholds across occasions were quite consistent among the three observers, each of them showed a huge range in thresholds to each compound, generally about 2000-fold in concentration. This demonstrates that there can be tremendous variability within an individual in measured olfactory sensitivity. Furthermore, it suggests that specific anosmia and other findings that demonstrate a wide ranges in olfactory sensitivity (e.g., Punter, 1983) may be due to *intra*individual variability. While the findings of Stevens et al. (1988) do not rule out the possibility of stable individual differences to other compounds that were not part of their experiment, they do suggest that a convincing demonstration of specific anosmia will require much more extensive testing of an individual than a single threshold assessment on one day. Of possible significance in this study was the finding that the odorant to which subjects were most sensitive, phenylethylmethylethyl carbinol (PEMEC), was one which did show a reliable difference among the three subjects. The two other (less potent) stimuli, butanol and pyridine, showed greater consistency. This suggests that wider individual differences may be observed in those compounds that have high odor impact.

Toward a Psychophysics of Food Perception

As stated in the introduction, a common opinion is that sensory evaluation is a collection of methods for testing products, rather than for understanding variables. However, understanding psychophysical relationships in a product system can have profound consequences. For example, knowing something about the discriminability of different concentration levels and understanding the nature of the dose-response curve for a given ingredient has many applications. The difference threshold tells us about recipe tolerance in manufacturing an quality control. The slope of the psychophysical function (or its curvature) tells us how fast responses change with concentration. It gives us a variable (perceived intensity) with which to model relationships to consumer acceptance. As noted above, acceptance as a direct function of ingredient concentrations can be a weak relationship filled with pitfalls if perceived intensity is not considered. Finally, such knowledge has applications in product research and development as well. The dose-response curve provides reference data for reformulations, e.g., ingredient reductions in the face of escalating costs, or even ingredient reductions to create a consumer-perceived "light" product. Knowledge of the dose-response curve (or of JNDs) can suggest to a formulator the degree of change in physical ingredient level that is needed to evoke a perceived change to the human observer.

At this time, it is difficult to conceive of a psychophysical compendium that would describe the behavior of every food ingredient under every possible condition of processing and in every possible combination with other ingredients. It remains to be seen whether sensory evaluation matures into a discipline with the quantitative precision of engineering and the cumulative knowledge base of science in general, or whether it remains a collection of technical methodologies. In the meantime, sensory scientists and evaluation practitioners must continue to explore links between scientific findings and practical product applications. Evaluation specialists should benefit from the development of more sensitive, more predictive test procedures which take into account the sensory and perceptual functioning of human observers. Conversely, sensory scientists may benefit from exposure to natural phenomena that suggest heretofore unknown perceptual principles or even underlying physiological mechanisms.

REFERENCES

Acree, T. E., Barnard, J., and Cunningham, D. G. 1984. A procedure for the sesory analysis of gas chromatographic effluents. *Food Chem.* 14: 273.

Amoore, J. E. 1977. Specific anosmia and the concept of primary odors. *Chem. Senses* 2: 267.

Amoore, J. E., Forrest, L. J., and Pelosi, P. 1976. Specific anosmia to isobutyraldehyde: The malty primary odor. *Chem. Senses Flav.* 2:17.

Amoore, J. E., Venstrom, D., and Davis, A. R. 1968. Measurement of specific anosmia. *Percep. Mot. Skills* 26: 3.

ASTM. 1978. *Compilation of Odor and Taste Threshold Values Data*. American Society for Testing and Materials, Philadelphia.

Azevedo, E., Krieger, H., Mi, M. P., and Morton, N. E. 1965. PTC taste sensitivity and endemic goiter in Brazil. *Am. J. Hum. Gen.* 17: 87.

Bartoshuk, L. M. 1968. Water taste in man. *Percep. Psychophys.* 3: 69.

Bartoshuk, L. M. 1979. Bitter taste of saccharin related to the genetic ability to taste the bitter substance 6-n-propylthiouracil. *Science* 205: 934.

Blakeslee, A. F. 1932. Genetics of sensory thresholds: Taste for phenylthiocarbamide. *Proc. Nat. Acad. Sci. USA* 18: 120.

Bodyfelt, F. W., Tobias, J., and Trout, G. M. 1988. *The Sensory Evaluation of Dairy Products,* 2nd edition. AVI/Van Nostrand Reinhold, New York.

Brennand, C. P., Ha, J. K., and Lindsay, R. C. 1989. Aroma properties and thresholds of some branched-chain and other minor volatile fatty acids occurring in milkfat and meat lipids. *J. Sensory Stud.* 4:105.

Brinberg, D. and MacGrath, J. E. 1985. *Validity and the Research Process*. Sage, Beverly Hills.

Buttery, R. G., Teranishi, R., Flath, R. A., and Ling, L. C. 1989. Fresh tomato volatiles. In *Flavor Chemistry,* ACS Symposium Series 388 (R. G. Buttery and R. Teranishi, Ed.), p. 213. American Chemical Society, Washington, DC.

Civille, G. V. and Lawless, H. T. 1986. The importance of language in describing perceptions. *J. Sensory Stud.* 1: 203.

Cohen, J. and Ogdon, D. P. 1949. Taste blindness to phenyl-thio-carbamide and related compounds. *Psych. Bull.* 46: 490.

Cox, E. P. 1980. The optimal number of response alternatives for a scale: A review. *J. Mkt. Res.* 17: 407.

Day, E. A., Lillard, D. A., and Montgomery, M. W. 1963. Autoxidation of milk lipids. III. Effect on flavor of the additive interactions of carbonyl compounds at subthreshold concentrations. *J. Dairy Sci.* 46: 291.

Desor, J. A. and Beauchamp, G. K. 1974. The human capacity to transmit olfactory information. *Percep. Psychophys.* 16: 551.

Dorries, K. M., Schmidt, H. J., Beauchamp, G. K., and Wysocki, C. J. 1989. Changes in sensitivity to the odor of androstenone during adolescence. *Develop. Psychobiol.* 22: 423.

Ennis, D. M. 1990. Relative power of difference testing methods in sensory evaluation. *Food Technol.* 44: 114.

Fischer, R. and Griffin, F. 1963. Quinine dimorphism: A cardinal determinant of taste sensitivity. *Nature* 200: 343.

Fischer, R., Griffin, F., England, S., and Garn, S. M. 1961. Taste thresholds and food dislikes. *Nature* 191: 1328.

Fox, A. L. 1932. The relationship between chemical constitution and taste. *Proc. Nat. Acad. Sci. USA* 18: 115.

Frank, R. A. and Byram, J. 1988. Taste-smell interactions are tastant and odorant dependent. *Chem. Senses* 13: 445.

Frank, R. A., Ducheny, K., and Mize, S. J. S. 1989. Strawberry odor, but not red color enhances the sweetness of sucrose solutions. *Chem Senses* 14: 371.

Frank, R. A., Wessel, N., and Shaffer, G. 1990. The enhancement of sweetness by strawberry odor is instruction-dependent. Paper presented at the 1990 Association for Chemoreception Sciences, Sarasota, FL.

Gent, J. F. and Bartoshuk, L. M. 1983. Sweetness of sucrose, neohesperidin dihydrochalcone and sacchain is related to genetic ability to taste the bitter substance 6-*n*-propylthiouracil. *Chem. Senses* 7: 265.

Greene, L. S. 1974. Physical growth and development, neurological maturation and behavioral functioning in two Ecuadorian Andean communities in which goiter is endemic. *Am. J. Phys. Anthrop.* 41:139.

Hall, M. L., Bartoshuk, L. M., Cain, W. S., and Stevens, J. C. 1975. PTC taste blindness and the taste of caffeine. *Nature* 253: 442.

Harris, H. and Kalmus, H. 1949. Chemical specificity in genetical differences of taste sensitivity. *Ann. Eugen.* 15: 32.

Harris, H., Kalmus, H., and Trotter, W. R. 1949. Taste sensitivity to phenyl-thiourea in goitre and diabetes. *Lancet* 257: 1038.

Hoover, E. F. 1956. Reliability of phenylthiocarbamide-sodium benzoate method of determining taste classifications. *J. Agr. Food Chem.* 4: 345.

Jones, L. V. and Thurstone, L. L. 1955. The psychophysics of semantics: An experimental investigation. *J. Appl. Psychol.* 39: 31.

Jones, L. V., Peryam, D. R., and Thurstone, L. L. 1955. Development of a scale for measuring soldier's food preferences. *Food Res.* 20: 512.

Lawless, H. T. 1977. The pleasantness of mixtures in taste and olfaction. *Sensory Proc* 1: 227.

Lawless, H. 1979a. The taste of creatine and creatinine. *Chem. Senses Flav.* 4: 249.

Lawless, H. 1979b. Evidence for neural inhibition in bittersweet taste mixtures. *J. Comp. Physiol. Psychol.* 93: 538.

Lawless, H. T. 1984. Flavor description of white wine by "expert" and non-expert wine consumers. *J. Food Sci.* 49: 120.

Lawless, H. T. 1986. Sensory interactions in mixtures. *J. Sensory Stud.* 1: 259.

Lawless, H. T. 1989. Logarithmic transformation of magnitude estimation data and comparisons of scaling methods. *J. Sensory Stud.* 4: 75.

Lawless, H. T. In press. The sense of smell in food quality and sensory evaluation. *J. Food Qual.*

Lawless, H. T. and Klein, B. P. 1989. Academic vs. industrial perspectives on sensory evaluation. *J. Sensory Stud.* 3: 205.

Lawless, H. T. and Malone, J. G. 1986a. The discriminative efficiency of common scaling methods. *J. Sensory Studies* 1: 85.

Lawless, H. T. and Malone, G. J. 1986b. A comparison of scaling methods: Sensitivity, replicates and relative measurement. *J. Sensory Studies* 1: 155.

Lawless, H. and Schlegel, M. P. 1984. Direct and indirect scaling of sensory differences in simple taste and odor mixtures. *J. Food Sci.* 49: 44.

Lyon, B. G. 1987. Development of chicken flavor descriptive attribute terms aided by multivariate statistical procedures. *J. Sensory Stud.* 2: 55.

Marlow, P. 1987. Qualitative research as a tool for product development. *Food Technol.* 41: 74.

Marin, A. B. 1989. Evaluating odor perception by charm and sensory analysis. Ph.D. thesis, Cornell University, Ithaca, NY.

Marin, A. B., Acree, T. E., Hotchkiss, J., and Nagy, S. In press. Interaction between limonene and odor active volatiles from orange juice by packaging materials. *J. Agr. Food Chem.*

Marshall, R. J. and Kirby, S. P. J. 1988. Sensory measurement of food texture by free-choice profiling. *J. Sensory Stud.* 3: 63.

McBurney, D. H. 1966. Magnitude estimation of the taste of sodium chloride after adaptation to sodium chloride. *J. Exp. Psychol.* 72: 869.

McBurney, D. H., Smith, D. V., and Shick, T. R. 1972. Gustatory cross-adaptation: Sourness and bitterness. *Percep. Psychophys.* 11: 228.

McEwan, J. A. and Thomson, D. M. H. 1988. An investigation of the factors influencing consumer acceptance of chocolate using the repertory grid method. In *Food Acceptability* (D. M. H. Thomson, Ed.), p. 347. Elsevier Applied Science, London.

McEwan, J. A., Colwill, J. S., and Thomson, D. M. H. 1989. The application of two free-choice profile methods to investigate the sensory characteristics of chocolate. *J. Sensory Stud.* 3: 271.

Meilgaard, M., Civille, G. V., and Carr, B. T. 1987. *Sensory Evaluation Techniques,* Vol. II, p. 1. CRC Press, Boca Raton, FL.

Meilgaard, M. C., Dalgliesh, C. E., and Clapperton, J. F. 1979. Beer flavor terminology. *Am. Soc. Brew, Chem. J.* 37: 47.

Meiselman, H. L. and Halpern, B. P. 1973. Enhancement of taste intensity through pulsatile stimulation. *Physiol. Behav.* 11: 713.

Miller, G. A. 1956. The magical number seven, plus or minus two: Some limitations on our capacity for processing information. *Psychol. Rev.* 63: 81.

Moskowitz, H. R. 1983. *Product Testing and Sensory Evaluation of Foods.* Food and Nutrition Press, Westport, CT.

Murphy, C. and Cain, W. S. 1980. Taste and olfaction: Independence vs. interaction. *Physiol. Behav.* 24: 601.

Murphy, C., Cain, W. S., and Bartoshuk, L. M. 1977. Mutual action of taste and olfaction. *Sensory Proc.* 1: 204.

O'Mahony, M. 1986. Sensory adaptation. *J. Sensory Stud.* 1: 237.

O'Mahony, M. and Odbert, N. 1985. A comparison of sensory difference testing procedures: Sequential sensitivity analysis and aspects of taste adaptation. *J. Food Sci.* 50: 1055.

O'Mahony, M. and Ishii, R. 1986. Umami taste concept: Implications for the dogma of four basic tastes. In *Umami: A Basic Taste* (Y. Kawamura and M. R. Kare, Ed.), p. 75. Marcel Dekker, New York.

O'Mahony, M. and Goldstein, L. 1987. Tasting successive salt and water stimuli: The roles of adaptation, variability in physical signal strength, learning, supra- and subadapting signal detectability. *Chem. Senses* 12: 425.

Parducci, A. 1965. Category judgment: A range-frequency model. *Psych. Rev.* 72: 407.

Pelosi, P. and Viti, R. 1978. Specific anosmia to l-carvone: The minty primary odour. *Chem. Senses Flav.* 3: 331.

Perng, C. M. and McDaniel, M. R. 1989. Optimization of a blackberry juice drink using response surface methodology. Presented at the 49th Annual Meeting of the Institute of Food Technologists, Chicago, IL.

Piggot, J. R. and Canaway, P. R. 1981. Finding the word for it: Methods and uses of descriptive sensory analysis. In *Flavour '81,* p. 33. Walter de Gruyter, Berlin.

Punter, P. H. 1983. Measurement of human olfactory thresholds for several groups of structurally related compounds. *Chem. Senses* 7: 215.

Rainey, B. A. 1986. Importance of reference standards in training panelists. *J. Sensory Stud.* 1: 149.

Rosenthal, R. 1987. *Judgment Studies: Design, Analysis and Meta-analysis.* Cambridge University Press, Cambridge (U.K.).

Rosenthal, R. and Rosnow, R. L. 1984. *Essentials of Behavioral Research: Methods and Data Analysis.* McGraw-Hill, New York.

Schutz, H. G. 1988. Beyond preference: Appropriateness as a measure of contextual acceptance of food. In *Food Acceptability* (D. M. H. Thomson, Ed.), p. 115. Elsevier Applied Science, London.

Shaffer, G. and Frank, R. A. 1990. An investigation of taste-smell interactions across four tastant and six odorants. Presented at the 1990 Association for Chemoreception Sciences, Sarasota, FL.

Shapiro, E. 1990. New products clog groceries. *New York Times,* May 29, D1, D17.

Steenkamp, J-B. E. M. and Van Trijp, H. C. M. 1988. Free-choice profiling in cognitive food acceptance research. In *Food Acceptability* (D. M. H. Thomson, Ed.), p. 363. Elsevier Applied Science, London.

Stevens, J. C. and Cain, W. S. 1987. Old-age deficits in the sense of smell as gauged by thresholds, magnitude matching and odor identification. *Psych. Aging* 2: 32.

Stevens, J. C., Cain, W. S., and Burke, R. J. 1988. Variability of olfactory thresholds. *Chem. Senses* 13: 643.

Stone, H. and Sidel, J. L. 1985. *Sensory Evaluation Practices.* Academic Press, New York.

Stone, H., Sidel, J., Oliver, S., Woolsey, A., and Singleton, R. C. 1974. Sensory evaluation by quantitative descriptive analysis. *Food Technol.* 28: 24.

Thompson, R. H. and Pearson, A. M. 1977. Quantitative determination of 5 androst-16-en-3-one by gas chromatography-mass spectrometry and its relationship to sex odor intensity of pork. *J. Agric. Food Chem.* 25: 1241.

Vie, A. and O'Mahony, M. 1989. Triangular difference testing: Refinements to sequential sensitivity analysis for predictions for individual triangles. *J. Sensory Stud.* 4: 87.

von Sydow, E., Moskowitz, H., Jacobs, H., and Meiselman, H. 1974. Odor-taste interactions in fruit juices. *Lebensm.-wiss. u. Technol.* 7: 18.

Williams, A. A. 1988. Procedures and problems in optimizing sensory and attitudinal characteristics in foods and beverages. In *Food Acceptability* (D. M. H. Thomson, Ed.), p. 297. Elsevier Applied Science Publishers, London.

Wiseman, J. J. and McDaniel, M. R. 1989. Modification of fruit flavors by aspartame and sucrose. Presented at the 49th Annual Meeting of the Institute of Food Technologists, Chicago, IL.

Wysocki, C. J. and Beauchamp, G. K. 1988. Ability to smell androstenone is genetically determined. *Proc. Nat. Acad. Sci. USA* 81: 4899.

Wysocki, C. J. and Labows, J. 1984. Individual differences in odor perception. *Perfum. Flavorist* 9: 21.

Wysocki, C. J., Dorries, K. M., and Beauchamp, G. K. 1989. Ability to perceive androstenone can be acquired by ostensibly anosmic people. *Proc. Nat. Acad. Sci. USA* 85: 7976.

2

More than Meets the Tongue: Temporal Characteristics of Taste Intensity and Quality

Bruce P. Halpern

Cornell University
Ithaca, New York

INTRODUCTION

The Study of Human Taste Judgments

Two quite different traditions have developed with regard to research on human taste judgments. One approach, frequently used in psychological and physiological studies, assumes that a single numerical description of the intensity of a taste stimulus or a single-word description of the quality of a taste stimulus will adequately, if not completely, represent all significant aspects of the perceived taste. This approach commonly uses stimuli consisting of one molecular species, in aqueous solution. Responses for taste quality are often restricted to the words "salty," "sweet," "sour," and "bitter," with "other" sometimes also permitted. Implicit assumptions of this approach are that: (1) a very close relationship exists between events at taste receptors and taste perception, with little contribution from postreceptor central nervous system processing; (2) events distributed over time are not significant for taste perception, except for reductions in perceived intensity and increases in threshold subsequent to the initial perception (Gent, 1979; Naim et al., 1986); and (3) taste perception can be understood by using a very limited set of single-component solutions that correspond to, and will consistently evoke, the limited set of permitted taste quality descriptions. This gustatory research tradition represents the application of classical experimental psychology paradigms to taste perception (e.g., Bartoshuk, 1988; Land and Shepherd, 1988). Stimuli that are thought to be "physically simple" are used to evoke responses that will be as consistent as possible, both within and between subjects.

A second approach to the study of human taste perception differs from that outlined above in several aspects. These differences include the response measures used, stimuli employed, explicit and implicit assumptions, and context from which the research tradition grew. This approach assumes that taste perception is an ongoing process that can, and probably should, be studied by obtaining successive response measures over time. Both taste intensity and taste quality are conceived of as having temporal aspects that are significant, and sometimes crucial, factors in perceived taste (e.g., Beck, 1975; Lawless and Skinner, 1979; Pangborn et al., 1983). Stimuli may be mixtures and are sometimes multiple component beverages or foods (Lee and Pangborn, 1986; Powers, 1988). The physiology of taste is often of limited interest, and few assumptions may be made concerning its nature. However, if physiological interpretations are made, they generally assume a close relationship between events at taste receptors and human taste perception (e.g., Birch et al., 1980; DuBois and Lee, 1983; Guinard et al., 1986b; Overbosch, 1986; O'Mahony and Wong, 1989).

In common with the first approach, taste quality responses are sometimes limited to the words "salty," "sweet," "sour," "bitter," and "other," but more extensive vocabularies, or nonverbal measures such as sorting, are also used. Intensity of taste perception may be measured by nonverbal tracking over time as well as by single-report numerical judgments. This taste research tradition developed from the emprical needs and insights of food scientists. Stimuli are, or are selected to resemble, foods or beverages that might be tasted and consumed by humans (e.g., Guinard et al., 1986a,b). Response measures focus on procedures that will characterize, or distinguish between, such foods and beverages. The use of multiple component, "physically complex" stimuli and relatively rich response measures causes this approach to be potentially compatible with cognitive emphases in psychology, with a focus on ecological relevance, and with investigations of the role of central nervous system processing in taste perception. However, because this tradition arose from empirical roots that are separate from cognitive or neural science, it is often unknown in the latter domains.

The present review will focus on the more ecologically relevant approach to human taste perception just outlined. Measurements of taste perception over time that have been gathered in several laboratories will be emphasized. Implications of these data for the nature of human gustatory perception and the apparent dominance of central nervous system events in human taste processing (Halpern, 1986b) will be noted.

Patterns and Principles

Before taste responses are considered in a later part of this chapter, we will provide very brief coverage of temporal factors in nongustatory systems. There are several reasons for addressing nongustatory systems. One purpose is to indicate that temporal patterns are found in most if not all sensory systems. Another goal of the nongustatory segment is to present at a general level several basic concepts that will subsequently be regularly employed (see Perceptual Responses). For example, a distinction is made between two broad categories of responses: discrete responses and tracking responses. This distinction will be used throughout the remainder of the chapter. Also introduced is a separation of perceptual judgments into the domains of intensity judgments and quality descriptions. It too will repeatedly reappear. Finally, an important psychophysical method, magnitude estimation, is first encountered in the section about Perceptual Responses. The method will be operationally defined using standard literature sources. This approach to relevant methodology will be continued when taste data that were obtained using other psychophysical methods are considered.

Representative Samples Versus Experts

Psychophysical procedures are designed to measure relationships between environmental energy patterns (stimuli) and behavioral responses to these stimuli (Engen, 1971a; Matlin, 1988; Levine and Shefner, 1991). When these procedures are applied to humans, the response may be linguistic, such as a numerical judgment or a description, or may be nonlinguistic motor activities, for example, moving a joystick in order to draw a picture of intensity over time.

These procedures may be used to obtain two quite different types of information about human judgments. One type of information represents the "typical," central tendency (e.g., mean or median) response of a specified general class of people. The class may be as broad as all humans or as narrow as individuals of a particular cultural background living in a certain neighborhood of a particular city. In order to obtain the desired information, one must measure responses from individuals who are representative of the population of humans who are of interest. Futhermore, measurements must be made from a relatively large number of these people, so that the final central tendency obtained will be likely to be a fair estimate of that population.

A different goal would be to attempt to study the most sensitive, rapid, or consistent responses that humans, or a specified class of humans, are capable of giving to the stimuli of interest. To accomplish this goal, some preliminary testing or screening will often be done. Only those individuals who meet specified performance criteria will be studied further. Those who meet these criteria will frequently receive additional training, and then may participate in the experiment for a long period of time. These individuals have become experts in using the specified psychophysical procedure to make judgments concerning the stimuli. Under these circumstances, a very small number of people are studied, sometimes as few as two or three. The results of such experiments indicate what is possible; they establish limits for humans or for a specified class of humans. One would expect that these judgments would be similar to those of few of the individuals who would be included in the type of representative sample outlined in the previous paragraph. However, the experts would represent one extreme when compared to the general population. Since the experts were not only selected for their judgment ability but also received extensive training, their responses may be outside the limits of any members of the untrained population.

The psychophysical data presented in this chapter are derived from experiments that used experts. In some cases, "expert" may be too strong a term, since the subjects were not screened for their perceptual ability and may have received only limited training. However, in almost all cases the experiments did not attempt to fairly represent the general characteristics of a specified population, did not test very large numbers of individuals, and did obtain many judgments from a few individuals. Consequently, as discussed above, these data should be taken as indications of what is possible for human observers, rather than as typical values that would often be found in a broad survey.

TEMPORAL FACTORS IN NONGUSTATORY SYSTEMS

Many investigations of temporal factors in neural or perceptual responses from nongustatory systems have been reported. This vast and interesting literature cannot be presented or adequately characterized in the present chapter. A few studies that indicate the breadth of effects will be mentioned. The primary purposes of this section are to demonstrate that study of the temporal aspects of sensory and perceptual phenomena has not been limited to taste, and to introduce some of the general concerns and concepts that will subsequently be used to organize gustatory observations.

In selecting experiments to consider, some emphasis was given to responses to cutaneous stimulation. This was done because cutaneous receptors and central nervous system pathways and processing appear to have a similarity to those in the gustatory systems of primates.

Neural Responses

The temporal characteristics of both stimuli and responses are of general importance across sensory systems. With regard to stimuli, rate of change in intensity is often a significant parameter. One reason is that sensory systems are especially sensitive to the dynamic aspects of inputs (Husmark and Ottoson, 1971; Pierau and Wurster, 1981; Hensel, 1982; Halpern, 1983; Pubols and Pubols, 1983; Shepherd, 1988). However, only a certain range of rates of change may be optimal, with both higher and lower rates producing either lesser responses or none at all (Darian-Smith, 1984; Iggo, 1984; Mountcastle, 1967; Pickles, 1988). In some instances, such as responses of certain central nervous system neurons to moving cutaneous stimuli (Whitsel et al., 1978), interactions between stimulus velocity and other parameters can be quite complex.

Stimulus duration is another powerful variable for neural responses in sensory systems. If intensity is held constant, sufficiently brief durations may evoke little or no response, while appreciably longer durations may produce a sizable response. This has been observed, for example, in peripheral visual responses over a 10-fold increase in duration (Fuortes, 1971).

Perceptual Responses

Discrete responses   Quantitative perceptual measurements most commonly involve single, discrete responses, each representing the total judgment for a particular trial. Use of single numerical, graphical, or descriptive responses to characterize stimuli assumes either that the stimuli are perceived as constant during a trial or that subjects will know what aspect of a changing perception is to be judged. Although stimuli are often presented and removed during a trial, thus ensuring change, the change is generally not mentioned in instructions to the subject. This practice may be fully justified in many auditory or visual experiments. For taste, however, change in concentration of flowing liquids at the human tongue requires 10 msec under the best conditions (Halpern, 1986a; Kelling and Halpern, 1986), 500 msec in many experiments. If taste perceptions are measured over time, they too exhibit measurable rises and falls, but with time courses generally much slower than those of chemical stimuli at the tongue.

When perceived intensity is directly measured, it is often scaled using Magnitude Estimation (Stevens, 1956; Bartoshuk and Marks, 1986). Because each observer reports the perceived magnitudes of stimuli by assigning numbers that they consider appropriate, magnitude estimation is called a direct psychophysical method. In principle, any positive number could be a response. Subjects are explicitly instructed to maintain constant ratio relationships between their responses. For example, if one intensity is judged to be three times more intense than another, responses of 100 and 300, 2 and 6, or 1 and 3 would all properly express the ratio of the perceived intensities. This characteristic is described as ratio scaling. It differs from equal interval or arithmetic relationships (Wyscezki, 1986, Land and Shepherd, 1988; Luce and Krumhansl, 1988).

Intensity. Both stimulus duration and rate of change of stimulus intensity can have substantial effects on human perceptual judgments. Rate of change of cutaneous thermal or mechanical stimuli affects threshold (Mountcastle, 1967; Hensel, 1982; Darian-Smith, 1984) and perceived intensity (Mountcastle, 1967; Burgess et al., 1983; Sherrick and Cholewiak, 1986). Only a certain range of rates of change may produce optimal responses; sufficiently slow or rapid rates may fail to elicit any responses.

Thresholds are decreased when the duration of fixed intensity cutaneous mechanical or thermal events increases (Sherrick and Cholewiak, 1986). The time over which cutaneous mechanical temporal summation occurs may be less than 10 msec, but threshold for temperature change appears to sum over 1.0–1.5 sec. These examples indicate temporal summation, but also demonstrate that the temporal limits of such summation can have large quantitative differences between perceptual systems. Detection of two separate skin mechanical stimulations requires an interval of about 6 msec, again suggesting temporal summation of cutaneous mechanical events.

Perceived cutaneous intensity cumulates with stimulus duration over relatively long time periods. For example, cutaneous cold intensity increases dramatically with stimulus duration over a span of many seconds (Sherrick and Cholewiak, 1986). The effects of cutaneous stimulus duration that have been noted are of particular relevance to temporal processes in taste, since gustatory receptor cells are modified cutaneous cells. Consequently, substantial temporal summations for thermal and mechanical stimulation of the skin (Sherrick and Cholewiak, 1986) suggest that the taste system is likely to also exhibit temporal summation. The locus of summation for cutaneous stimulation is apparently within the central nervous system, and this appears to also be the case for gustatory temporal cumulation (Halpern, 1986b; Kelling and Halpern, 1988).

In the auditory and visual systems, thresholds can be a joint function of stimulus duration and intensity [e.g., Bloch's Law in vision (Levine and Shefner, 1991; Watson, 1986)]. For both vision and hearing, these time versus stimulus-intensity trade-offs generally are restricted to relative brief durations, such as 100–200 msec (Thurlow, 1971; McBurney and Collings, 1984; Hood and Finkelstein, 1986). However, the temporal separation necessary to detect two separate clicks is approximately 10 μsec, which is about a thousand times briefer than that indicated for the skin. On the other hand, two flashes of light must be separated by about 25 msec to be reliably distinguished (Sherrick and Cholewiak, 1986). Thus, *although various aspects of temporal summation occur in most sensory systems, the quantitative parameters differ widely.*

Perceived loudness increases with sound duration over a range similar to that noted above for visual time-intensity trade-offs (Scharf and Houtsma, 1986; Luce and Krumhansl, 1988; Algom et al., 1989), with the central nervous system taken to be the site of temporal integration. The maximum duration of these duration-loudness interactions is largely independent of stimulus intensity. However, sound frequency must be held constant in such studies, since loudness varies nonmonotonically with frequency. The latter constraint indicates that *if stimuli are changed such that perceived quality is altered, complex differences in intensity-duration relationships may occur.* This may also apply to gustatory time-intensity interactions.

In similar fashion, and over a similar duration range, ability to discriminate between two tones improves with stimulus duration (Scharf and Buus, 1986). Furthermore, for a tone rather than a click to be perceived, a minimum stimulus duration must be presented (Gulick et al., 1989). The latter observation may have a conceptual relationship to reports discussed later in this chapter that the durations of tracked taste quality descriptions differ from durations of tracked taste intensity.

In the visual domain, the range of durations over which judgments of brightness increase with stimulus duration depends upon stimulus luminance (Luce and Krumhansl, 1988). At low luminances, stimulus durations up to 500 msec yield larger estimates of brightness. For relatively intense flashes, this occurs only up to 15 msec. This is a specific illustration of the general observation that *the intensity range in which studies are done will often have profound effects on perceptual sensitivity to other stimulus parameters.*

Rate of modulation of light intensity is also important for visual perception. For example, the dependence of visual flicker thresholds and contrast sensitivity upon the rate of change of luminance is well known (Watson, 1986). Major parameters in modulation studies are the amount of change in stimulus intensity (depth of modulation), the average stimulus intensity around which intensity is modulated, the rate at which maximum and minimum intensity are reached, the number of times per second that complete cycles of intensity increase and decrease can occur (frequency), and the overall stimulus duration (train length). Many of these parameters have been used in experiments on the effects of temporal modulation of taste stimulus concentration on perceived taste intensity, and the detectability of such modulation.

In general, the judged magnitude (e.g., brightness, loudness, strength) of nongustatory perception grows as stimulus duration increases over some specified range of duration and stimulus intensity. Other stimulus properties, such as wavelength, will often affect the stimulus duration-judged intensity interactions. Similar relationships are found in taste. Judged taste intensity tends to rise as stimulus duration lengthens, with the concentration and chemical identity of the stimulus modifying the pattern.

Quality.   Rate of change and duration of cutaneous thermal or mechanical stimuli affect the perceived pattern (Whitsel et al., 1978; Darian-Smith, 1984; Sathian, 1989). Cutaneous stimulus durations in the 5–20 msec range result in poor tactual character recognition (Loomis and Lederman, 1986), but obvious improvement occurs with 100 msec durations, while performance asymptotes near 500 msec durations. This approximately 500 msec range over which qualitative descriptions of skin stimulation improve is much broader than the range of about 10 msec in which perceived strength of skin stimulation grows as a function of stimulus duration (Sherrick and Cholewiak, 1986). It appears that *the effect of stimulus duration on judged intensity may operate in a range quite different from that in which differences in perceived quality occur.*

In hearing, relations between sound frequency (changes per sec in sound pressure level) and perceived auditory pattern (pitch) are well known (Scharf and Buus, 1986; Matlin, 1988). Pitch also changes as a function of sound intensity. The relationship is nonlinear and nonmonotonic. An increase in intensity may result in an increased, a decreased, or an unchanged pitch (Scharf and Houtsma, 1986).

In similar fashion, perceived hue can undergo substantial change with sufficient increase or decrease in photopic stimulus luminance (Bezold-Brücke effect) (Wyszecki, 1986). However, these changes in hue occur only for certain colors. Of particular interest to our consideration of temporal aspects of taste quality is the importance of stimulus duration for these changes in hue as luminance is altered. Present gustatory data indicate that for some stimulus solutions, two separate taste quality descriptors may be reported, with each quality duration different than stimulus duration (Zwillinger and Halpern, 1991). However, effects of concentration or duration have not been systematically explored.

Tracking

Direct approaches to quantitatively measuring perception over time can provide the times of appearance and disappearance, the rates of increase and decrease, the extent of plateaus, time-lags between stimulus and response changes, tracking errors, and total duration of the phenomena. Judged intensity is most commonly measured. In this case, intensity tracking is done. "In all tracking tasks, the subject manipulates a control device so as to minimize the difference between the input signal and the output that is produced" (Pew and Rosenbaum, 1988).

Measurements of change in the loudness of long auditory stimuli or the brightness of prolonged visual stimuli are not often done. However, some laboratory data do exist. For example, small changes in the tracked loudness of a constant intensity tone have been reported (Harris and Pikler, 1960). The task was to maintain constant the loudness of a tone that varied in intensity. Intervals of constant intensity, with durations of 9–15 sec, were separated by epochs of increasing and decreasing intensity. Interestingly, tracking was relatively slow when intensity was changed, with a time-lag of about 3 sec. For longer durations, sustained tones near threshold were reported to disappear after about 30 sec (Scharf and Buus, 1986). Somewhat more intense tones decreased about 25% in loudness over 30 sec, with an asymptote at about a 65% decrease after 150 sec (Scharf and Houtsma, 1986). This time course is similar to some descriptions of decreased judged taste intensity during sustained stimulation (Halpern, 1985).

Perceived duration is one of the aspects of a stimulus that is normally provided by tracking measures. For visual stimuli, perceived duration has been studied indirectly with a different method. The technique was a cross-modal reaction time procedure. The subject adjusted a brief click to be coincident with the onset or the offset of a visual stimulus (Haber and Standing, 1970). Under conditions of dark adaptation, perceived duration exceeded flash duration through 1000 msec flashes. The difference between flash duration and perceived duration was greatest for brief stimuli, reaching a maximum, 400 msec, for a 10 msec flash. These results are of particular interest because tracked durations of taste stimulus intensity exceeded stimulus duration when gustatory stimuli of 100 msec through 2000 msec duration were used.

TEMPORAL FACTORS IN TASTE

Neural Responses

Human neurophysiological taste responses   Few neurophysiological data exist on human gustatory neural activity. Experiments in which chemosensory responses were recorded from the human chorda tympani nerve were pursued by Diamant, Zotterman, and their colleagues for approximately a decade (Diamant and Zotterman, 1959; Diamant et al., 1963). Such neurophysiological experiments were discontinued after adverse effects of preparing the nerve for recording became apparent (Diamant, 1968; Schuknecht, 1974, Oakley, 1985). Consequently, the available data, although they are based upon relatively few individuals and limited experimental designs, are very important. They are our only direct information on the characteristics of human peripheral taste responses. Because of the unique nature of these data, as well as the somewhat scattered and occasionally complex articles in which they are presented, a relatively detailed summary will be provided. For similar reasons, psychophysical judgments made by individuals from whom gustatory neural recording were subsequently made will be carefully examined (Fig. 2.1, Table 2.1). These psychophysical data were not gathered with the rigor normally expected in taste psychophysics. However, since no other parallel psychophysical and neural data presently exist for taste, these limited observations deserve attention.

The chorda tympani nerve, which is a branch of the seventh cranial nerve (the facial nerve), connects with taste buds on the front (anterodorsal region) of the human tongue (Halpern, 1977). This nerve passes through the middle ear space. It was accessed during necessary procedures for middle ear surgery (Zotterman, 1967). The nature of the surgical procedure used for exposure of the middle ear cavity required that the patients receive general anesthesia. In addition, a drug that prevents muscular activity was administered to reduce spurious electrical events during the experiment (Zotterman, 1967). Consequently, no taste judgments could be obtained while the recordings were being made.

The neural response measure was the maximum magnitude of electronically processed activity from many axons in the chorda tympani nerve. In order to obtain recordings, it was necessary to cut the connection between the chorda tympani nerve and the central nervous system before placing the nerve on a recording electrode. Attempts to record from individual axons of the human chorda tympani nerve were not successful (Zotterman, 1967).

During the neurophysiological experiments, stimulus solutions were delivered to the tongue at approximately 30°C. Each stimulus had a volume of 15 ml and flowed onto the tongue from a plastic tube. The solutions were prepared in distilled water, which was also used as the solution removal liquid. A gravity flow system, originating in a reservoir about 25 cm above the tongue, delivered the liquids when a stopcock was turned (Diamant et al., 1963,1965).

Sometimes psychophysical judgments of taste intensity were obtained from a patient a few days before a successful chorda tympani nerve recording session (Diamant et al., 1965; Borg et al., 1967; Diamant, 1968; Oakley, 1985). Magnitude estimates of taste intensity were compared with neural responses to the same stimulus solutions, presented in the same order. Both the neural and the psychophysical measures of taste intensity could be fitted with power functions. For single patients, the combined neural and psychophysical data after log-log

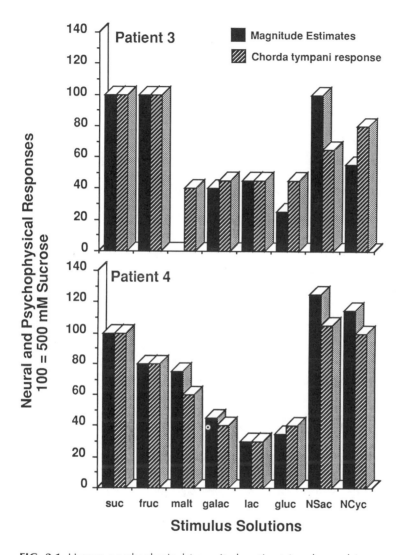

FIG. 2.1 Human psychophysical (magnitude estimate) and neural (summated chorda tympani nerve) responses from two patients to a series of sweetening agents. The test solutions were 500 mM sucrose (suc), fructose (fruc), maltose (malt), galactose (galac), lactose (lac), and glucose (gluc), 4 mM sodium saccharin (NSac), and 30 mM sodium cyclamate (NCyc). No magnitude estimates for maltose were obtained from patient 3. *(From Diamant et al., 1965, Table 1.)*

TABLE 2.1   Relative Psychophysical and
Gustatory Neural Responses of Two Patients
to Three Stimulus Solutions

Patient No.	200 mM NaCl	20 mM citric acid	500 mM sucrose
		Magnitude Estimates	
	400	200	100
3			
		Neural Response	
	150	60	100
		Magnitude Estimates	
	200	240	100
4			
		Neural Response	
	50	100	100

The sucrose solution was assigned the value of 100.

conversion had a slope of 0.85 for citric acid; 1.1 for sucrose (Borg et al., 1967). It should be noted that the neural response magnitudes could be fit equally well, and perhaps better, with linear functions (Diamant, 1968; B. Oakley, personal communication, 1990).

A series of sweetening agents was used with two patients for both psychophysical and neural measurements (Fig. 2.1). For one patient ("Patient 3") stimuli in the psychophysical experiment were applied by cotton swab to the anterodorsal region of the tongue on the same side as (i.e., ipsilateral to) the chorda tympani nerve from which recordings were subsequently made. For the other patient ("patient 4") stimuli were delivered as 5 ml sips and spits in the psychophysical portion of the experiment. Patients made two magnitude estimates of the intensity of each sweetening agent. These two judgments were averaged. During the chorda tympani nerve-recording session, a single neural response was recorded for each solution, except for repeated responses to 500 mM sucrose. All results were expressed as percentages with respect to the neural response or mean magnitude estimate for 500 mM sucrose, rounded to the nearest 5%. Within each patient, there was considerable similarity between the relative response magnitudes of the magnitude estimates and the maximum summated peripheral nerve responses, but also some notable differences (e.g., glucose and sodium saccharin for patient 3). Between the two patients, the eight sweetening agents apparently produced different patterns (Fig. 2.1). Since no repetitions of the neural responses were done, and only two magnitude estimates, it is difficult to know if specific differences are reliable. Of course, this consideration must also apply to the responses that appear to not differ.

However, when relative gustatory neural and psychophysical response magnitudes were compared across stimulus solutions that contained sufficiently diverse molecular species, psychophysical and neural data were not congruent (Table 2.1). Based upon these observations, it has been suggested that differential central nervous system processing of peripheral gustatory responses is indicated when stimuli elicit tastes that are qualitatively distinct (Oakley, 1985).

The time of stimulus solution arrival at the tongue was not precisely measured in the recording experiments, since stimulus delivery was indicated by a switch on the stopcock that controlled stimulus delivery (Diamant et al., 1963). This precluded accurate evaluation of the time interval between the beginning of taste stimulation and the start of the human chorda tympani nerve gustatory neural response (response latency). In addition, electronic processing (summation) of the tape-recorded responses was produced by a resistance-capacitance integration network with rise and fall times of 1.5 sec (Diamant et al., 1965). This processing device was a running averager that provided an analog approximation of the electrical energy in the original record (see Halpern, 1973). Rise and fall times of 1.5 sec will cause the time course of the processed neural responses to be much slower than the original activity if any rapid changes are present. With these limitations in mind, the time course of the processed responses can be examined. Responses to 500 mM NaCl, 500 mM sucrose, 2 mM sodium saccharin (NaSac), 20 mM quinine sulfate, and 200 mM citric acid all rose from baseline to maximum in less than 1 sec (Zotterman, 1967). The initial slope of the rising response was quite steep. For example, processed human neural responses to both 500 mM NaCl and 2 mM NaSac reached 66% of maximum in 200 msec.

The processed human chorda tympani nerve responses also fell relatively quickly. For example, for patient 3 the response to a continuous flow of 200 mM NaCl decreased to approximately 34% of maximum within 15 sec after the response had begun, and had returned to prestimulus baseline in about 71 sec after response onset (Diamant et al., 1965). This same patient "indicated in the psychophysical test that the salt taste disappeared after 90 sec. Patient no. 4 indicated that he could no longer taste salt after 79 sec, which corresponded to a 95% reduction in the magnitude of his neural response" (Diamant et al., 1965). The psychophysical test was not described, but may have consisted of holding 5 ml of 200 mM NaCl solution in the mouth with minimal motion, and indicating manually when a salt taste was no longer perceived (B. Oakley, personal communication, 1990).

The data reviewed in this section indicate that human peripheral gustatory neural responses rapidly increase in magnitude once they begin. They also decline relatively rapidly. These observations suggest that humans might be able to respond to relatively brief taste stimuli, and that human taste may be oriented towards change in stimulation rather than steady states.

Nonhuman neurophysiological taste responses   In contrast to the relatively detailed review of human gustatory neural responses and associated psychophysical judgments that was presented in the previous section, nonhuman gustatory neural responses will receive only a very brief summary, and nonhuman taste-dependent behavior will not be considered at all. The nonhuman taste literature is extensive but, as noted below, often unrelated to questions of temporal patterns. The references in the present section can lead interested readers to relevant literature.

Temporal characteristics of mammalian gustatory neural responses have not been a major focus of most researchers. Neither the time of stimulus arrival nor stimulus duration has been accurately or precisely measured in typical studies. In addition, the time intervals over which gustatory neural responses have been grouped before analysis were generally 5–10 sec. Such intervals are very long compared with the temporal patterns of neural responses and the time required for nonhuman species to make taste-dependent decisions (Halpern, 1985, 1986b). Time-dependent taste events have not been considered sufficiently important to include in compendia on chemosensory function (Finger and Silver, 1987).

Useful data on temporal properties of mammalian gustatory neural responses and a related set of interpretations do exist. Response latencies in the chorda tympani nerve are generally in the 30–100 msec range, although latencies of 300–400 msec are sometimes reported for certain sugars and amino acids (see Halpern, 1985, 1986b, for references).

Nonhuman mammalian chorda tympani nerve gustatory neural responses often increase in magnitude rapidly, reaching maximum within 200–300 msec after stimulus arrival at the tongue. Following this maximum, the response usually quickly decreases in magnitude. This rapid rise and fall at the start of responses is known as the *phasic* component of the response. It appears that the phasic component can be sufficient for identification of taste stimuli by laboratory animals. After the phasic component, the peripheral nerve gustatory neural responses generally continue for many seconds, and sometimes minutes, at a relatively low steady-state level. This sustained neural activity is known as the *tonic* component of the response.

The time to maximum, maximum magnitude, and overall temporal envelope of nonhuman mammalian peripheral nerve gustatory neural responses can change as a function of stimulus concentration, stimulus molecule, and stimulus duration. Briefer durations lead to shorter times to maximum, and smaller maximum magnitude. The tonic component of the response is also sensitive to the concentration, duration, and composition of the stimulus, but to a lesser degree than the phasic component (Halpern, 1985,1986b).

Most mammalian gustatory neural data are from acute studies using anesthetized laboratory rats, hamsters, or mice, with lesser amounts from rabbits, dogs, and cats. Relatively few data are presently available from awake preparations or from nonhuman primates. The large and poorly understood behavioral and neural differences between animals in gustatory responses and the effects of anesthetics on central nervous system function suggest that data from unanesthetized nonhuman primates are necessary to explain some aspects of human taste. Studies in which gustatory responses were recorded from electrodes implanted in the central nervous system of unanesthetized nonhuman primates may be especially useful (e.g., Rolls et al., 1985; Scott et al., 1986a,b). Such experiments are particularly valuable when their design incorporates investigation of temporal parameters, and the stimuli are selected to include substances related to the natural habits and evolutionary development of the animals under study. It should be noted that a substantial body of data exists on peripheral gustatory neural responses from nonmammalian vertebrates and from invertebrates. This information is useful not only for understanding these animals themselves, but also for providing insights into basic and general mechanisms of chemosensory function.

Perceptual Responses

In common with the majority of neurophysiological studies of taste, temporal parameters have been of little or no interest to most students of gustatory perception who applied standard psychophysical methods (Halpern, 1986a,b). Indeed, consideration of the onset characteristics of perceived taste intensity or quality, the extent of plateaus, and time-lags between stimulus and response changes, as well as analyses of effects of stimulus duration on taste judgments, are absent from several treatments of chemosensory perception (Meilgaard et al., 1987; Bartoshuk, 1988; Land and Shepherd, 1988). Some temporal properties of taste were examined by McBurney in a comprehensive review of sensory discrimination in taste and olfaction. He wrote, "The temporal properties of taste and smell have received very little attention. Indeed, many investigators do not consider the duration of stimulus presentation worthy of note" (McBurney, 1984).

Adaptation   Taste adaptation, often defined as a decrease in perceived intensity and increase in threshold with prolonged constant stimulation (Matlin, 1988), is invariably included in chapters on taste (Bartoshuk, 1985, Bartoshuk et al., 1985; Lawless, 1987). Because taste adaptation has been extensively treated elsewhere (Bartoshuk, 1971,1980; Pfaffmann et al., 1971; Gent, 1979; O'M-ahony, 1979; McBurney, 1984; Overbosch, 1986; Overbosch and Soeting, 1989), it will receive only brief coverage here. Phenomena that are related to adaptation, specifically the effects of temporal modulation of taste stimulus concentration on perceived intensity and the detectability of such modulation, are discussed in a later section.

Taste adaptation was described in the previous paragraph as a decrease in perceived intensity and increase in threshold following prolonged constant stimulation. The elevation in threshold of course indicates that a stimulus concentration higher than the normal threshold value will be necessary for consistent detection of the stimulus to occur. However, Bartoshuk (1971,1980) has pointed out that the taste-detection threshold situation is more complicated than just suggested. This is the case because concentrations sufficiently *lower* than the adapting concentration will also be detected as different from the adapting stimulus. The taste quality descriptions that would be elicited by these lower concentration stimuli might well be different from those given to detect-able stimuli at concentrations above the adapting concentration. This difference in descriptors may not be noticed because procedures for measurement of a detection threshold normally permit only two responses: one denoting detection, the other absence of detection (e.g., yes or no; 1st or 2nd) (Engen, 1971a; McBurney and Collings, 1984). The fundamental problem is that the detection threshold concept (also called the stimulus threshold or the absolute threshold) assumes that the relevant sensory system will be adapted to zero intensity, and that only increases in stimulus intensity will be possible (Luce and Krumhansl, 1988). A parallel difficulty occurs for the claim that taste adaptation will produce a decrease in judged intensity. This is certainly the case for concentrations close to the adapting concentration. However, as the concentration becomes suf-ficiently higher *or lower* than the adapting concentration, the perceived intensity will increase (Bartoshuk, 1971, 1980; Pfaffmann et al., 1971; McBurney, 1984).

The general depiction of taste adaptation as only a reduction in responsiveness is an incomplete rendition. Taste adaptation's decrease in absolute sensitivity is accompanied by increased sensitivity to small changes away from the adapting stimulus (Pfaffmann et al., 1971). Improved differential taste sensitivity after adaptation has been reported in studies that have measured gustatory just notice-able differences (also called difference thresholds) (Engen, 1971a,b), which are the smallest consistently detectable differences from a specified reference con-centration. There are two interesting implications of the observation that differ-ence thresholds are reduced by taste adaptation. First, a central nervous system component of the phenomena may be indicated. Second, the observation sug-gests that a changing taste stimulus should exhibit less reduction in perceived intensity or threshold over time, since in effect a type of just noticeable differ-ence is being measured. This latter possibility is considered in the next section.

Application of the term adaptation to many different stimulus and measurement conditions may itself be a problem (Bartley, 1959; Bartoshuk, 1971). At its broadest, adaptation has been said to include all outcomes produced by repeated or steady presentation of a stimulus to an organism (McBurney and Collings, 1984). A global definition of this sort includes increases and decreases in responsiveness, and encompasses central nervous system and receptor level processes. However, narrower definitions of adaptation are commonly used. Often, only changes in responsiveness that can be attributed to events at sensory receptors are considered adaptation. If a central nervous system locus rather than peripheral effects is indicated for decreased responsiveness after repeated peripheral stimulation, the term *habituation* is frequently employed by researchers in taste (Kroeze, 1982) and other areas (Hearst, 1988; Matlin, 1988; Thompson et al., 1988).

Discrete Responses

Temporal Modulation. The increased sensitivity to small changes away from an adapting taste stimulus that was discussed in the previous section on adaptation suggests that a changing taste stimulus should elicit less reduction in perceived intensity and less increase in threshold than a constant stimulus. Effects on judged taste intensity or taste sensitivity of prolonged, intermittently fluctuating taste stimuli were examined in two series of experiments. In both, stimuli were delivered by an open taste stimulus delivery system, and stimulus concentration at the tongue was not measured. The use of an open flow system precluded rapid removal of taste stimuli from the tongue (Kelling and Halpern, 1986). Consequently, with a open delivery system, an intended square wave stimulus train of alternate 1 sec flows of stimulus solution and water (Meiselman and Halpern, 1973; Halpern et al., 1986) was actually a continuous modulation of concentration (Halpern, 1986a). When a train consisting of briefer pulses was attempted with an open taste stimulus delivery system (McBurney, 1976), the result was necessarily a shallow modulation of concentration. Therefore, the actual changes in concentration that were delivered in these experiments were often much less than those intended.

One set of studies measured taste intensity during approximately 1 min long stimulus trains of NaCl (50–500 mM) at intended pulse durations of 1 or 2 sec (Meiselman and Halpern, 1973; Halpern et al., 1986). These NaCl pulses alternated with 1, 2, or 3 sec intended duration pulses of water or an artificial saliva containing 36 mM Na^+ (similar to natural human saliva). Judged intensity was measured by magnitude estimation starting several seconds after flow onset. Magnitude estimates were obtained every 5–12 sec. No attempt was made to measure initial or initially rising intensity. Instead, the initial intensity was taken as the standard against which subsequent intensities were to be compared.

No decrease in judged intensity occurred with any of the NaCl–water trains or the NaCl–artificial saliva train. For some of the conditions, the magnitude estimations increased during the trial. However, if NaCl pulses were alternated with identical NaCl pulses or were alternated with periods of no liquid flow, intensity gradually decreased. These results confirmed the earlier supposition (Pfaffmann et al., 1971) that stimulus change over time is important in taste. Relatively small, slow (every 2 sec) modulations, such as the intended 28% change between 50 mM NaCl and the 36 mM Na^+ artificial saliva, were sufficient to prevent any decrease in judged intensity.

A conceptually more sophisticated set of studies on gustatory temporal modulation was reported by McBurney (1976, 1978). He asked subjects to indicate whether or not a flowing liquid was changing in intensity, or the direction of change, during 30 sec presentations. Solutions of NaCl, citric acid, HCl, and quinine HCl were used. Quite small changes, $\leq 15\%$ modulation of concentration, were reliably detected, even though some of the changes were so slow that about 17 min would have been required to present a full cycle. Thus, both this and the previous experiment indicated that the human taste system is very responsive to temporal aspects of stimulation.

Duration

Reaction time. Bujas (1934) observed a trade-off between taste stimulus concentration and the stimulus duration that was necessary before humans detected the taste stimulus (see Halpern 1986b for graphed data and details). With sufficiently low concentrations, such as 153 μM quinine dihydrochloride or 35 mM NaCl, necessary taste stimulus durations and associated human taste reaction times were very long (e.g., ≥ 4080 msec). This indicates substantial temporal summation in the gustatory system, because a prolonged stimulus can be effective only if the stimulation is summed in some fashion. Since stimulus access to the taste receptors and subsequent peripheral neural events could account for only a small part of the necessary stimulus durations, summation of peripheral taste responses in the central nervous system was occurring (Halpern, 1986b). Reaction times decreased appreciably with small increases in concentration. A large change in response for a small change in stimulation demonstrates that the overall human gustatory process is nonlinear, but does not identify the nonlinear loci.

Stimulus duration has been directly varied in a series of studies on taste reaction time (Lester and Halpern, 1979; Kelling and Halpern, 1983,1987a, 1988). A closed flow taste delivery apparatus (Kelling and Halpern, 1986) was developed to permit reliable taste stimulation with durations as brief as 50 msec, and to provide measurement at the tongue of the wave form and concentration of the flowing liquid (Halpern, 1986a). The time required to detect any change in taste, that is, simple taste reaction time, differed between 50 and

2000 msec stimulus durations for NaSac, $MgSO_4$, HCl, and monosodium gluta-mate (MSG). For some stimuli, a 50 msec increase in duration, from 50 to 100 msec, was sufficient to decrease simple taste reaction time. In other cases, a 200 msec increase in duration was necessary for a decrease in reaction time. However, simple taste reaction times to 1000 msec durations were never different from those to 2000 msec stimuli.

Effects of duration on taste quality identification (i.e., descriptor) reaction time were also measured in this series of experiments (Kelling and Halpern, 1983,1988). These complex taste reaction times were approximately 200 msec longer than the corresponding simple reaction times. Surprisingly, the quality identification (taste descriptor) reaction times did not change with stimulus duration.

Overall, these reaction time studies provided two observations that are especially relevant to an analysis of temporal factors in human gustation. First, human taste responses occurred to very brief stimuli, such as a single 50 msec solution flow. This duration is described as very brief because it is much shorter than the normal range of human taste stimulation during drinking. A sip and swallow will have a duration of about 1 sec, during which liquid is transferred into the mouth, with the swallow coming about 500 msec later (Halpern, 1985) if no taste-dependent decision is required. Thus the entire sequence would last about 1500 msec. Responses to 50 msec taste stimuli indicate a sensitivity to gustatory events at the tongue with durations that are only 3% of the sips that humans typically produce (Halpern, 1977,1985). It follows that some aspects of human taste-dependent behavior only require, and may normally only use, the initial portions of responses to stimulation. Second, reaction time changes occurred with stimulus duration for simple taste reaction times but apparently not for identification (taste descriptor) reaction times. Perhaps the central nervous system processing time required for the identification reaction times becomes the predominant factor (Halpern, 1986b). It should be noted that quality (taste descriptor) identification was defined in those studies as an unrestricted vocal (spoken) description of a stimulus liquid, responding to the question "What was the taste?" The reaction time of the identification response was the time interval from solution arrival at the tongue until the beginning of the vocal response.

Intensity

Continuous Stimulation.   Increases in taste intensity as a function of taste stimulus duration have been observed in several laboratories. Two early but rarely cited studies that used discrete responses were done by Bujas and Ostojcic (1939) and by Bekesy (1964b). The experimenters presented liquid stimuli using closed flow delivery systems (see Halpern, 1986a, and Kelling and Halpern, 1986, for discussions), but measures of concentration or stimulus duration at the tongue were not available. Both experiments scaled taste intensity with a form of the method of constant stimuli. The method of constant stimuli is an indirect, classical psychophysical technique (Engen, 1986). In this procedure, the experimenter selects about seven (± two) discrete stimuli that will be presented to the subject "The middle concentration is designated as the standard (St) and is compared in a random order with each of the other stimuli, the comparison

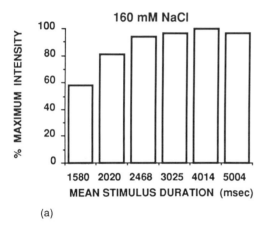

(a)

FIG. 2.2 Psychophysical judgments by one subject, based upon the method of constant stimuli, of the intensity of taste solutions flowed over 1.4 cm² of the annterodorsal region of the subject's tongue. Responses are shown as percentages of the maximum response magnitude for each of three stimulus solution: 160 mM NaCl (a), 560 mM NaCl (b), or 40 mM sucrose (c). The NaCl concentrations were 4 (160 mM) and 14 (560 mM) times this subject's absolute threshold; the sucrose concentration, 4 times the absolute threshold. Stimulus durations are means of the durations presented on 12 individual trials. The set of 8 comparison stimuli for each standard stimulus duration and concentration were flowed for 6 to 13.5 sec through a second closed taste stimulus delivery system. The tongue was rinsed for 7 sec outside the closed systems between standard and comparison stimulus flows. *(From Bujas and Ostojcic, 1939, Tables 1 and 5.)*

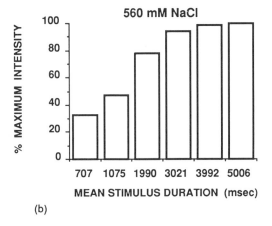

(b)

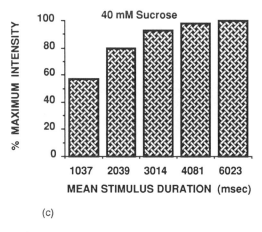

(c)

FIG. 2.2 *(cont.)*

stimuli. The subject's task is to judge for each pair whether the second stimulus has more or less of a perceived attribute . . ." (Engen, 1986).

Bujas and Ostojcic (1939) found that longer stimulus durations produced greater judged taste intensities. This relationship was observed for aqueous solutions of NaCl, sucrose, or quinine dihydrochloride (Fig. 2.2, quinine hydrochloride responses not shown). These experimenters observed that total taste intensity initially increased rapidly. In addition, maximum intensity was greater for higher concentrations (for example, 560 mM NaCl at a duration of 4–5 sec produced a response 3.6 times as large as that yielded by 160 mM NaCl; both responses were maxima). Thus, there was an interaction between stimulus duration, concentration and intensity. For the higher NaCl concentration, a 707 msec stimulus duration gave a response about one-third of maximum intensity. However, this duration was not an effective stimulus at the lower concentration. Instead, with 160 mM NaCl a much longer duration (1255 msec) (not shown in Fig. 2.2) gave about half of the maximum for that concentration of NaCl. For all NaCl and sucrose solutions, stimulus durations of approximately 2 sec gave at least 71% of maximum intensity (Fig. 2.2), but similar durations of quinine dihydrochloride solutions yielded only $\geq 60\%$ of maximum intensity. However, even for the quinine dihydrochloride solutions, the initial rise was relatively rapid. For example, about 1 sec of 1.2 mM quinine dihydrochloride gave 56% of maximum.

Higher judged taste intensity with increased stimulus duration was also reported by Bekesy (1964b). He presented a salt or acid solution to " . . . a small, constant area of the tongue. . . ." Judged intensity increased rapidly for stimulus durations of 1 sec or less, reaching approximately 87% of maximum intensity for about 1 sec of stimulation (see Bekesy 1964b, Figure 9). The subsequent slope of the intensity-duration function was relatively flat. Thus, a 2 sec stimulus duration gave 91% of maximum intensity. The general pattern of these observations is similar to the more detailed report made 30 years earlier by Bujas and Ostojcic (1939).

Overall, the Bujas and Ostojcic (1939) and Bekesy (1964b) studies found that less than 1 sec of taste stimulus flow produced a substantial proportion of the maximum obtainable intensity. Durations of 1–2 sec always gave more than half of maximum intensity, and sometimes 80–90%. However, durations briefer than 707 msec were not reported by Bujas and Ostojcic (1939), while 500 msec was probably the shortest duration used by Bekesy (1964b).

A series of experiments reported 20 years after Bekesy's study used magnitude estimation to evaluate the effects of taste stimulus duration on gustatory intensity (Kelling and Halpern, 1983,1988). This direct scaling method was combined with a closed taste stimulus delivery apparatus that permitted consistent stimulus durations down to 50 msec and provided measurement of duration and concentration at the tongue (Halpern, 1986a; Kelling and Halpern, 1986). For NaCl, NaSac, HCl, and $MgSO_4$ solutions, judged intensity significantly changed with stimulus duration (Fig. 2.3). The increases in intensity were quite large. For example, the intensity for 300 msec durations was more than three times (230% increase) that for 50 msec durations. A high sensitivity to small increases in duration was also found. Thus, a 50 msec increase in duration, from 50 to 100 msec, produced a significant change in intensity for all solutions tested. The asymptote for the duration-intensity function was not identified. That is, the longest duration used, 2000 msec, produced intensities that differed significantly for all solutions from those elicited with a 1000 msec stimulus. However, as the Bujas and Ostojcic (1939) and Bekesy (1964b) data suggested,

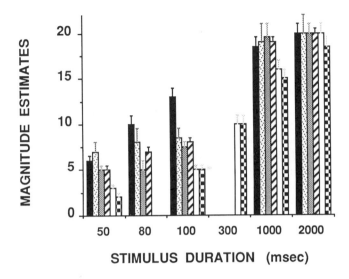

FIG. 2.3 Taste intensity judgments (medians and standard errors of the median) made as vocal magnitude estimations for six stimulus solutions. Magnitude estimates for a duration of the stimulus solution were made with reference to identified (as the standard) 2000 msec duration stimulus solution presentations that were assigned an intensity of 20. The stimulus solutions were 250 mM $MgSO_4$ (solid bar), 500 mM $MgSO_4$ (speckled bar), 3.2 mM HCl (dotted bar), 10 mM HCl (diagonal bar), 500 mM NaCl (clear bar), and 2 mM NaSac (black and white bar). Only the NaCl and NaSac were used at the 300 msec duration; they were not used at the 80 msec duration. A single stimulus solution, randomized with distilled water control stimuli, was presented twice at each duration, in random order, in five data collecting sessions. Medians are based on at least five practiced subjects. See Halpern, 1986b, and Kelling and Halpern, 1988, for procedural details.

the duration-intensity function was clearly not linear. For example, intensities for 1000 msec durations were only about 150% of those to 300 msec stimuli, while the 2000 msec durations gave intensity responses that differed from those to the 1000 msec stimuli by no more than 25%.

These observations on duration-judged intensity interactions (Kelling and Halpern, 1983,1988) supported and extended those provided by earlier experimenters. Over at least a 200-fold range, duration of taste stimulation was processed in a cumulative, nonlinear manner into taste intensity. This cumulation function started at stimulus durations much shorter than those produced by human drinking. It extended into stimulus durations that would occur during normal sipping and swallowing. For very brief durations, small increments in duration lead to large increases in intensity, but at long durations, much longer stimuli provided only slight increases in intensity. There is little doubt that the locus of this cumulative intensity processing over time is the human central nervous system (Kelling and Halpern, 1988). This is compatible with the major central nervous system component that had been suggested earlier for taste reaction times (Halpern, 1986b; Kelling and Halpern, 1987).

The duration-intensity studies just discussed employed a throat microphone to record the start of spoken magnitude estimates (Kelling and Halpern, 1988). This provided the time at which each magnitude estimate began. In the same experiments, a flow-through conductivity cell in the closed stimulus delivery system (Kelling and Halpern, 1986) indicated time of stimulus arrival at the tongue. Consequently, the time interval between arrival of the taste stimulus at the tongue and the magnitude estimate response was known. This time interval was the magnitude estimation reaction time.

The magnitude estimation reaction times to 50–2000 msec duration taste stimuli were very long. Their medians ranged from 1639 to 2106 msec. For the same stimuli, simple taste reaction times had been 518–818 msec (Kelling and Halpern, 1987a), while taste quality identification (descriptor) reaction times had been 761–1048 msec. Neither the taste quality identifications nor their reaction times had changed significantly with stimulus duration. Therefore, human taste *intensity* processing has three interesting characteristics that are not found in human taste *quality* processing: (1) taste intensity processing converts a temporal characteristic of stimuli, duration, into intensity; (2) a wide range of different stimulus durations, from 50 msec to above 1000 msec, yield different intensities; and (3) taste intensity processing is cumulative over time.

Ongoing cumulative processing of sustained gustatory stimulation, with a long delayed judgmental output for magnitude estimates (i.e., long magnitude estimation reaction times), strongly distinguishes taste intensity processing from taste quality processing. It was noted earlier that taste quality processing may require only the initial component of peripheral taste responses. An intriguing aspect of the intensity processing is the great length of magnitude estimation reaction times, about 900 msec longer than simple taste reaction times. In contrast, taste quality identification (descriptor) reaction times were only 200 msec longer than simple taste reaction times. Perhaps the magnitude estimation judgments were made only after the duration-to-intensity cumulation process had reached its maximum value or had started to decrease. If so, an interesting question arises: Are subjects aware that taste intensity is increasing during this cumulation period, or do they have no intensity information until the cumulation process has peaked, and the conditions for a magnitude estimate are reached?

One approach to answering this question would be to signal subjects to report taste intensity at times earlier than the normal magnitude estimation reaction time. The earliest reasonable time would be simple taste reaction time. At times longer than the simple reaction time but less than the magnitude estimation reaction time, one might expect greater and greater taste intensities to be reported. The general procedure of signaling subjects when to make a judgment rather than allowing them to make the judgment when they wish is called cued reaction time (Posner, 1986). This procedure was applied by Pantzer et al. (1987) to magnitude estimations of 300 msec duration 2 mM NaSac or 214 mM MSG, flowed over the anterodorsal tongue with a closed-flow taste stimulus apparatus. A 300 msec stimulus duration was used so that the stimulus solution at the tongue would be removed before the earliest possible responses. Since the vocal simple taste reaction time for 300 msec 2 mM NaSac was 730 msec, for 214 mM MSG, 585 msec (Kelling and Halpern, 1987a), all cued magnitude estimations were expected to have longer reaction times than these values. This approach ensured that all cued responses would be based upon a fixed duration input to the central nervous system from the stimulated receptors. Auditory cueing signals sounded between 730 and 1640 msec after stimulus arrival at the tongue. These times spanned the range from simple taste reaction times to uncued magnitude estimation times.

The experimenters found that the earliest *cued* magnitude estimation responses for 300 msec duration 2 mM NaSac were at 1138 ± 15 msec after stimulus arrival; for 214 mM MSG, at 1086 ± 34 msec. The *cued* MSG magnitude estimates at these times were 100% of the uncued magnitude estimates; the *cued* NaSac estimates, 70% of the uncued estimates. There were no significant changes in the perceived intensity represented by *cued* magnitude estimates, across the five cueing times. Thus, the subjects were unable to provide cued magnitude estimates with reaction times less than 1000 msec. This was surprising because a judgment that might be expected to require considerable central nervous system processing, taste quality (descriptor) identifications, had had reaction times less than 1000 msec (Kelling and Halpern, 1988). The cued magnitude estimates made between 1086 msec and the normal uncued magnitude estimation reaction time were similar in judged intensity to those that would have been made voluntarily at the full time. This result might mean that subjects do not have any intensity information until the duration-to-intensity cumulation process has peaked. Another possibility is that the linguistic nature of the magnitude estimation response is the problem (Carr, 1986). That is, perhaps intensity information is available, but it cannot be expressed as numbers until the cumulation process has reached or passed its maximum. If this is the case, a nonnumerical form of taste intensity tracking should be successful. This approach is discussed below.

Intermittent Stimulation.   Experiments that used continuous stimulus solution flows at various durations have led to the conclusion that a cumulation process located in the central nervous system accounts for the greater taste intensity that is perceived with longer taste stimulus durations. Intermittent stimulation in which flows of solution and solvent alternate might also elicit greater taste intensity when the duration of presentation is increased. For intermittent stimuli, stimulus train duration, generally expressed as train duration, specifies the total time of stimulus presentation. If judged intensity does become greater for increased train durations, the implication is that the intensity-duration function cumulates the separated responses that are elicited by the train. This would be especially strong evidence of a central nervous system locus for the taste stimulus duration-to-taste intensity conversion process.

Some methodological details are necessary at this point. A liquid square wave train exists when the durations of the solution and solvent flows of a train are equal, and the transitions between them are rapid. The time required for one solution flow and one alternate liquid flow is the period of the train, while the number of such complete transitions per second (cycles) is the frequency, in hertz (Hz).

Magnitude estimates of 5 Hz and 10 Hz square wave trains of 2 mM aqueous NaSac increased with train duration for 100, 200, 500, 1000, and 2000 msec durations (Kelling and Halpern, 1987b). The two frequencies produced the same train duration-intensity function. Each cycle of the 10 Hz train would ideally consists of a 50 msec NaSac flow and a 50 msec water flow. This was only approximated by the closed taste stimulus delivery apparatus, since the water to NaSac transition had a rise time of 10 msec, while the NaSac to water transition had a 23 msec rise time (Halpern, 1986a). Nonetheless, these rise times (10 to 90% of final concentration) were sufficiently rapid to provide a series of separated stimulations, and, presumably, separate receptor responses.

It is possible that central nervous system cumulation of responses to square wave trains of gustatory stimuli would result in lower intensities than would continuous stimulus solution flow. This might be expected because of the substantial time intervals during which no stimulus solution is being delivered to the tongue (approximately half of each cycle). During these nonstimulation (and presumably nonresponse) intervals, some of the cumulated receptor input might "dissipate." However, a direct comparison between intensities to the same train and continuous solution flow durations cannot be made because for equal stimulus durations, twice the solution is presented under continuous solution flow conditions. An appropriate approach is to compare intensities for equivalent total solution presentation durations. For example, intensity for a 2000 msec duration continuous solution flow is compared with that for a 4000 msec duration square wave train. When this was done, the 2 mM NaSac trains were at least as effective as continuous solution flow (Kelling and Halpern, 1987b). These observations suggested that the duration-to-intensity cumulation process has a time constant such that a 25 msec interval without stimulation leads to no loss of accumulated input.

A confounding factor in the comparison between responses to square wave stimuli and continuous solution flows is the repeated stimulus presentations that occur during a square wave train. Gustatory responses are often greater at the start of stimulus presentation than later during ongoing stimulation (Halpern, 1985). If taste responses to stimulus onsets in a train follow this pattern, the resulting series of large initial responses might compensate for loss of cumulated input during intervals of nonstimulation. It could then be expected that a higher frequency train would elicit greater intensities, since the number of stimulus onsets for a given train duration would be larger than that with a lower frequency train. This was not observed for the 5 and 10 Hz 2 mM NaSac square wave trains. One plausible interpretation is that initial transients diminish when stimuli are repeatedly presented in these square wave trains. Alternatively, the 5 Hz train, in which 100 msec of water flow separates every 100 msec of 2 mM NaSac flow, may provide more complete recovery from the previous stimulation (i.e., less adaptation) than the 10 Hz train, and thus balance the doubling of transients that the 10 Hz train gives.

A noteworthy general outcome of these intermittent stimulation studies was the observation that greater square wave train length resulted in greater judged intensity. This finding went beyond the limited confines of an examination of adaptation and its prevention by alternation of stimulus solution and solvent (e.g., Halpern et al., 1986). It also supplemented the understanding that could have been achieved by using only continuous solution stimulation. Two conclusions may be stated: (1) peripheral gustatory input is cumulated over time such that longer peripheral stimulation yields increased perceived intensity; and (2) the cumulation process is tolerant of brief interruptions in input, and interruptions approaching 100 msec can be accepted.

Tracking   Single, discrete measures of the perceived intensity or quality of taste stimuli necessarily provide very limited and generally incomplete information on temporal patterns (Pangborn et al., 1983). The response to each presentation of a particular stimulus duration is typically one scaling judgment or one description. Occasionally, repeated judgments are obtained (Meiselman and Halpern, 1970, 1973, 1980; Lawless and Skinner, 1979; Kennedy and Halpern, 1980; Halpern et al., 1986), especially in studies of adaptation (Gent, 1979). However, time intervals between judgments are 5–15 sec, and sometimes as long as 1 min.

Multiple judgments during a single stimulus presentation can be obtained using tracking. The subject is asked to continuously indicate taste intensity or quality. Responses are not made by writing or speaking numbers or words, or by pointing at previously written discrete linguistic or numerical items. Instead, subjects move their fingers or limbs such that the position of a response device is changed. This change in position of the response device denotes change in total taste intensity or change in the intensity of a specified taste quality or descriptor (Pew and Rosenbaum, 1988).

When tracking is done, the scaling can be considered a form of cross-modal matching (Luce and Krumhansl, 1988). Cross-modal or cross-modality matching eliminates the use of numbers (or other linguistic components) in the response (Engen, 1971b). Instead, the perceived taste stimuli are matched by visual or auditory stimuli. Gustatory cross-modal matching is most commonly applied to discrete taste stimuli which are matched individually (e.g., Weiffenbach et al., 1986a,b). For measurements over time, this method is converted into a cross-modal tracking task (Overbosch et al., 1986). Cross-modal tracking has been employed in a number of contexts in addition to taste research. For example, it has received considerable use in studies of kinesthesia (Clark and Horch, 1986).

Tracking of taste intensity or quality requires stimulus presentation to gustatory receptors. Because measurements will be taken over time, stimulus duration can be brief or prolonged. A limited region of taste receptors or the entire oral cavity can be involved. One approach to specifying liquid taste stimulus delivery in tracking studies is to observe the manner in which humans consume beverages. Drinking by humans often consists of a series of sips and swallows. During this process, suction produced inside the mouth transfers liquid into the oral cavity (Halpern, 1977). The transfer per se continues for approximately 1 sec (Halpern, 1985). Subsequent to the transfer or ingestion phase, the liquid bolus is held in the oral cavity for 0.5–4 sec, and then swallowed. Alternatively, if rejection occurs after intraoral ingestion, much of the bolus is spit out, although a substantial concentration remains in the mouth for some seconds (Halpern, 1986a).

This sequence, intraoral ingestion by suction drinking and then a swallow or spit, brings the ingested liquid into contact with many of the oral gustatory receptor populations. It is the most common means for acquiring liquids (Halpern, 1977). Consequently, it is not surprising that many gustatory time-intensity studies present stimuli in this manner. However, there are some disadvantages. One is that responses cannot be related to a particular receptor population or region of the oral cavity; another, that salivary dilution may alter the stimulus over time; and a third, that strict control of stimulus duration is not available. Consequently, a few time-intensity investigations have restricted stimulation to the anterodorsal tongue. They are discussed below.

Prolonged, Anterodorsal Tongue Stimulation. Perhaps the simplest form of tracking is done using a pencil and graph paper. This approach was employed by Holway and Hurvich (1937) in a study of taste intensity. One drop of 0.5, 1.0, 2.0, 3.0, or 4.0 M aqueous NaCl was placed onto the tongue. The authors stated " . . . we undertook to obtain records of the way that the intensity of the salt-process varies as a function of time during a period of 10 sec. The technique employed was quite simple. The O was given a pencil and a sheet of graph paper. The abscissa represented the time axis; the ordinate, the degree of saltiness. The O was instructed to place the point of the pencil on the intersection of the two axes, and to trace the intensity of the process on graph paper. At the end of the 10-sec presentation, he was stopped abruptly" (Holway and Hurvich, 1937). One result was that " . . . maximum intensity is reached later, for a concentration equal to 1.0 mol., than for a concentration equal to 0.5 mol.—although the initial slope of the former is apparently greater than that for the latter" (Fig. 2.4). Maximum intensity also increased with concentration. The authors concluded "Ordinarily the concentration of the gustatory substances is regarded as the stimulus to intensity. A glance at any one of the intensitive diagrams shows, however, that, while the concentration originally placed on the tongue is 'fixed', the intensity varies in a definite manner from moment to moment. Saline intensity depends upon time as well as concentration" (Holway and Hurvich, 1937).

These observations and conclusions apply to all subsequent studies of changes in taste intensity over time that include the onset phases: (1) higher stimulus

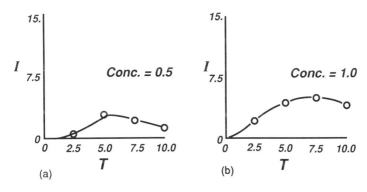

FIG. 2.4 Time-intensity tracking of the saltiness intensity of NaCl solutions. Data are shown for 500 mM NaCl (a) and 1000 mM NaCl (b). Subjects refrained from smoking on testing days. Subjects' nostrils were blocked with cotton; they steadied their heads with chin rests and extended their tongues outside the mouth. Intensity was tracked with a pencil and graph paper for 10 sec after a 8.2 mm^3 drop of solution was released onto the protruded anterodorsal region of the tongue from a burette. A 10 sec water rinse and a 75 sec interstimulus interval followed each stimulus presentation. Each plotted point is the mean of 20 response measurements (10 from each of two practiced subjects). The horizontal axis is time (T) in seconds; the vertical axis, graphed intensity in cm. *(From Holway and Hurvich, 1937.)*

concentrations tend to evoke intensity functions with steeper initial slopes; and (2) maximum intensity is reached later for higher concentrations, and is greater when it is reached. Similar time-intensity patterns, obtained from measurements of interactions between stimulus duration and taste intensity, were described in a previous section.

The authors (Holway and Hurvich, 1937) pointed out that they had provided no latency (reaction time) measurements. They suggested, "Latency could be measured, however, if the pencil were placed in a straight, narrow, vertical groove, and the graph paper were moved horizontally past the groove at a constant rate by means of a revolving drum. In this way more reliable results would be obtained since the O would have to record only the vertical component of the intensitive-temporal complex" (Holway and Hurvich, 1937). This suggestion was adopted for gustatory time-intensity measurements approximately 40 years later (Larson-Powers and Pangborn, 1978).

Reaction time measurements during time-intensity tracking would require both a sufficiently rapid and precise response device and an accurate indication of stimulus arrival at the gustatory receptor surface (Halpern, 1973,1986a,b). Most gustatory time-intensity studies provided neither, but the necessary conditions have been met for a series of experiments discussed below.

More than 40 years after the Holway and Hurvich paper, Lawless and Skinner (1979) reported another time-intensity investigation that restricted stimulation to the anterodorsal tongue. A gravity-driven open stimulus system (see Kelling and Halpern, 1986) delivered 100, 180, 320, or 560 mM aqueous sucrose solutions (3 ml/sec flow rate) over the extended tongue, after a 10 sec water rinse. In order to respond, " . . . the subjects moved a pointer along a line scale 26 cm in length with a zero mark at one end. They were instructed to move the pointer to mark off a distance from zero proportional to the taste intensity. They were told to estimate ratios, so that if the taste intensity doubled from some level, the pointer would be moved twice as far from zero" (Lawless and Skinner, 1979). This ratio scaling procedure is a form of magnitude production, which has considerable similarity to magnitude estimation (Wyszecki, 1986). The position of the pointer was recorded on moving chart paper. Each stimulus solution flowed and was tracked for 2 min unless 15–20 sec of zero intensity was indicated first. Intensity values were obtained every 3 sec from the chart record (H. T. Lawless, personal communication, 1990). No measurements of time to maximum intensity or total response duration were reported because of a 3 sec response latency, which was thought to obscure response onset. Time from maximum intensity to zero intensity was analyzed in several ways.

The authors compared the maximum intensities and postmaximum rate constants obtained using the just described ratio time-intensity tracking approach with those from a 10 ml sip and spit (after 10 sec) procedure in which category judgments were made on a 11-point verbally anchored scale. This psychometric method can be considered a rating procedure in which categorical judgments are made. The name "method of successive categories" is sometimes applied. "Essentially, the experimental operation is that of judging each of several stimuli as belonging in one of a limited number of categories differing quantitatively along a defined continuum. . . . It is assumed only that the categories are in correct rank order and that their boundary lines are stable . . . " (Guilford, 1954). "Category scale data are generally considered to be at least ordinal level data. They do not generally provide values which measure the degree (how much) one sample is more than another" (Meilgaard et al., 1987).

In the Lawless and Skinner (1979) study under discussion, category judgments were written every 5 sec through 25 sec, then every 10 sec until 150 sec. Two other conditions were also used. In one, dorsal flow stimulus presentation was combined with category tracking, so that subjects indicated intensity by moving a pointer along a 11-point scale. In this case, measurements from the chart were made every 5 or 10 sec, as in the written category judgment approach. For the fourth condition, stimulus presentation was by sip and spit, while intensity was indicated by the ratio tracking procedure.

For all four methods, "The taste of sucrose rose rapidly, peaked within 10 sec, and then declined . . . stronger concentrations were judged to have greater peak intensities and to last longer" (Lawless and Skinner, 1979). Sip and spit responses showed greater differences in maximum intensity as a function of concentration and had longer times to zero. Equations fitted to the data from the four procedures were more similar to those from previous experiments when area under the decreasing intensity curve, rather than just maximum intensity values, were used. This was taken to indicate that taste intensity over time normally is used in making single-value intensity judgments.

The written repeated category judgments procedure appeared to provide data that were as precise and accurate as those obtained by the continuous tracking response methods. (Lawless and Skinner, 1979). Since continuous tracking required extraction of numbers from the graphed record, while the written category judgments gave numbers immediately (and needed no chart recorder), the repeated written judgments approach seems superior when measurements are to be made every 5–10 sec. Of course, if greater temporal resolution were needed, written responses would not be satisfactory. This is particularly pertinent to instances in which the rising phase of intensity is of interest.

In defense of their continuous tracking method, Lawless and Skinner noted, "The ratio scaling was coupled to the chart recorder, and therefore produced a more detailed account" (Lawless and Skinner, 1979). However, this was not apparent in the reported data. Investment in the apparatus and techniques required for continuous tracking appears justified only when a temporal resolution superior to that of repeated written judgments is both available and utilized. This has rarely been the case in the time-intensity literature.

Prolonged, Whole Mouth Stimulation. Most investigations of gustatory time-intensity interactions have used similar stimulation procedures and response rules: The stimulus was generally a 5–15 ml bolus of aqueous solution that was sipped, held in the mouth for 5 sec to 2 min, and then either spit out or swallowed. Nonaqueous liquids and gels have also been used (Lee and Pangborn, 1986; Overbosch et al., 1986). In a few experiments, additional sips were taken at specified intervals (Guinard et al., 1986a,b). Intensity of a specified taste quality or descriptor—most often bitterness, sweetness, saltiness, or astringency—was tracked by moving a writing instrument or response device. Tracking studies done since the late 1970s have restricted movement of the response or writing device to a single axis. Sometimes rotation replaced linear motion (Birch and Munton, 1981). The subject's task was to record the taste intensity over time. Once a change in response had occurred, subjects were not permitted to view earlier components of their responses during a trial. Tracking began around the time that the subject sipped the stimulus liquid. Tracking continued beyond a spit or swallow until the intensity of the specified taste quality or descriptor was perceived to be zero or, sometimes, "faint" (DuBois et al., 1981a,b) or was constant for at least 1 minute (Leach and Noble, 1986). Occasionally, tracking was done for a specified period of time—1–5 min (O'Mahony and Wong, 1989). In some studies, a tracking pause of 6 sec was inserted when the subject first perceived maximum intensity (e.g., DuBois and Lee, 1983).

A numerical or descriptive scale was placed next to the response device or writing instrument. Reference solutions for one or more of the scale positions were usually provided. In some instances, subjects were first given extensive practice in assigning particular scale values to specified concentrations (O'Mahony and Wong, 1989), with the intent of securing time-intensity measurements based upon a calibrated scale (O'Mahony and Wong, 1989). Tracked intensity was generally measured every 1–6 sec, even when position of the response device was digitized at 500 msec intervals (Guinard et al., 1985, 1986a,b). Many of the studies that have followed this approach, and their stimulus solutions, are listed in an article on time-intensity techniques and applications (Lee and Pangborn, 1986).

A much simpler procedure has been followed in some studies. "For the time/concentration measurements, subjects used stopwatches to record reaction time, time to maximum intensity, time to end of maximum intensity and total persistence time on separate occasions" (Birch et al., 1980). Timing by stopwatch might be considered a rather primitive technique. However, when prolonged stimulation is used, the accuracy of stopwatch timing for specifying time to maximum intensity and time until zero intensity (extinction) is probably comparable to, and conceivably better than, the more elaborate approaches described above, in which tracked intensity is measured every 1–6 sec (see Birch and Munton, 1981). In similar fashion, a response device (joystick) that was directly input to a digital computer did not provide time-intensity data that were superior in accuracy to those obtained with a writing instrument and moving paper, when measurements were made every 5 sec (Guinard et al., 1985).

Reaction time is, of course, a different matter. As already noted earlier, no useful measures of taste reaction time were obtained in any of the time-intensity investigations that have thus far been discussed. Consequently, although the subjects' task might have been tracking the intensity of a specified taste quality or descriptor, reports of reaction time in these studies (e.g., Birch et al., 1980) should not be considered indications of taste quality reaction time. Precise and accurate knowledge of the time of stimulus arrival is necessary for reaction time (Halpern, 1986a; Kelling and Halpern, 1988), as well as sufficiently frequently measurement of the position of the response device (Yoshida, 1986).

Results of these time-intensity tracking studies are generally shown as graphs and expressed as various specific time intervals. Common temporal values that are provided are *time to maximum intensity* [sometimes referred to as taste onset or taste appearance time (DuBois and Lee, 1983)], *time to extinction,* and *total persistence time of the specified taste quality or descriptor.* The first quantity, time to maximum intensity, is the interval from the beginning of the time-intensity response function until maximum intensity is reached (Birch et al., 1980). It is not a reaction time, since no measure of stimulus arrival is involved. Time to extinction is often taken as the time interval from the beginning of decline in intensity until tracked intensity reaches zero or some very low level (Birch et al., 1980; DuBois and Lee, 1983). Finally, total persistence time or total duration is the time from the beginning to the end of the time-intensity function (Birch et al., 1980). A related measure, area under the time-intensity curve (Harrison and Bernhard, 1984), combines and confounds intensity and duration. All the "time to" values are limited by the temporal resolution of the measurement and analytical procedures. In normal usage the temporal resolution is, at best, ± 1 seconds, and often ± 5 seconds.

The time to maximum intensity is a key event in most of these time-intensity measures. It has several problems. First, the temporal locus of maximum intensity becomes complex if swallowing is done, because tracked intensity tends to increase at the time of swallowing (Guinard et al., 1986a,b). A more serious problem is use of maximum intensity per se. As previously noted, early studies by Bujas and Ostojcic (1939) and Bekesy (1964b) showed that maximum taste intensity is approached asymptotically over time. These observations were confirmed in a later experiment (Kelling and Halpern, 1988). A major observation of these discrete response studies was that 60–93% of maximum taste intensity was reached with a 2 sec stimulus duration, but maximum intensity could require an additional 2–4 sec of stimulation (Figs. 2.2, 2.3). For example, the taste onset time of 5.1 sec reported for 146 mM sucrose by DuBois and Lee (1983) was about the same as the stimulus duration that Bujas and Ostojcic (1939) found necessary for a similar concentration of sucrose to give maximum intensity. However, the latter authors also noted that 72% of maximum was produced by a 2 sec stimulus duration.

Since maxima are approached asymptotically in many systems, rise time is often used instead of time to maximum. Rise time is the time needed for the output of a system to change from a specified small percentage (usually 5 or 10%) of its steady-state level to a specified large percentage (usually 90 or 95%) (Lapedes, 1974). Perhaps this concept can be useful in time-intensity studies. The time required to reach 50, 75, or 90% of maximum intensity may be much more relevant to perceptual judgments of foods than the time required to asymptote at maximum intensity. For example, Beck noted, "Both sucrose and cyclamate-saccharin in combination have above moderate sweet taste at 2 to 3 seconds" (Beck, 1975). Time to fall to 50% of maximum has been measured in some studies (Lawless and Skinner, 1979).

Time-Limited Anterodorsal Tongue Stimulation

(1) intensity. Little information on the onset phase of taste judgments has been provided by most time-intensity studies. The problem is, as indicated above, the low temporal resolution that is utilized. Since a substantial proportion of total intensity is achieved within the first second or two of intensity tracking, even a measurement every second (Yoshida, 1986) will give only one or two data points during this important onset component of taste perception. Much more frequent measurements, e.g., 10 per second, are needed in order to adequately characterize the development of taste intensity perception.

A suitable stimulus presentation procedure is also necessary if the onset of perceived taste intensity is to be studied. High temporal resolution tracking of judged taste intensity may be incompatible with sip and spit or swallow stimulus delivery. A subject's involvement with sipping is likely to make careful control of the response device difficult during the first second or two. The difficulty will reappear at the time of spit or swallow. This suggests that high temporal resolution tracking of taste intensity should use a stimulus flow delivery technique. A closed stimulus delivery apparatus (Kelling and Halpern, 1986) is one suitable approach.

Another unavailable datum from gustatory time-intensity investigations has been the reaction time for the start of tracking. This information is lacking because stimulus arrival times were not generally known. If measurement of stimulus arrival at the tongue were combined with high temporal resolution tracking, plus stimulus delivery through a closed system, the full time course of taste intensity could be measured. This measurement would include the time interval between stimulus delivery to the tongue and the start of intensity tracking (reaction time) as well as detailed information on the onset component that precedes maximum intensity. Time-intensity tracking reaction time is of special interest because, as discussed earlier, the reaction times of magnitude estimates of taste intensity were very long, almost 2 sec (Kelling and Halpern, 1988). Perhaps reaction time measurements based on a nonlinguistic form of taste intensity tracking (Guinard et al., 1985, 1986a,b) would give different results? It might be that taste intensity information is available quite early, but cannot be expressed at that time in numerical, linguistic responses.

Stimulus duration is another factor to consider. Studies discussed in earlier sections demonstrated that judged intensity increased with the duration of either continuous or intermittent taste stimuli. For durations less than 1.5 sec, magnitude estimates of intensity were made long after the stimulus solution had been removed from the tongue (Kelling and Halpern, 1988). This suggests that some portion of tracked taste intensity probably reflects much earlier stimulation at the tongue. Tracking of stimuli with known brief durations is required to provide answers.

Continuous Stimulation.   A series of gustatory time-intensity studies that met the above criteria has been reported (Kelling et al., 1988, 1989). The tongue was stimulated using an automated closed-flow taste stimulus delivery apparatus (Kelling and Halpern, 1986;1987b), which was calibrated by a flow-through conductivity cell in the liquid delivery tube (Kelling and Halpern, 1986). Solution duration and wave form were monitored during data collection trials by a conductivity cell in the liquid discard channel. Each trial consisted of a 10 sec distilled water flow through the liquid delivery tube and across a 39.3 mm^2 area of the anterodorsal tip region of the tongue immediately followed by a stimulus liquid presentation, and then a 15 sec distilled water flow (Kelling and Halpern, 1987a, 1988). The time interval between the trials of a session was 60 seconds.

Each session began with five identified practice trials. For the data collection trials the stimuli (distilled water control stimuli and aqueous stimulus solutions), which were presented for a predetermined range of stimulus durations, were randomized.

A nonlinguistic, high-resolution, visually guided (Pew and Rosenbaum, 1988) taste intensity-tracking procedure provided continuous measures of taste intensity with 100 msec resolution. Subjects indicated taste intensity by the vertical position of a moving visual display on a monochrome video monitor located in front of them. The video monitor's screen was centered on their chin rest. Liquid arrival at the tongue was simultaneous with the beginning of the display. The angular position of a single axis joystick determined vertical location of the display. Subjects were instructed that their task was to track any change in intensity of the taste of a liquid flowing over their tongue, using the joystick. They were told to indicate taste intensity with the vertical position of a bright line image on the screen before them, and that the joystick controlled the vertical position of that image.

At the upper and lower limits of the display on the video monitor were horizontal dotted lines that corresponded to the maximum and minimum positions of the joystick. These lines appeared simultaneously with the beginning of the display. The viewable region of the video monitor screen was delimited at top and bottom by opaque white areas adjacent to the horizontal dotted lines. Just above the beginning of the upper white area the words "Maximum Taste Intensity" were centered; just below the beginning of the lower white area the words "Minimum Taste Intensity" were centered. A horizontal white line 36 mm above the bottom of the viewing area (69% of the available vertical excursion) indicated the taste intensity of a standard stimulus.

Subjects were instructed that their responses should be made in reference to the intensity of a standard which would be given six times during a session, and that they would be notified before each such presentation. They were told that they should indicate the maximum taste intensity that they would perceive during a standard trial by moving the joystick so that the bright line image on the screen arrived at the standard position, which was represented on the screen by the fixed horizontal white line. Subjects were informed that if the taste became less intense, they should show this on the screen by moving the bright line back toward the lower dotted line, which represented minimum taste intensity. They were asked to try to make their graph of the intensity of the taste as accurate as possible, tracking the taste during its entire duration.

With reference to all trials *other than* the standard trials, subjects were instructed to track the intensity of the stimulus during its entire duration in proportion to the standard. They were told that if the intensity of the liquid flowing over their tongue at any moment was half of the standard, they should position the bright line image on the screen equally between the standard line and the lower dotted line. Subjects were informed that the perceived intensity of stimuli other than the standard might fall anywhere on the graph, above or below the standard line, and that if they did not notice a change in the intensity of the taste, the bright line image should remain positioned along the lower dotted line, which corresponded to the lowest position of the joystick. After each trial, they were asked to reposition the joystick to its lowest position.

Horizontal movement of the bright line image of the display occurred every 100 msec across the 80 "columns" of the video monitor, with the previous horizontal and vertical positions of the graphic display retained and visible to the subject during a trial. Position of the joystick was digitized every 100 msec by a microcomputer, processed to produce the next vertical coordinate of the line-segment graphic display, stored in random-access memory, and written to disk at the end of each trial under program control. After the 100 time-intensity values (joystick position for 10 sec from the beginning of a trial) and a trial identification code had been written to disk, the display disappeared for 60 sec until the beginning of the next trial.

Taste intensity tracking using joystick or mouse input to a computer had previously been described (Guinard et al., 1985, 1986a,b; Yoshida, 1986). However, the digitization rate of those studies never exceeded twice per second, and their time-intensity values were never analyzed more frequently that once per second. Consequently, the temporal resolution of the experiments now under discussion (Kelling et al., 1988, 1989) was 10 times greater than that of previous gustatory tracking studies.

Earlier experiments that used joystick tracking of taste intensity presented stimuli by sip and spit or swallow methods. As discussed above, this precluded precise control of stimulus duration and required manually coordinated oral ingestion activity by the subject. These motor activities may have been in competition with tracking. Finally, no measure of stimulus arrival at the tongue was available, thus preventing an accurate indication of reaction time.

For the present experiments, stimulus solutions were aqueous 2 mM NaSac or 10 mM and 30 mM citric acid (Kelling et al., 1988, 1989). Five practice sessions preceded 8 or 10 data collection sessions. In each session, stimulus durations of 100, 500, 1000, and 2000 msec (and sometimes 300 msec) (NaSac) or 800, 1000, and 2000 msec (citric acid) were randomized. The standard stimulus, a 2000 msec flow of 2 mM NaSac or 30 mM citric acid, was presented in an identified trial after every three data collection trials. During each session, the control stimulus (distilled water) and stimulus solution(s) were presented once at each of the durations. Quantitative results are given as medians.

The authors reported that the intensity-tracking reaction time for 2 mM NaSac was 700 msec after stimulus arrival at the tongue for all stimulus durations (Fig. 2.5). This latency was similar to the 760–756 msec simple taste reaction time for the same duration range of 2 mM NaSac (Kelling and Halpern, 1986b) but very different from (722–1306 msec faster than) magnitude estimation reaction times (Kelling and Halpern, 1988). In other words, they found that the initiation of taste intensity tracking was much earlier than magnitude estimates. This indicated that intensity information could be expressed as quickly as simple taste reaction time (an indication that a change in taste had occurred) if the intensity response was tracking rather than a linguistic response. Previous experiments had shown that taste quality vocal reaction times were < 1 sec (Kelling and Halpern, 1988). Thus, within less than 1 sec after taste stimulus onset, a subject is aware of and can report not only the taste quality but also information on taste intensity.

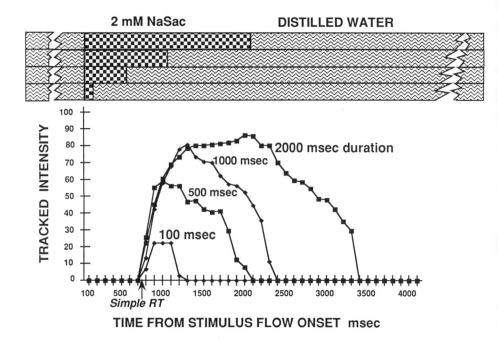

FIG. 2.5 Visually guided high resolution time-intensity tracking of four durations of 2 mM NaSac. The angular position of a single-axis joystick determined the vertical location of a moving display on a computer monitor every 100 msec. A closed stimulus delivery apparatus flowed liquids over a 39.3 mm² area near the tip of the anterodorsal region of the tongue (Kelling and Halpern, 1986). In the figure, the rectangular bar above the graph indicates a 10 sec distilled water prestimulus rinse (wave pattern bars) immediately followed by a 100, 500, 1000, or 2000 msec duration continuous solution flow of 2 mM NaSac (black and white bars) that was immediately followed by a 5 sec distilled water stimulus removal rinse. Data points are the medians of 50 measurements (10 from each of five practiced subjects). Stimulus durations were presented in random order; solution and distilled water control trials were randomized. No intensity changes from zero were indicated on control trials. *(From Kelling et al., 1988.)*

    A different approach to analyzing these data would be to compare the time course of taste stimulation at the tongue with that of the tracking responses. When this was done for the three shortest stimulus durations of 2 mM Na-Sac (100, 300 [not shown in figure], 500 msec), tracking did not start until hundreds of msec after the stimulus solution had been removed from the flowing liquid at the tongue (Fig. 2.5). This substantial difference between the temporal parameters of stimulus events at the tongue and perceived intensity fits with data discussed earlier. Those data had indicated that a central nervous system cumulation process is the basis for the taste stimulus duration-perceived taste intensity relationship that has been often encountered in this chapter.

Maximum taste intensity for 2 mM NaSac was reached 1000 msec after *stimulus* onset for the 100–500 msec durations; 1300–2000 msec after *stimulus* onset, for the 1000 and 2000 msec stimulus durations (Fig. 2.5). Once again the time course of perceived taste intensity had a time course quite different from stimulus events at the tongue.

In agreement with earlier studies that used nontracking measures, maximum intensity increased with stimulus solution duration. However, on control stimulus trials (distilled water), tracked intensity always remained at its lowest level (the lower dotted line of the display) throughout the trial. This indicated no change in intensity from the 10 sec distilled water flow that preceded every stimulus presentation.

The total duration of perceived intensity, often called total persistence time, also increased with 2 mM NaSac stimulus duration. Total persistence time ranged from 500 msec with the 100 msec stimulus duration to 2600 msec for the 2000 msec duration (Fig. 2.5). In every case total persistence time was at least 400 msec longer than stimulus duration. This is yet another taste intensity tracking measure for which the time course of perceived intensity continued well after stimulus events at the tongue had ended.

For the times to maximum intensity discussed above, zero time was stimulus arrival at the tongue. This is not the usual approach in tracking studies. It was indicated earlier that the common practice in taste intensity tracking studies is to measure from the beginning of *tracking* to maximum intensity (Birch et al., 1980; DuBois and Lee, 1983), that is, from the beginning of an indication of nonzero intensity. If that procedure were applied to these data (Kelling et al., 1988), the taste onset or taste appearance time would range from 200 to 1200 msec for 2 mM NaSac, increasing as a function of stimulus duration. Effective stimuli for these rapid taste onset times would include the 1000–2000 msec duration stimuli, which are in the range of normal sips and swallow sequences. This reveals a potential problem of some importance, because intensity tracking studies that make measurements every 3–5 sec would miss much of the onset component of tracking responses. If stimuli with the duration of a nonjudgmental sip and swallow were used (1500–2000 msec), such time-intensity studies might fail to capture any of the tracking response.

Stimulation with 10 mM and 30 mM citric acid solutions (Kelling et al., 1989) produced high temporal resolution taste intensity tracking patterns that were similar to those for 2 mM NaSac (Fig. 2.6). Tracking latency did not

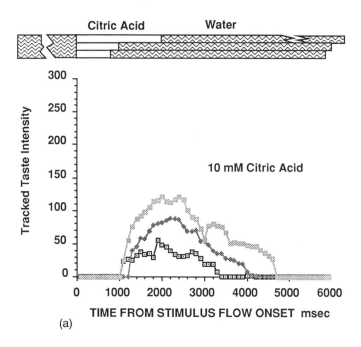

(a)

FIG. 2.6 Visually guided high resolution time-intensity tracking of three durations of 10 mM and 30 mM citic acid. The angular position of a single-axis joystick determined the vertical location of a moving display on a computer monitor every 100 msec. A closed stimulus delivery apparatus flowed liquids over a 39.3 mm^2 are near the tip of the anterodorsal region of the tongue. In the figure, the rectangular bar above the graph indicates a 10 sec distilled water prestimulus rinse (wave pattern bar) immediately followed by a 800, 1000, or 2000 msec duration (clear bars) continuous solution flow of 10 mM citric acid (a) or 30 mM citric acid (b) that was immediately followed by a 5 sec distilled water stimulus removal rinse. Data points are the medians of 40 measurements (8 from each of five practiced subjects). Stimulus durations were presented in random order; solution and distilled water control trails were randomized. Tracked intensity of 800 msec solution flows is shown as open squares with a centered dot; 1000 msec, filled diamonds; 2000 msec, filled squares. No intensity changes from zero were indicated on control trials. *(From Kelling et al. 1989.)*

covary with stimulus duration but was less for 30 mM than for 10 mM concentrations (Fig. 2.7). Time from *stimulus* onset to maximum intensity increased as 10 or 30 mM citric acid stimulus duration lengthened from 800 msec through 1000 msec to 2000 msec.

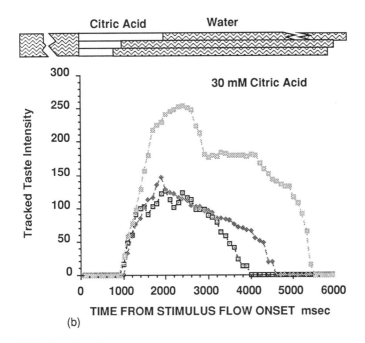

FIG. 2.6 *(cont.)*

Finally, total persistence time also increased with stimulus duration (Fig. 2.6, 2.7). Thus, as with 2 mM NaSac, total persistence time (from the beginning to the end of intensity tracking) increased with stimulus duration, but was always much longer than stimulus duration. For citric acid, the total duration of perceived intensity exceeded stimulus duration by at least 500 msec and as much as 2500 msec (for 2000 msec 30 mM citric acid). Both concentration and duration were significant variables for total persistence time. As expected, maximum tracked intensity for citric acid increased with stimulus duration and with concentration. As in the NaSac experiment, on control stimulus trials (distilled water) tracked intensity always remained at its lowest level (the lower dotted line on the display), indicating no change in intensity from the 10 sec distilled water flow that preceded every stimulus presentation.

For these NaSac and citric acid stimuli, some aspects of the data that were obtained with high temporal resolution taste intensity tracking and controlled stimulus durations corresponded to qualitative descriptions that would also fit low resolution tracking of prolonged stimuli:

1. Intensity increased rapidly to a maximum, and then decreased more slowly.

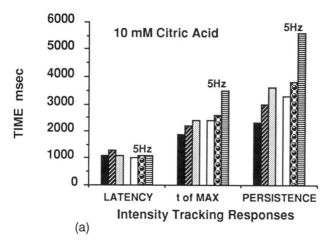

(a)

FIG. 2.7 Results from visually guided high resolution time-intensity tracking of three durations of continuous solution flow and 5 Hz square wave train 10 mM (a) and 30 mM (b) citric acid. Shown are LATENCY (median time from stimulus solution arrival at the tongue to the beginning of the intensity-tracking response), t of MAX (median time from stimulus solution arrival at the tongue to the maximum tracked intensity), and PERSISTENCE (median time from the start to the end of the intensity-tracking response). Each column is the median of 40 measurements (8 from each of five practiced subjects). Stimulus durations were presented in random order; continuous and 5 Hz presentations and solution and distilled water control trails were randomized. The left-hand triad of each group of six columns gives results for continuous solution flow; the right-hand triad, for 5 Hz trains. Solid columns represent median results for 800 msec continuous solution flow; diagonal pattern columns, 1000 msec; dotted columns, 2000 msec. Clear columns represents median results for 1600 msec duration 5 Hz train (800 msec cumulative solution duration); columns with solid spheres, 2000 msec duration 5 Hz train (1000 msec cumulative solution duration); horizontal pattern columns, 4000 msec duration 5 Hz train (2000 msec cumulative solution duration). No intensity changes from zero were indicated on control trials. *(From Kelling et al., 1989).*

    2. Larger maximum intensity and greater total persistence time were produced by a higher stimulus concentration.

    However, there were important quantitative characteristics of these data that required high temporal resolution taste intensity tracking using controlled stimulus durations and a known stimulus arrival time:

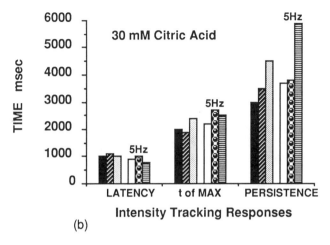

(b)

FIG. 2.7 *(cont.)*

3. Intensity tracking started ≤1 sec after stimulus arrival at the tongue.

4. For sip and swallow duration stimuli (1000–2000 msec), maximum intensity was reached within 2.5 sec after stimulus arrival. This translates into a taste onset time of less than 1700 msec. In general, the onset and rise-times of taste intensity were considerably more rapid than previous tracking studies had suggested.

5. Total persistence time was always much longer than stimulus solution duration at the tongue. It could be more than twice the stimulus duration.

6. Perceived taste intensity had a temporal pattern that differed substantially from the time course of stimulus solution flow at the tongue. Nonetheless, knowledge of stimulus time course, together with information on stimulus molecule and concentration, would allow perceived intensity to be predicted.

Intermittent Stimulation. Square wave trains of aqueous 2 mM NaSac or 10 mM and 30 mM citric acid solutions, and distilled water control stimulus trains, were randomized with continuous solution flows and control stimuli presentations (Kelling et al., 1988, 1989). Taste intensity tracking with 100 msec resolution, controlled stimulus durations, and known stimulus arrival times at the tongue was used as described above. Quantitative results for square wave train stimuli are given below as medians.

As noted earlier, train duration specifies the total time of stimulus presentation. The square wave trains had equal intervals of solution and distilled water flows, and rapid transitions between them. The time required for one solution flow and one distilled water flow is the period of the train, while the number of such complete transitions per second (cycles) is the frequency, in hertz (Hz).

Train durations were 200, 600, 1000, 2000, and 4000 msec, at 5 Hz, for 2 mM
NaSac. Citric acid square wave trains had durations of 1600, 2000, and 4000
msec, at 2.5 Hz and 5 Hz, for both 10 mM and 30 mM concentrations.

Many of the intensity tracking results for square wave trains were similar to
those for continuous solution flow stimuli. For example, intensity-tracking
reaction times (i.e., latencies) for 2 mM NaSac trains (not shown in a figure) at 5
Hz were 800–900 msec; for 10 mM and 30 mM citric acid trains at 2.5 Hz (not
shown in figure) and 5 Hz, 800–1100 msec (Fig. 2.7). Intensity-tracking latency
did not consistently increase or decrease with stimulus concentration, duration,
or frequency, or between continuous and intermittent presentations. On control
train trials (distilled water), tracked intensity always remained at its lowest level,
indicating no change in intensity from the 10 sec distilled water flow that
preceded every stimulus presentation.

Time to maximum intensity tended to increase with train duration (Fig. 2.7).
This relationship had been expected from continuous solution flow data.
Nonetheless, a consistent difference between train and continuous solution flows
times was found. Time to maximum intensity was usually greater by 100–500
msec for trains than for equal duration constant solution flows. The 2 mM NaSac
5 Hz 200–2000 msec duration trains had maxima at 800–2500 msec after
stimulus arrival. With 10 mM and 30 mM citric acid, stimulus durations of 1600
and 2000 msec (cumulative solution durations of 800 and 1000 msec) produced
intensity tracking maxima at 2100 and 2500 msec, respectively (Fig. 2.7).

However, one aspect of the cumulation process began to asymptote at the
longest train duration. With 2 mM NaSac, the tracked intensity of 2000 msec and
4000 msec 5 Hz trains reached maxima at 2500 msec and 2700 msec, respectively. The latter maximum occurred during the train (at approximately 68% of the
train duration) rather than after stimulation had ended. For 10 mM and 30 mM
citric acid 5 Hz trains, the 4000 msec train durations (2000 msec cumulative
solution durations) produced intensity tracking maxima at 3500 msec and 2500
msec after stimulus onset (Fig. 2.7); for 10 mM 2.5 Hz trains, at 2800 msec. In
all these cases, tracked intensity for a 4000 msec duration train was reached
before stimulus presentation ended. One exception was the 4000 msec train of 30
mM citric acid at 2.5 Hz, which had its intensity maximum at 4600 msec. A time
to maximum tracked intensity that exceeded stimulus duration even for a 4000
msec train length may indicate that the frequency and concentration combination
produced near-optimum input to the cumulation process.

Maximum intensity generally increased with train duration and solution concentration, but not with frequency. As noted earlier, a direct comparison between
tracked intensity for square wave train and continuous solution flows of the same
duration cannot be made because for equal stimulus durations, twice the solution
is presented under continuous flow conditions. An appropriate approach is to
compare equivalent total solution presentation durations. When this comparison

was made, NaSac and citric acid gave different results. For 2 mM NaSac, the 5 Hz trains always had lower maximum tracked intensities than comparable continuous solution flows, except with 1000 msec trains (compared to 500 msec constant solution flows). In contrast, for 10 mM citric acid at all durations and 30 mM citric acid except for 4000 msec train durations, trains always had greater maximum intensity (Fig. 2.8). An explanation for this difference is not presently available.

Total persistence times for trains also increased with stimulus duration, generally exceeding stimulus duration by at least 600 msec for the 5 Hz NaSac trains; at least 900 msec, for the citric acid trains (Fig. 2.7). The longest total persistence time, 6300 msec, was for 4000 msec 10 mM citric acid trains at 2.5 Hz. When equal stimulus durations (not total solution durations) were compared, persistence times for trains were similar to those for continuous solution flow presentations.

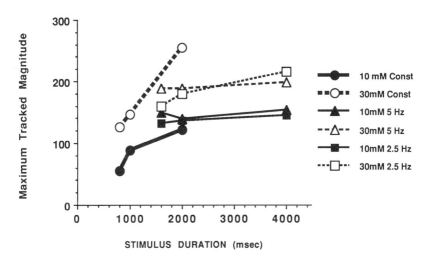

FIG. 2.8 Maximum tracked taste intensity from visually guided high resolution time-intensity tracking of continuous solution flow (circles), 2.5 Hz square wave train (square symbols) and 5 Hz square wave train (triangular symbols), 10 mM (filled symbols and solid lines) and 30 mM (open symbols and broken lines) citric acid. Each data point is the median of 40 measurements (8 from each of five practiced subjects). Stimulus durations for continuous solution flows were 800 msec, 1000 msec, and 2000 msec; for square wave trains, 1600 msec, 2000 msec, and 4000 msec. The latter provide 800 msec, 1000 msec, and 2000 msec cumulative *solution* duration. Stimulus durations were presented in random order; continuous, 2.5 Hz and 5 Hz presentations and solutions and distilled water control trails were randomized. No intensity changes from zero were indicated on control trials. *(From Kelling et al., 1989).*

This series of square wave stimulation experiments with citric acid and NaSac solutions found several similarities between time-intensity tracking of intermittent and continuous stimulus flow:

1. Tracking reaction times for stimulus solution trains were approximately the same as those for continuous stimulus flow.
2. On control train (distilled water) trials, tracked intensity always remained at zero.
3. Time to maximum tracked intensity increased with train duration.
4. Maximum tracked intensity increased with train duration.
5. Total persistence time increased with train duration; it always exceeded train duration.

These observation suggests that 2.5 Hz and 5 Hz trains produced many of the perceptual consequences of continuous taste solution flows. The similarities are important because they provide insights into the parameters of the underlying processes. Stimulus trains had repeated 200 msec (2.5 Hz trains) or 100 msec (5 Hz trains) intervals during which distilled water rather than solution flowed over the tongue. It appears that a relatively long time constant characterizes the central nervous system cumulation process that converts taste stimulus duration into perceived intensity.

Qualities or Descriptors

Single-Quality Intensity Tracking.   Gustatory time-intensity tracking studies have often specified that a particular taste quality or descriptor was to be tracked, rather than total intensity. This approach must assume: (1) more than one taste quality or descriptor will be present during the tracking interval; and (2) subjects can selectively attend to, and quantitatively scale, a specified taste quality or descriptor. To the extent that these two assumptions are correct, these experiments can provide information on the time course of the selected taste perception.

The linguistic descriptions of the taste qualities or descriptors that were to be tracked have been quite broad. Generally, a single quality or descriptor was specified. For example, subjects were instructed to track the intensity of sweetness (Birch and Munton, 1981; Harrison and Bernhard, 1984; Liu and MacFie, 1990; Yoshida, 1986; Yoshida and Mochizuki, 1987, 1988), bitterness (Pangborn et al., 1983; Guinard et al., 1985, 1986a; Leach and Noble, 1986), saltiness (Holway and Hurvich, 1937; O'Mahony and Wong, 1989), sourness (Yoshida and Mochizuki, 1989), umami (O'Mahony and Wong, 1989), or astringency (Guinard et al., 1986b). Intensity tracking of four quality categories (sweetness, bitterness, sourness, and saltiness) has also been done (Meiselman and Halpern,

1970). Special problems may exist for bitterness and sweetness intensity tracking. Behavioral and neurophysiological data have indicated that neither one is a unitary category (Faurion, 1987; Faurion et al., 1980; Lawless and Stevens, 1983; McBurney, 1984; Herness and Pfaffmann, 1986; Schiffman et al., 1986).

Standards were sometimes used to provide references for the ends or intermediate locations of a taste intensity tracking scale. These standards may have helped obtain reliable intensity tracking. They may also have complicated or confused tracking of the intensity of a specified taste quality or descriptor. This may have been the case because the standards were generally solutions of one molecular species, but the solutions tracked in a single experiment often had several different compounds as solutes. Subjects may have taken the standard to represent not only a given intensity but also the taste quality or descriptor that was to be tracked. When subjects subsequently tracked the sweetness or bitterness intensity of unknown solutions containing different molecules, did they try to track only the type of sweetness or bitterness produced by the standard, and attempt to reject all others? This is a potentially serious problem. Use of several different sweeteners or bitter-tasting substances as standards in single-quality intensity tracking studies might resolve the issue.

Perhaps standards could represent solely taste intensity if their other characteristics were sufficiently different from the test stimuli for which single-quality intensity tracking was to be done. This approach has been followed by Yoshida and Mochizuki (1989). They used three sucrose solutions as standards for time-intensity tracking of the sourness of six organic acids (acetic, succinic, lactic, maronic [*sic,* perhaps malonic acid], tartaric, and citric) and three "umami substances" (MSG, inocinic acid [*sic,* perhaps inosinic acid], and sodium guanylate). General procedures were as in Yoshida, 1986. The authors compared the time-intensity curves for the "umami substances" with data from a previous study in which sweetness had been tracked. The sourness curves for the three "umami substances" differed from the sweetness time-intensity curve. This indicates that instructions to track a particular taste quality or descriptor can have consequences.

Multiple-Quality Tracking.   A taste quality tracking method that allowed subjects to specify what taste qualities or descriptors were perceived, and their time course, has been reported (Zwillinger et al., 1990). Changes in the presence or absence of taste descriptors or qualities over time were measured using a 23-item single-letter computer keyboard code. In contrast to all other taste tracking procedures that have been discussed in this chapter, intensity per se was not scaled.

Three aqueous stimulus solutions and distilled water were used in this multiple taste quality-tracking experiment. The solutions were 2 mM NaSac, 214 mM MSG, and a mixture, 10 mM citric acid in 2 mM NaSac. The authors designated this mixture Artificial Lemonade (ArtLem). Each subject participated in 4 practice and 12 data-collecting sessions. In each session, two of the three solutions and distilled water were presented six times each, in random order, after three practice trials. The practice trials were two identified (as having a taste) solution practice trials and one identified (as not having a taste) distilled water practice trial. They were also in random order.

Liquids were presented with a closed taste delivery apparatus (Kelling and Halpern, 1986). A microprocessor controlled a four-way Teflon chromatography valve which determined whether distilled water or a stimulus solution flowed through a polypropylene delivery tube. A subject sealed an elliptical opening (10 × 5 mm, 39.3 mm^2) in the delivery tube with the anterior dorsal surface of their tongue. Distilled water flowing at 3 ml/sec from a separate source rinsed areas of the tongue not sealing the opening. Data-collecting trials were monitored with a flow-through conductivity cell in the waste channel of the Teflon valve. Each trial consisted of a 10 sec distilled water flow through the delivery tube followed by a 1000 msec stimulus delivery and a 5 sec water flow immediately following (for stimulus delivery procedure see Kelling and Halpern, 1986, 1987b). Liquid flow rate was 10 ml/sec. There was a 90 sec interval between trials.

Subjects were first trained and tested with a typing program until they could touch-type at least 45 words per minute on a computer keyboard with at least 95% accuracy. They then learned a 23-item key press code which represented taste descriptor words, by memorizing a list of words and the associated letter code. Next, they were tested for touch-typing accuracy and speed in response to displays of the taste descriptor words on a computer monitor with the subject positioned in the taste stimulus delivery setup. When subjects were 100% correct on key presses, and all key press reaction times were ≤550 msec with liquid flow, practice sessions in the multiple quality tracking experiment began.

Practice sessions continued until five performance criteria were met. These criteria assured that subjects could indicate when they noticed a taste beginning by pressing a key, could press the keys that corresponded to the tastes they perceived, could indicate when they no longer noticed a taste by pressing the space bar (the space bar corresponded to the "no taste" descriptor of the code), could switch from one key to another if they noticed a change in taste during a trial, and would notice no change in taste during distilled water control stimulus trials.

Each subject's task was to track the taste of the liquid flowing over her tongue. Subjects entered their responses by pressing a key on the keyboard of a computer. For each trial, 80 successive taste quality judgments were recorded. Two-hundred msec after stimulus arrival at the subject's tongue, an asterisk denoting "no taste" appeared on the computer monitor in front of her, and a "no taste" response for the preceding 200 msec was stored in the computer's memory. Since 200 msec is much shorter than the fastest taste quality reaction times

(Kelling and Halpern, 1988), an automatic initial "no taste" response was considered to be appropriate on all trials. Unless and until the first key press (other than the space bar ["no taste"]) was made after trial initiation, another asterisk was displayed every 200 msec in successive columns of the monitor, for 16 seconds, and another "no taste" response was stored in the computer's memory. When and if the first key had been pressed, the letter pressed would appear on the display at the appropriate successive column location, or, if it was an illegal key (one not in the 23 item descriptor code), a "?" would appear on the display. The letter pressed and the time of that key press would be stored in memory. For the remainder of the 80 response measurements, the first legal key press was repeatedly stored in memory and displayed at 100 msec intervals, unless either the space bar was pressed, denoting "no taste," or a new legal key was pressed. Each such legal key press or space bar press was recorded at the time at which it occurred and was displayed on the computer monitor screen. A space bar press caused the "no taste" symbol, *, to be recorded and displayed every 200 msec unless and until another key (other than the space bar) was pressed.

The 23 descriptors that comprised the keyboard code were derived from spoken taste quality descriptions that other subjects had given in previous experiments (Halpern, 1987; Kelling and Halpern, 1988). They were: "Acid, Bitter, boUillon, breaD, Chocolate, citRus, Egg, fIsh, Fruity, Honey, Lemon, meatY, Metallic, MilK, Salty, SacchariN, soaPy, sOur, suGar, sWeet, Vegetable, Tomato," and "no taste" (space bar) (the capitalized letter in each descriptor is its key press code).

The authors (Zwillinger et al., 1990; Zwillinger and Halpern, in press) reported that the most frequently selected descriptors for the stimuli corresponded to results from earlier nontracking experiments. For example, sweet (a W key press) was the response on 75% of the trials in which NaSac was the stimulus, while sugar (a G key press) was the response on 6.5% of the 2 mM NaSac trials (Fig. 2.9). In similar fashion, sour (an O key press), citrus (a C key press), or sugar represented 43, 20, and 11% of the responses to ArtLem. As expected (see Halpern, 1987; Kelling and Halpern, 1988) many different descriptors were used for 214 mM MSG, with no one descriptor accounting for more than 28% of the responses (Fig. 2.9). Distilled water control stimulus trials were always "no taste" for the entire trial. On the other hand, "no taste" was never the descriptor throughout an ArtLem trial, and was the sole descriptor on only 0.5% of the 214 mM MSG and the 2mM NaSac trials.

Several new temporal patterns of taste quality were revealed by the quality-tracking technique. For example, two descriptors during a single trial, separated by a "no taste" (space bar key press) interval, was a relatively frequent response to MSG (23% of all 214 mM MSG trials). The most common combinations of this sort were soapy (a P key press), then "no taste," then bitter (a B key press),

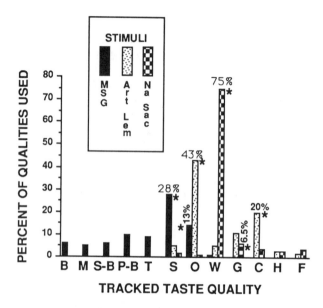

FIG. 2.9 Percentages of taste quality descriptors selected during taste quality track-ing of stimulus liquids. Percentages were calculated from responses made by five practiced subjects to 48 presentations of each solution. Subjects indicated with a computer keyboard the nature, times of occurrence, and durations of the taste quality categories (descriptors) evoked by stimulus liquids flowed through a closed delivery system over 39.3 mm^2 of the anterodorsal tongue tip region for a duration of 1000 msec. Each stimulus presentation was preceded by a 10 sec distilled water flow and immediately followed by 5 sec flow of distilled water. Subjects used touch-typing to press the single keys that corresponded to the taste words of a previously learned and practiced 23 item code. Responses could indicate a single taste quality category (descriptor) for a trial, multiple taste quality categories (a change in taste from one descriptor to another), and no taste. The 12 most frequently used taste quality descriptors (or temporal sequence of descriptors) across the three stimulus solutions are represented on the horizontal axis by the keyboard letter or letters used to indicate that quality descriptor or sequence of descriptors. B = bitter; C = citrus; F = fruity; G = sugar; H = honey; M = metallic; O = sour; S = salty; T = tomato; W = sweet; P-B = soapy-*gap*-bitter; S-B = salty-*gap*-bitter. The stimulus solutions were 214 mM MSG (MSG; black columns), 10 mM citric acid in 2 mM NaSac (ArtLem; dotted columns), and 2 mM NaSac (black and white columns). These solutions were randomized with distilled water control stimuli, which received "no taste" on all trials. For each stimulus solution, the two most frequently selected quality de-scriptors are marked with asterisks. The percentage represented by an asterisk-marked column is indicated at the top of that column. *(From Zwillinger et al., 1990, and Zwillinger and Halpern, in press.)*

which occurred on 10% of all MSG trials, and salty (an S key press), then "no taste," then bitter, on 6% of all MSG trials (Fig. 2.9). The "no taste" interval between the other two descriptors for 214 mM MSG had a median duration of 843 msec. It is important to recall that the "no taste" tracking response required an active key press (the space bar) once any other key press had been made. In contrast to these results with MSG, this type of pattern, descriptor #1—"no taste"—descriptor #2, was very rare for 2 mM NaSac (1.0% of all NaSac trials) and ArtLem (1.5% of all ArtLem trials).

Other plausible quality-tracking patterns were infrequent for all stimuli. Thus, two different successive descriptors *without* an intervening interval of "no taste" was never indicated for 2 mM NaSac, was rarely given for MSG (1.5% of all MSG trials), and was uncommon (4%) for ArtLem. Equally uncommon were tracking responses for which two descriptors were perceived to be present *at the same time* (never for 2 mM NaSac, 0.5% of trials for MSG, and 2.5% of trials for ArtLem).

Reaction times to the 1st descriptor key press during quality tracking seemed long (Fig. 2.10). The quality-tracking reaction time for MSG was 1353 ± 60 msec (median ± standard error of the median); for NaSac and ArtLem, 1145 ± 0 msec. In a previous experiments (Kelling and Halpern, 1988), vocal taste quality identification reaction times for 1000 msec stimulus durations had been 886 ± 44 msec for MSG, 768 ± 37 msec for ArtLem, and 923 ± 41 msec for 2 mM NaSac. A comparison between the latter taste quality reaction times and the quality-tracking results suggests that quality-tracking reaction times were approximately 200–500 msec slower than vocal quality identification reaction times. Differences of this magnitude might indicate that the quality-tracking task was very difficult for subject. The multiple item key press code and the key press responses themselves could be sources of such difficulty.

However, in the taste quality identification component of the cited experiment (Kelling and Halpern, 1988), two solutions (e.g., ArtLem and MSG), plus distilled water, were repeatedly presented across sessions. The practiced subjects may have developed a strategy of not responding if a change in taste did not occur (distilled water trials) and responding in one of two ways (an ArtLem or a MSG response) if there was a change in taste. When the taste change wasn't ArtLem, then a MSG-type response was always correct. A response strategy of this type could permit relatively brief reaction times.

If a larger number of stimuli were used from which a random subset were presented in any one session, longer taste quality identification reaction times might occur. This would be expected because the procedure would prevent highly practiced "A" or "B" responses. Such a design was used in another taste quality identification study (Halpern, 1987). In each session, three out of four stimulus solutions were randomly selected for presentation. Under these circumstances, 1000 msec flows of 2 mM NaSac or 214 mM MSG had quality

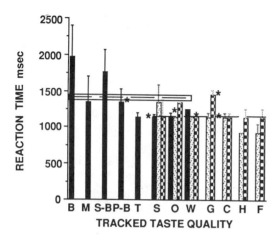

FIG. 2.10 Reaction times, in msec, from stimulus liquid arrival at the tongue to 1st taste quality descriptor key press during taste quality tracking of three stimulus solutions. Five practiced subjects tracked each solution a total of 48 times across 12 sessions. Subjects indicated with a computer keyboard the nature, times of occurrence, and durations of the taste quality categories (descriptors) evoked by stimulus liquids flowed through a closed delivery system over 39.3 mm^2 of the anterodorsal tongue tip region for a duration of 1000 msec. Columns are median reaction times, and the standard error of the median, of the 12 most frequently used taste quality descriptors (or temporal sequence of descriptors) across the three stimulus solutions. The 12 descriptors are represented on the horizontal axis by the keyboard letter or letters used to indicate that quality descriptor or sequence of descriptors. The stimulus solutions were 214 mM MSG (MSG; black columns), 10 mM citric acid in 2 mM NaSac (Art ArtLem; dotted columns), and 2 mM NaSac (black and white columns). These solutions were randomized with distilled water control stimuli, which received "no taste" on all trials. For each stimulus solution, the columns representing the two most frequently selected quality descriptors are marked with asterisks. The left-hand dashed horizontal line at 1353 msec, and the rectangle surrounding it, represent the overall median reaction time to the 1st taste quality descriptor key press for MSG, ± the standard error of the median. The right-hand continuous horizontal line at 1145 msec indicates for NaSac and for Art ArtLem the overall median reaction time to the first taste quality descriptor key press for these solutions, ± the standard error of the median. See text and Fig. 2.9 legend for procedural details. *(From Zwillinger et al., 1990, and Zwillinger and Halpern, in press.*

identification reaction time of 1117–1121 msec ± 52–68 msec. The latter vocal quality identification reaction times are relatively similar to the quality-tracking experiment reaction times. Of course, only three solutions, of which two were randomly selected for a session, were used in the quality-tracking experiment (Zwillinger et al., 1990; Zwillinger and Halpern, in press). Nonetheless, the lack of stimulus predictability may have been an important factor in the length of the quality-tracking reaction times.

The multiple descriptor quality-tracking responses were especially interesting because they occurred after the end of stimulus presentation at the tongue. That is, all stimuli had 1000 msec durations, while the quality-tracking reaction times for the initial key press were ≥1145 msec (Fig. 2.10). Since the closed stimulus delivery apparatus had a removal time (90 to 10% of full concentration) of 23 msec (Halpern, 1986a; Kelling and Halpern, 1986), the solution concentration of the flowing liquid at the tongue was below threshold even before the first quality tracking responses began. The second descriptor key presses that followed a "no taste" interval started much longer than 1000 msec after the end of stimulation at the tongue. This suggests that taste quality processing in the central nervous system can operate in a discrete, serial manner. Once processing for a taste quality is complete, the same input can apparently yield a second taste quality. The second processing time was often substantial, since the median "no taste" interval was 843 msec. Perhaps much of this interval is required for the second bout of processing? This possibility was supported by the virtual absence of responses that would have represented two contiguous descriptors.

Simultaneous perception of the taste qualities or descriptors represented by two descriptors was very rare, and, as noted above, two different but contiguous descriptors without an intervening "no taste" interval was quite uncommon. These observations could suggest that taste quality perceptions are actually always simple and can appear complex only because successive perceptions are combined into a single linguistic report. However, some readers might assert that certain of the individual descriptors already combined several taste qualities or descriptors. This might include citrus, which was a common descriptor for ArtLem (20% of trials), as well as metallic, soapy, and tomato, which were used with 214 mM MSG (5.5%, 11%, and 9% of trials) (Fig. 2.9). Thus, single descriptors that may represent complex perceptions were directly selected. If these key press selections for MSG were accidental, similar percentages would be expected across all stimulus solutions. This was not the case. Metallic or soapy were never used for NaSac or ArtLem, and tomato was rare (≤1%) for these two stimuli. In similar fashion, citrus was never used for MSG (Fig. 2.9).

The duration of any one tracked taste quality or descriptor was surprisingly brief. The most common descriptors indicated for the three stimulus solutions had median durations ≤650 msec (Fig. 2.11). For 2 mM NaSac in particular,

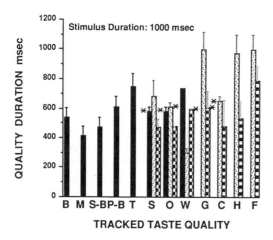

FIG. 2.11 Durations of 1st taste quality, in msec, during taste quality tracking of three stimulus solutions. Five practiced subjects tracked each solution a total of 48 times across 12 sessions. Subjects indicated with a computer keyboard the nature, times of occurrence, and durations of the taste quality categories (descriptors) evoked by stimulus liquids flowed through a closed delivery system over 39.3 mm² of the anterodorsal tongue tip region. The duration began when the 1st taste quality descriptor key (other than "no taste") was pressed, and continued, *without a need for a prolonged key press,* until a different key press was made. That is, the "no taste" tracking response which generally ended a taste quality descriptor duration required an *active* key press (the space bar) once any other key press had been made. Columns are median durations, + the standard error of the median, of the 12 most frequently used taste quality descriptors (or temporal sequence of descriptors) across the three stimulus solutions. The 12 descriptors are represented on the horizontal axis by the keyboard letter or letters used to indicate that quality descriptor or sequence of descriptors. The stimulus solutions were 214 mM MSG (MSG; black columns), 10 mM citric acid in 2 mM NaSac (ArtLem; dotted columns), and 2 mM NaSac (black and white columns). These solutions were randomized with distilled water control stimuli, which received "no taste" on all trials. For each stimulus solution, the columns representing the two most frequently selected quality descriptors are marked with asterisks. See text and Fig. 2.9 legend for procedural details. *(From Zwillinger et al., 1990, and Zwillinger and Halpern, in press.)*

duration across all descriptors (other than "no taste") was 594 ± 11 msec (median ± standard error of the median). This was unexpected. Tracking of 2 mM NaSac total intensity, using the same stimulus delivery system, tongue area, and 1000 msec stimulus duration (Fig. 2.12), had found a total duration (persistence time) of 1600 msec (median) (Kelling et al., 1988). The relatively brief durations of quality-tracking descriptors could not be due to inattention by

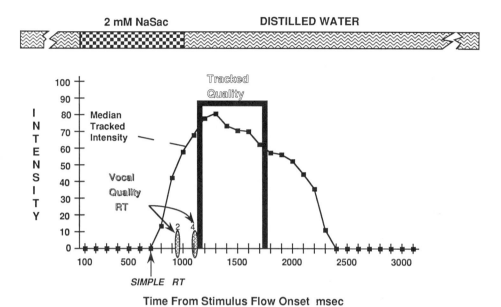

FIG. 2.12 A comparison of the temporal patterns of tracked taste intensity and tracked taste quality in response to stimulation of the anterodorsal tongue tip region with 1000 msec duration 2 mM NaSac. In the figure, the rectangular bar above the graph indicates a 10 sec distilled water prestimulus rinse (wave pattern bars) immediately followed by a 1000 msec duration continuous solution flow of 2 mM NaSac (black and white bar) that was immediately followed by a 5 sec distilled water stimulus removal rinse (wave pattern bars). In the graph, the continuous line and filled squares represent the median taste intensity values produced by visually guided high resolution time-intensity tracking of 2 mM NaSac, as described for Fig. 2.5 (Kelling et al., 1988). The rectangle labeled Tracked Quality represents the median reaction time (latency) and duration of the 1st taste quality tracking key press response for 2 mM NaSac, as described for Figs. 2.9, 2.10, and 2.11 (Zwillinger et al., 1990; Zwillinger and Halpern, in press). The arrow at 700 msec denotes the simple taste reaction time to 2 mM NaSac (time from solution arrival at the tongue to detection of any change in taste) (Kelling and Halpern, 1987a). The cross-hatched ellipse at 923 msec indicates the median vocal taste quality identification reaction time for 2 mM NaSac, ± the standard error of the median, from an experiment in which subjects spoke descriptions of 2 stimulus solutions and distilled water (Kelling and Halpern, 1988). The cross-hatched ellipse at 1120 msec indicates the median vocal taste quality identification reaction time for 2 mM NaSac, ± the standard error of the median, from an experiment in which subjects spoke descriptions of four stimulus solutions and distilled water (Halpern, 1987).

subjects. A key press on a different key was required to indicate that a descriptor was no longer perceived. In the absence of a second key press, a descriptor, once selected, would continue to be repeatedly displayed on the computer monitor and recorded as present for the remainder of the trial.

Total intensity durations that exceed taste quality durations by approximately 1000 msec raise interesting questions (Fig. 2.12). How can there be a taste intensity in the absence of taste quality? Do comparable phenomena exist in other sensory domains? If so, what explanations have been offered?

One possibly related observation is the absence of perceived color under conditions of dim illumination (Hood and Finkelstein, 1986). Taking this approach for single descriptor quality-tracking responses, one might suggest that the taste stimulus intensity became too low to support taste quality processing, but was still sufficient to permit cumulative intensity processing. This "explanation" is totally ad hoc. Even worse, it does not comfortably fit the pattern in which two different descriptors were selected, separated by a "no taste" interval. Under the latter circumstances, the second descriptor is likely to be perceived at the time when tracked taste intensity is at or near zero.

Auditory and somesthetic perception also contain low stimulus intensity situations in which the presence of a stimulus is detected, but no reliable description of stimulus pitch or vibration rate (Mountcastle et al., 1990) is possible. For hearing, this concept is represented by the difference between audibility and tonality thresholds. When an audibility threshold is measured, subjects are instructed to respond to " . . . something that is different from silence," but for a tonality threshold subjects are told to respond if they hear " . . . a tone with characteristic pitch" (Hirsh, 1952). The difference between these two threshold criteria is approximately 7 db. The relevant intensity range is sometimes called the atonal interval or atonal gap. A perhaps related but larger threshold difference is that between the speech detectability threshold and the speech intelligibility threshold. In this case, the thresholds are separated by 12 db (Hirsh, 1952). Once again, the descriptor #1—"no taste"— descriptor #2 pattern found for MSG quality tracking is a problem, since it cannot be explained simply by differences between a taste detectability threshold and a taste quality (descriptor) threshold. This pattern appears similar to the common experience in which one hears something spoken, attaches a meaning, and then decides that a second meaning, different from the first, is or may be correct. The spoken utterance and the peripheral processes that it produced are over before the second meaning develops. Of course, an analogy of this sort is not a substitute for a testable hypothesis. Further investigations are necessary to verify, and to help to explain, the apparent brief taste quality or descriptor durations, and the multiple descriptor patterns, that have been reported during taste-quality tracking.

SUMMARY

Many different approaches have been useful in attempts to understand the temporal characteristics of human taste intensity and taste quality judgments. Studies on gustatory temporal modulation indicated that the human taste system is very responsive to temporal aspects of stimulation. Although taste adaptation occurs with prolonged constant stimulation, it is accompanied by increased sensitivity to small changes away from the adapting stimulus. On the other hand, relatively small, slow changes, such as a 28% modulation between 50 mM NaCl and the 36 mM Na^+ artificial saliva, are sufficient to maintain constant judged intensity.

Investigations using intermittent stimulation demonstrated that greater square wave train length resulted in greater judged intensity. These data led to two conclusions: (1) peripheral gustatory input is cumulated over time such that longer peripheral stimulation yields increased perceived intensity; and (2) the cumulation process is tolerant of brief interruptions in input. Interruptions approaching 200 msec can be accepted.

Taste quality (descriptor) reaction time studies provided two observations that are especially relevant to an analysis of temporal factors. First, human taste responses can occur to very brief stimuli, such as a single 50 msec solution pulse. This indicates a sensitivity to gustatory events at the tongue with durations only 0.5% of the sips that human typically produce. It follows that some aspects of human taste-dependent behavior only require, and may normally only use, the initial portions of responses to stimulation. Second, reaction time changes occur with stimulus duration for simple reaction times but apparently not for taste quality (descriptor) identification reaction times. Perhaps the central nervous system processing time required for the identification reaction times becomes the predominant factor.

A relationship between judged taste intensity and taste stimulus duration has been repeatedly observed. Over at least a 200–fold range, duration of taste stimulation is processed in a cumulative but nonlinear manner into taste intensity. This cumulation function starts at stimulus durations much shorter than those produced by human drinking. It extends into stimulus durations that occur during normal sipping and swallowing. For very brief durations, small increments in duration lead to large increases in intensity, but at long durations, much longer stimuli provide only slight increases in intensity. There is little doubt that the locus of this cumulative intensity processing over time is the human central nervous system. This ongoing processing of sustained gustatory stimulation had long magnitude estimation reaction times. These data clearly distinguished taste intensity processing from taste quality (descriptor) processing.

Cued magnitude estimation experiments on taste intensity found that subjects were unable to provide cued magnitude estimates that had reaction times less than 1000 msec. This was surprising because a judgment that might be expected to require considerable central nervous system processing, taste quality (descriptor) identifications, had had reaction times less than 1000 msec under some circumstances. The cued magnitude estimates made between 1086 msec and the normal uncued magnitude estimation reaction time were similar in judged intensity to those that would have been made voluntarily at the full time. This result might have meant that subjects did not have any intensity information until the duration-to-intensity cumulation process had peaked. Another possibility was that intensity information was available, but could not be expressed as numbers until the cumulation process had reached or passed its maximum. If this were the case, a nonnumerical form of taste intensity tracking should be successful.

In 1937, gustatory time-intensity tracking data derived from graph paper and a pencil were reported. Two general observations from that experiment can be applied to all subsequent studies of changes in taste intensity over time which measured the onset phase: (1) higher stimulus concentrations tended to evoke intensity functions with steeper initial slopes; and (2) maximum intensity was reached later for higher concentrations, and was greater when it was reached.

The apparatus used for time-intensity tracking has become increasingly complex. A stopwatch or pencil and graph paper were replaced with a moving chart on which a manually held or remotely controlled pen wrote. In many instances, the latter has given way to response devices such as a mouse or joystick that inputs directly to a digital computer. Investigators often assumed that the more complex apparatus was necessarily better. However, elaborate equipment was not always advantageous for gustatory tracking research. When low temporal resolution tracking was done, with measurement every 3–5 sec, written judgments were as useful as time-intensity tracking recorded on a remote chart or in a digital computer.

On the other hand, high temporal resolution gustatory tracking, with measurements every 100 msec, did require direct input of joystick position or keyboard key presses into a digital computer. This type of measurement was necessary when the rising phase of taste intensity was of interest, when relationships between taste stimulus duration and tracking responses were investigated, when the reaction time of gustatory tracking was under study, or when the temporal pattern of taste quality (descriptors) tracking was measured. For these questions, tracking with high temporal resolution was combined with detection of the time of stimulus arrival at the tongue. Stimulus delivery was done using an automated closed stimulus system, which eliminated the possibility of attention being divided between ingestion or swallowing versus tracking actions. This stimulus presentation system also provided control of stimulus duration.

Data from the high temporal resolution taste-intensity tracking studies permitted a series of conclusions:

1. Taste intensity increased rapidly to a maximum, and then decreased more slowly.

2. Larger maximum taste intensity and greater total persistence time were produced by a higher stimulus concentration.

3. Taste intensity tracking started ≤1 sec after stimulus arrival at the tongue.

4. For sip and swallow duration stimuli (1000–2000 msec), maximum taste intensity was reached within 2.5 seconds after stimulus arrival. This translates into a taste onset time of about 1700 msec. The onset and rise times of taste intensity were considerably more rapid than had been suggested by low temporal resolution tracking studies.

5. Total persistence time was always much longer than stimulus duration. It could be double the stimulus duration.

6. Perceived taste intensity had a temporal pattern that differed substantially from the time course of stimulation at the tongue.

Taste-quality tracking produced both expected and novel results. As expected, the most frequently selected descriptors for the stimuli corresponded to results from earlier nontracking experiments. However, a previously unreported temporal pattern of taste quality was also revealed by the quality-tracking technique: two descriptors during a single trial, separated by a "no taste" interval, occurred on almost one-quarter of MSG trials. This was not found for other stimuli. On the other hand, two different successive descriptors *without* an intervening interval of "no taste," or tracking responses for which two descriptors were perceived to be present *at the same time,* rarely occurred for any solution. Another unexpected finding was the duration of taste quality-tracking responses. Subjects indicated that taste quality ended in about 650 msec. This was very surprising, because the total duration (persistence time) of taste intensity was almost three times as long.

In general, humans perceived tastes as changing experiences originating in the mouth, which normally existed for a limited time and then either subsided or transformed into qualitatively different gustatory perceptions. Taste experiences did not begin at the moment of stimulus arrival in the mouth, did not suddenly appear at full intensity, were influenced by the pattern of taste stimulation, and often continued well beyond stimulus removal. These temporal patterns of perceived intensity differed, sometimes substantially, as a function of stimulus molecule. In many cases, direct measurements of stimulus concentration at the tongue and control of stimulus duration were necessary for these generalizations.

ACKNOWLEDGMENTS

I thank Herbert N. Wright for a valuable reference, Carol Krumhansl for discussions of psychophysics and acoustics, Bruce Oakley for a critical reading of the section on human neurophysiological taste responses, and Kathleen M. Dorries, Pauline A. Halpern, and the editors of this volume for critical readings and helpful comments on the entire chapter. Original data were collected with support from the United States National Science Foundation and from the Umami Manufacturers Association of Japan.

REFERENCES

Algom, D., Rubin, A., and Cohen-Raz, L. (1989). Binaural and temporal integration of the loudness of tones and noises. *Perception and Psychophysics.* 46: 155–166.

Bartley, S. H. (1959). Organ systems in adaptation: certain special senses. In *Handbook of Physiology. Section 4: Adaptation to the Environment* (J. Field, Ed.). American Physiological Society, Washington, DC, pp. 91–107.

Bartoshuk, L. M. (1971). The chemical senses. I. Taste. In *Woodworth and Schlosberg's Experimental Psychology*. Third Edition (J. W. Kling and L. A. Riggs, Eds.), Holt, Rinehart, Winston, New York, pp. 169–191.

Bartoshuk, L. M. (1980). Sensory analysis of the taste of NaCl. In *Biological and Behavioral Aspects of Salt Intake,* M. R. Kare, M. J. Fregley, R. A. Bernard (Eds.), Academic Press, New York, pp. 83–98.

Bartoshuk, L. M. (1985). Chemical sensation: Taste. In *Nutrition in Oral Health and Disease* (R. L. Pollack and E. Kravitz, Eds.). Lea and Febiger, Philadelphia, pp. 53–67.

Bartoshuk, L. M. (1988). Taste. In *Stevens' Handbook of Experimental Psychology*. 2nd Edition. Volume 1: *Perception and Motivation* (R. C. Atkinson, R. J. Herrnstein, G. Lindzey, and D. Luce, Eds.). John Wiley & Sons, New York, pp. 461–499.

Bartoshuk, L. M., Cain, W. S., and Pfaffmann, C. (1985). Taste and olfaction. In *Topics in the History of Psychology*. Volume 1 (G. A. Kimble and K. Schlesinger, Ed. Lawrence Erlbaum Associates, Hillsdale, NJ, pp. 221–260.

Bartoshuk, L. M. and Marks, L. E. (1986). Ratio scaling. In *Clinical Measurement of Taste and Smell* (H. L. Meiselman and R. S. Rivlin, Eds.). Macmillan Publishing Co., New York, pp. 39–65.

Beck, K. M. (1975). Practical considerations for synthetic sweeteners. *Food Product Development,* 9: 47–50.

Bekesy, G. von (1963). Interaction of paired sensory stimuli and conduction in peripheral nerves. *J. Appl. Physiol.*, 18: 1276–1284.

Bekesy, G. von (1964a). Rhythmical variations accompanying gustatory stimulation observed by means of localization phenomena. *J. Gen. Physiol.*, 47: 809–825.

Bekesy, G. von (1964b). Sweetness produced electrically on the tongue and its relation to taste theories. *J. Appl. Physiol.*, 19: 1105–1113.

Birch, G. G., Latymer, Z., and Hollaway, M. (1980). Intensity/time relationships in sweetness: Evidence for a queue hypothesis in taste chemoreception. *Chem. Senses*, 5: 63–78.

Birch, G. G. and Munton, S. L. (1981). Use of the 'SMURF' in taste analysis. *Chem. Senses*, 6: 45–52.

Borg, G., Diamant, H., Ström, L., and Zotterman, Y. (1967). The relation between neural and perceptual intensity: a comparative study on the neural and psychophysical response to taste stimuli. *J. Physiology (London)*, 192: 13–20.

Bujas, Z. (1934). Le temps d'action des stimuli de la sensibilité gustative. *C. R. Seances Soc. Biol.*, 116: 1307–1309.

Bujas, Z. and Ostojcic, A. (1939). L'évolution de la sensation gustative en fonction du temps d'excitation, *Acta Instituti Psychologici Universitatis Zagrebensis*, 3: 1–24.

Burgess, P. R., Horch, K. W., and Tuckett, R. P. (1983). Boring's formulation: A scheme for identifying functional neurons in a sensory system. *Federation Proceedings*, 42: 2521–2527.

Carr, T. H. (1986). Perceiving visual language. *Handbook of Perception and Human Performance. Volume II. Cognitive Processes and Performance* (K. R. Boff, L. Kaufman, and J. P. Thomas, Eds.). Wiley, New York, pp. 29-1–29-92.

Clark, F. J. and Horch, K. W. (1986). Kinesthesia. In *Handbook of Perception and Human Performance. Volume I. Sensory Processes and Perception* (K. R. Boff, L. Kaufman, and J. P. Thomas, Eds.). Wiley, New York, pp. 13-1–13-62.

Darian-Smith, I. (1984). The sense of touch: Performance and peripheral neural processing. In *Handbook of Physiology—The Nervous System. III. Sensory Processes* (J. M. Brookhart and V. B. Mountcastle, Eds.). Williams and Wilkins, Baltimore, pp. 739–788.

Diamant, H. (1968). Comparison between neural and psychophysical recordings of taste stimulations. *Acta Oto-laryngologica*, 65: 51–54.

Diamant, H. and Zotterman, Y. (1959). Has water a specific taste? *Nature.* 183: 191–192.

Diamant, H., Funakoshi, M., Ström, L., and Zotterman, Y. (1963). Electrophysiological studies on human taste nerves. In *Olfaction and Taste* (Y. Zotterman, Ed.). Pergamon Press, Oxford, England, pp. 193–203.

Diamant, H., Oakley, B., Ström, L., Wells, C., and Zotterman, Y. (1965). A comparison of neural and psychophysical responses to taste stimuli in man. *Acta Physiologica Scandinavica*, 64: 67–74.

DuBois, G. E., Crosby, G. A., and Stephenson, R. A. (1981a). Dihydrochalcone sweeteners. A study of the atypical temporal phenomena. *J. Medicinal Chem.*, 24: 408–428.

DuBois, G. E., Crosby, G. A., Lee, J. F., Stephenson, R. A., and Wang, P. C. (1981b). Dihydrochalcone sweeteners. Synthesis and sensory evaluation of a homoserine-dihydrochalcone conjugate with low aftertaste, sucrose-like organoleptic properties. *J. Ag. Food Chem.*, 29: 1269–1276.

DuBois, G. E. and Lee, J. F. (1983). A simple technique for the evaluation of temporal taste properties. *Chem. Senses*, 7: 237–246.

Engen, T. (1971a) Psychophysics. I. Discrimination and detection. In *Woodworth and Schlosberg's Experimental Psychology*. Third Edition (J. W. Kling and L. A. Riggs, Eds.). Holt, Rinehart, and Winston, New York, pp. 11–46.

Engen, T. (1971b) Psychophysics. II. Scaling methods. In *Woodworth and Schlosberg's Experimental Psychology*. Third Edition (J. W. Kling and L. A. Riggs, Eds.), Holt, Rinehart, and Winston, New York, pp. 47–86.

Engen, T. (1986). Classical psychophysics: humans as sensors. In *Clinical Measurement of Taste and Smell*, H. L. Meiselman and R. S. Rivlin (Eds.), Macmillan Publishing Co., New York, pp. 39–49.

Faurion, A. (1987). Physiology of the sweet taste. *Progress in Sensory Physiology*, 8: 129–201.

Faurion, A., Saito, S., and MacLeod, P. (1980). Sweet taste involves several distinct receptor mechanisms. *Chem. Senses*, 5: 107–121.

Finger, T. E. and Silver, W. L. (1987). *Neurobiology of Taste and Smell*. Wiley-Interscience, New York.

Fuortes, M. G. F. (1971). Generation of responses in receptor. In *Handbook of Sensory Physiology*. Volume 1. *Principles of Sensory Physiology* (W. R. Loewenstein, Ed.), Springer-Verlag, Berlin, pp. 243–268.

Gent, J. F. (1979). An exponential model for adaptation in taste. *Sensory Proc.*, 3: 303–316.

Guilford, J. P. (1954). *Psychometric Methods*. McGraw-Hill Book Co., New York.

Guinard, J-A., Pangborn, R. M., and Shoemaker, C. F. (1985). Computerized procedure for time-intensity sensory measurements. *J. Food Sci.*, 50: 543–546.

Guinard, J-A., Pangborn, R. M., and Lewis, M. J. (1986a). Effect of repeated ingestion on temporal perception of bitterness in beer. *Am. Soc. Brew. Chem. J.*, 44: 28–32.

Guinard, J-A., Pangborn, R. M., and Lewis, M. J. (1986b). The time-course of astringency in wine upon repeated ingestion. *Am. J. Enol. Vitic.*, 37: 184–189.

Gulick, W. L., Gescheider, G. A., and Frisina, R. D. (1989). *Hearing: Physiological Acoustics. Neural Coding, and Psychoacoustics.* Oxford University Press, New York, pp. 216–282.

Haber, R. N. and Standing, L. G. (1970). Direct estimates of the apparent duration of a flash. *Canad. J. Psych.*, 24: 216–229.

Halpern, B. P. (1973). The use of vertebrate laboratory animals in research on taste. In *Methods of Animal Experimentation.* Vol. 4 (W. I. Gay, Ed.). Academic Press, New York, pp. 225–362.

Halpern, B. P. (1977). Functional anatomy of the tongue and mouth of mammals. In *Drinking Behavior: Oral Stimulation, Reinforcement, and Preference* (J. A. W. M. Weijnen and J. Mendelson, Eds.). Plenum Press, New York, pp. 1–92.

Halpern, B. P. (1983). Tasting and smelling as active, exploratory sensory processes. *Am. J. Otolaryng.*, 4: 246–249.

Halpern, B. P. (1985). Time as a factor in gustation: temporal patterns of taste stimulation and reponse. In *Taste, Olfaction, and the Central Nervous System* (D. W. Pfaff, Ed.). The Rockefeller University Press, New York, pp. 181–209.

Halpern, B. P. (1986a). What to control in studies of taste. In *Clinical Measurement of Taste and Smell* (H. L. Meiselman and R. S. Rivlin, Eds.), Macmillan Publishing Co., New York, pp. 125–153.

Halpern, B. P. (1986b). Constraints imposed on taste physiology by human taste reaction time data. *Neurosci. Biobehav. Rev.*, 10: 135–151.

Halpern, B. P. (1987). Human judgements of MSG taste: Quality and reaction time. In *Umami: a Basic Taste* (Y. Kawamura and M. R. Kare, eds.), Marcel Dekker, New York, pp. 327–354.

Halpern, B. P., Kelling, S. T., and Meiselman, H. L. (1986). An analysis of the role of stimulus removal in taste adaptation by means of simulated drinking. *Physiology and Behavior,* 36: 925–928.

Halpern, B. P. and Meiselman, H. L. (1980). Taste psychophysics based on a simulation of human drinking. *Chem. Senses,* 5: 279–294.

Harris, D. and Pikler, A. G. (1960). The stability of a standard of loudness as measured by compensatory tracking. *Am. J. Psych.,* 73: 573–580.

Harrison, S. K. and Bernhard, R. A. (1984). Time-intensity sensory characteristics of saccharin, xylitol and galactose and their effect on the sweetness of lactose. *J. Food Sci.,* 49: 780–786.

Hearst, E. (1988). Fundamentals of learning and conditioning. In *Stevens' Handbook of Experimental Psychology*. 2nd Ed. Volume 2: *Learning and Cognition* (R. C. Atkinson, R. J. Herrnstein, G. Lindzey, and D. Luce, eds.), John Wiley & Sons, New York, pp. 3–109.

Hensel, H. (1982). *Thermal Sensations and Thermoreceptors in Man*. Charles C. Thomas, Springfield, IL.

Herness, M. S. and Pfaffmann, C. (1986). Generalization of conditioned taste aversion in hamsters: evidence for multiple bitter receptor sites. *Chemical Senses,* 11: 347–360.

Hirsh, I. J. (1952). *The Measurement of Hearing*. McGraw-Hill, New York, 1952.

Holway, A. H. and Hurvich, L. M. (1937). Differential gustatory sensitivity to salt. *Amer. J. Psych.,* 49: 37–48.

Hood, D. C. and Finkelstein M. A. (1986). Sensitivity to light. In *Handbook of Perception and Human Performance*. Volume I. *Sensory Processes and Perception* (K. R. Boff, L. Kaufman, and J. P. Thomas, eds.), Wiley, New York, pp. 5-1–5-66.

Husmark, I. and Ottoson, D. (1971). The contribution of mechanical factors to the early adaptation of the spindle response. *Acta Physiol. Scand.,* 82: 545–554.

Iggo, A. (1984). Cutaneous sensory mechanisms. In *The Senses* (H. B. Barlow and J. D. Mollon, eds.), Cambridge University Press, Cambridge, pp. 369–408.

Kelling, S. T. and Halpern, B. P. (1983). Taste flashes: Reaction times, intensity, and quality. *Science,* 219: 412–414.

Kelling, S. T. and Halpern, B. P. (1986). Physical characteristics of open flow and closed flow taste stimulus delivery apparatus. *Chem. Senses,* 11: 89–104.

Kelling, S. T. and Halpern, B. P. (1987a). Taste judgments and gustatory stimulus duration: simple taste reaction times. *Chem. Senses,* 12: 543–562.

Kelling, S. T. and Halpern, B. P. (1987b). Comparisons between human gustatory judgments of square wave trains and continuous pulse presentations of aqueous stimuli: Reaction times and magnitude estimates. *ISOT IX. Ann. N. Y. Acad. Sci.,* 510: 406–408.

Kelling, S. T. and Halpern, B. P. (1987c). Taste intensity tracking of aqueous square wave trains. *Chem. Senses,* 12: 670–671 (abstract).

Kelling, S. T. and Halpern, B. P. (1988). Taste judgments and gustatory stimulus duration: Taste quality, taste intensity, and reaction time. *Chem. Senses,* 13: 559–586.

Kelling, S. T., Maney, D. L., and Halpern, B. P. (1988). Continuous and pulsatile (5 hz train) taste stimulation: Differing effects on judged maximum intensity but similar adaptation. *Chem. Senses,* 13: 702–703 (abstract).

Kelling, S. T., Lyke, K., and Halpern, B. P. (1989). Temporal patterns of citric acid taste intensity. *Chem. Senses*, 14: 716–717 (abstract).

Kennedy, L. M. and Halpern, B. P. (1980). Extraction, purification, and characterization of a sweetness-modifying component from *Ziziphus jujuba*. *Chem. Senses*, 5: 123–147.

Kroeze, J. H. A. (1982). Reduced sweetness and saltiness judgements of NaCl-sucrose mixtures depend on a central inhibiting mechanism. In *Determination of Behaviour by Chemical Stimuli* (J. E. Steiner and J. R. Ganchrow, eds.), IRL Press Ltd., London, England, pp. 161–174.

Land, D. G. and Shepherd, R. (1988). Scaling and ranking methods. In *Sensory Analysis of Foods*, second ed. (J. R. Piggott, ed.), Elsevier Applied Science, London, pp. 155–185.

Lapedes, D. N. (1974). *McGraw-Hill Dictionary of Scientific and Technical Terms*, McGraw-Hill Book Co., New York, p. 1270.

Larson-Powers, N. and Pangborn, R. M. (1978). Paired comparison and time-intensity measurements of the sensory properties of beverages and gelatins containing sucrose or synthetic sweeteners. *J. Food Sci.* 43: 41–46.

Lawless, H. T. Gustatory psychophysics. (1987). In *Neurobiology of Taste and Smell* (T. E. Finger and W. L. Silver, eds.), John Wiley and Sons, New York, pp. 401–420.

Lawless, H. T. and Skinner, E. Z. (1979). The duration and perceived intensity of sucrose taste. *Perception & Psychophysics*, 25: 180–184.

Lawless, H. T. and Stevens, D. A. (1983). Cross adaptation of sucrose and intensive sweeteners. *Chemical Senses*, 7: 309–315.

Leach, E. J. and Noble, A. C. (1986). Comparison of bitterness of caffeine and quinine by a time-intensity procedure. *Chemical Senses*, 11: 339–345.

Lee, W. E. III and Pangborn, R. M. (1986). Time-intensity: The temporal aspects of sensory perception. *Food Technology*, 40: 71–78.

Lester, B. and Halpern, B. P. (1979). Effect of stimulus presentation duration on gustatory reaction time. *Physiol. Behav.*, 22: 319–324.

Levine, M. W. and Shefner, J. M. (1991). *Fundamentals of Sensation and Perception*. Second Edition. Brooks/Cole Publishing Company, Pacific Grove, CA.

Liu, Y.-H. and MacFie, H. J. H. (1990). Methods for averaging time-intensity curves. *Chem. Senses*, 15: 471–484.

Loomis, J. M. and Lederman, S. J. (1986). Tactile perception. In *Handbook of Perception and Human Performance*, Volume II. *Cognitive Processes and Performance* (K. R. Boff, L. Kaufman, and J. P. Thomas, eds.), Wiley, New York, pp. 31-1–31-41.

Luce, R. D. and Krumhansl, C. L. (1988). Measurement, scaling, and psychophysics. In *Stevens' Handbook of Experimental Psychology,* 2nd Ed. Volume 1: *Perception and Motivation* (R. C. Atkinson, R. J. Herrnstein, G. Lindzey, and D. Luce, eds.), John Wiley & Sons, New York, pp. 3–74.

Matlin, M. W. (1988). *Sensation and Perception,* Second Edition. Allyn and Bacon, Inc., Boston.

McBurney, D. H. (1976). Temporal properties of the human taste system. *Sensory Processes,* 1: 150–162.

McBurney, D. H. (1978). Psychological dimensions and perceptual analyses of taste. In *Handbook of Perception,* Volume VIA, *Tasting and Smelling* (E. C. Carterette and M. P. Friedman, eds.), Academic Press, New York, pp. 125–155.

McBurney, D. H. (1984). Taste and olfaction: sensory discrimination. In *Handbook of Physiology—The Nervous System III* (J. Brookhart and V. B. Mountcastle, eds.), Williams and Wilkins, Baltimore, pp. 1067–1086.

McBurney, D. H. and Collings, V. B. (1984). *Introduction to Sensation/ Perception.* Prentice-Hall, Inc., Englewood Cliffs, NJ, pp. 136–138.

Meilgaard, M., Civille, G. V., and Carr, B. T. (1987). *Sensory Evaluation Techniques,* Volume I. CRC Press, Boca Raton, FL.

Meiselman, H. L. and Halpern, B. P. (1970). Human judgments of *Gymnema sylvestre* and sucrose mixtures. *Physiology and Behavior,* 5: 945–948.

Meiselman, H. L. and Halpern, B. P. (1973). Enhancement of taste intensity through pulsatile stimulation. *Physiol. Behav.,* 11: 713–716.

Mountcastle, V. B. (1967). The problem of sensing and the neural coding of sensory events. In *The Neurosciences, A Study Program* (G. C. Quarton, T. Melnechuk, and F. O. Schmitt, eds.), The Rockefeller University Press, New York, pp. 393–408.

Mountcastle, V. B., Steinmetz, M. A., and Romo, R. (1990). Frequency discrimination in the sense of flutter: psychophysical measurements correlated with postcentral events in behaving monkeys. *Journal of Neuroscience,* 10:3032–3044.

Naim, M., Dukan, E., Yaron, L., Levinson, M., and Zehavi, U. (1986). Effects of the bitter additives naringin and sucrose octa-acetate on sweet persistence and sweet quality of neohesperidin dihydrochloride. *Chem. Senses,* 11: 471–483.

Oakley, B. (1985). Taste response of the human *chorda tympani* nerve. *Chem. Senses,* 10: 469–481.

O'Mahony, M. and Wong, S-Y. (1989). Time-intensity scaling with judges trained to use a calibrated scale: Adaptation, salty, and umami tastes. *J. Sensory Studies,* 3: 217–236.

Overbosch, P. (1986). A theoretical model for perceived intensity in human taste and smell as a function of time. *Chem. Senses,* 11: 315–329.

Overbosch, P., van den Enden, J. C., and Keur, B. M. (1986). An improved method for measuring perceived intensity/time relationships in human taste and smell. *Chem. Senses,* 11: 331–338.

Overbosch, P. and Soeting, W. J. (1989). Temporal aspects of flavoring. In *Flavor Chemistry: Trends and Development. ACS Symposium Series 388* (R. Teranishi, R. G. Buttery, and F. Shahidi, eds.), American Chemical Society, Washington, DC., pp. 138–150.

Pangborn, R. M., Lewis, M. J., and Yamashita, J. F. (1983). Comparison of time-intensity with category scaling of bitterness of iso-alpha-acids in model systems and in beer. *J. Inst. Brew.,* 89: 349–355.

Pantzer, T., Kelling, S. T., and Halpern, B. P. (1987). Brief taste stimuli: cued and uncued magnitude estimates and their reaction times. In: ISOT IX, (S. D. Roper and J. Atema, Eds.), *Ann. N. Y. Acad. Sci.* 510: 541–543.

Pew, R. W. and Rosenbaum, D. A. (1988). Human movement control: Computation, representation, and implementation. In *Stevens' Handbook of Experimental Psychology,* 2nd Ed. Volume 2: *Learning and Cognition* (R. C. Atkinson, R. J. Herrnstein, G. Lindzey, and D. Luce, eds.), John Wiley & Sons, New York, pp. 473–509.

Pfaffmann, C., Bartoshuk, L. M., and McBurney, D. H. (1971). Taste psychophysics. In *Taste. Handbook of Sensory Physiology,* Volume IV. *Chemical Senses,* Part 2 (L. M. Beidler, (ed.), Springer-Verlag, Berlin, pp. 75–101.

Pickles, J. A. (1988). *An Introduction to the Physiology of Hearing,* second edition. Academic Press, London, pp. 258–269.

Pierau, F.-K. and Wurster, R. D. (1981). Primary afferent input from cutaneous thermoreceptors. *Federation Proceedings,* 40: 2819–2824.

Posner, M. I. (1986). Overview. In *Handbook of Perception and Human Performance. Volume II. Cognitive Processes and Performance,* K. R. Boff, L. Kaufman, and J. P. Thomas (Eds.), Wiley, New York, pp. V-3–V-10.

Powers, J. J. (1988). Current practices and applications of descriptive methods. In *Sensory Analysis of Foods,* Second Edition (J. R. Piggott, ed.), Elsevier Applier Science, London, pp. 187–266.

Pubols, Jr., B. H. and Pubols, L. M. (1983). Tactile receptor discharge and mechanical properties of glabrous skin. *Federation Proceedings,* 42: 2528–2535.

Rolls, E. T., Yaxley, S., Sienkiewicz, Z. J., and Scott, T. R. (1985). Gustatory responses of single neurons in the orbitofrontal cortex of the macaque monkey. *Presented at the 7th Annual Meeting of the Association for Chemoreception Sciences,* Sarasota, FL.

Sathian, S. (1989). Tactile sensing of surface features. *Trends in Neurosciences,* 2: 513–519.

Scharf, B. and Buus, S. (1986). Audition I. Stimulus, physiology, thresholds. In *Handbook of Perception and Human Performance*. Volume I. *Sensory Processes and Perception* (K. R. Boff, L. Kaufman, and J. P. Thomas, eds.), Wiley, New York, pp. 14-1–14-71.

Scharf, B. and Houtsma, A. J. M. (1986). Audition II. Loudness, pitch, localization, aural distortion, pathology. In *Handbook of Perception and Human Performance*, Volume I. *Sensory Processes and Perception* (K. R. Boff, L. Kaufman, and J. P. Thomas, eds.), Wiley, New York, pp. 15-1–15-60.

Schiffman, S. S., Hopfinger, A. J., and Mazur, R. H. (1986). The search for receptors that mediate sweetness. In *The Receptors*, Vol. IV (M. Conn, ed.), Academic Press, New York, pp. 315–377.

Schuknecht, H. F. (1974). *Pathology of the Ear*. Harvard University Press, Cambridge, MA, pp. 343–350.

Scott, T. R., Yaxley, S., Sienkiewicz, Z. J., and Rolls, E. T. (1986a). Gustatory responses in the nucleus tractus solitarius of the alert cynomologous monkey. *J. Neurophysiol.*, 55: 182–200.

Scott, T. R., Yaxley, S., Sienkiewicz, Z. J., and Rolls, E. T. (1986b). Gustatory responses in the frontal opercular cortex of the alert cynomologous monkey. *J. Neurophysiol.*, 56: 876–890.

Shepherd, G. M. (1988). *Neurobiology*, 2nd Edition. Oxford University Press, New York, p. 279.

Sherrick, C. E. and Cholewiak, R. W. (1986). Cutaneous sensitivity. In *Handbook of Perception and Human Performance, Volume I. Sensory Processes and Perception*. K. R. Boff, L. Kaufman, and J. P. Thomas (Eds.), Wiley, New York, pp. 12-1–12-58.

Stevens, S. S. (1956). The direct estimation of sensory magnitude-loudness. *Am. J. Psychol.*, 69: 1–25.

Thompson, R. F., Donegan, N. H., and Lavond, D. G. (1988). The psychobiology of learning and memory. In *Stevens' Handbook of Experimental Psychology*, 2nd Ed. Volume 2: *Learning and Cognition* (R. C. Atkinson, R. J. Herrnstein, G. Lindzey, and D. Luce, eds.), John Wiley & Sons, New York, pp. 245–347.

Thurlow, W. R. (1971). Audition. In *Woodworth and Schlosberg's Experimental Psychology*, Third Edition (J. W. Kling and L. A. Riggs, eds.), Holt, Rinehart, Winston, New York, pp. 223–271.

Watson, A. B. (1986). Temporal sensitivity. In *Handbook of Perception and Human Performance*, Volume I. Sensory Processes and Perception (K. R. Boff, L. Kaufman, and J. P. Thomas, eds.), Wiley, New York, pp. 6-1–6-43.

Weiffenbach, J. M., Cowart, B. J., and Baum, B. J. (1986a). Taste intensity perception in aging. *J. Gerentol.*, 41: 460–468.

Weiffenbach, J. M., Fox, P. C., and Baum, B. J. (1986b). Taste and salivary function. *Proc. Natl. Acad. Sci. USA,* 83: 6103–6106.

Whitsel, B. L., Dreyer, D. A., and Hollins, M. (1978). Representation of moving stimuli by somatosensory neurons. *Federation Proceedings,* 37: 2223–2227.

Wyscezki, G. (1986). Color appearance. In *Handbook of Perception and Human Performance,* Volume I. *Sensory Processes and Perception* (K. R. Boff, L. Kaufman, and J. P. Thomas, eds.), Wiley, New York, pp. 9-1–9-57.

Yoshida, M. (1986). A microcomputer (PC 9801/MS mouse) system to record and analyze time-intensity curves of sweetness. *Chem. Senses,* 11:105–118.

Yoshida, M. and Mochizuki, K. (1987). Time-intensity curves of taste, magnitude estimation with a microcomputer system. *Bulletin of the Faculty of Science and Engineering, Chuo University,* 30: 321–340 (in Japanese, with English abstract).

Yoshida, M. and Mochizuki, K. (1988). Time-intensity curves of taste, magnitude estimation with a microcomputer system. *Bulletin of the Faculty of Science and Engineering, Chuo University,* 31: 215–230 (In Japanese, with English abstract).

Yoshida, M. and Mochizuki, K. (1989). Time-intensity curves of sourness, magnitude estimation with a microcomputer system, NEC 9801. *Bulletin of the Faculty of Science and Engineering, Chuo University,* 32: 135–146 (in Japanese, with English abstract).

Zotterman, Y. (1967). The neural mechanisms of taste. *Sensory Mechanisms. Progress in Brain Research* (Y. Zotterman, ed.), Elsevier, Amsterdam, pp. 139–154.

Zwillinger, S. A., Kelling, S. T., and Halpern, B. P. (1990). Time-quality tracking: Temporal patterns of taste quality. *Chem. Senses,* 15: 657 (abstract).

Zwillinger, S. A. and Halpern, B. P. (in press). Time-quality tracking of monosodium glutamate, sodium saccharin, and a citric acid-saccharin mixture. *Physiology and Behavior.*

3

Oral Chemesthesis: The Importance of Time and Temperature for the Perception of Chemical Irritants

Barry G. Green

Monell Chemical Senses Center
Philadelphia, Pennsylvania

INTRODUCTION

In addition to taste and smell, the chemical attributes of things we consume are also sensed by a third modality, which has historically been referred to as the "common chemical sense" (Parker, 1912; Silver, 1987). The sensory end organs that serve the latter modality, which are thought to arise primarily from the trigeminal and vagus nerves, reside in the mucosal skin that lines the nose, mouth, and anterior reaches of the alimentary canal. Although the exact identity of these fibers remains to be established, there is mounting evidence that many and perhaps all of them belong to either the thermal senses or the nociceptive (pain) sense (e.g., Hensel and Zotterman, 1951; Okuni, 1978; Konietzny and Hensel, 1983; Harada et al., 1987; Szolcsányi, 1990). It is therefore likely that cutaneous sensory fibers that respond to chemicals also respond either to moderate changes in temperature or to potentially damaging levels of thermal or mechanical stimulation (e.g., burning or crushing). Accordingly, the term "common chemical sense," which implies the existence of a distinct anatomical system devoted to detection of chemicals, no longer seems appropriate to describe the physiological basis for the sensitivity to chemical irritants. Instead, the term "chemesthesis" has been suggested as an alternative (Green et al., 1990; Green and Lawless, in press). The latter term refers to a chemical *sensibility* rather than to a chemical sense, and communicates the close anatomical and functional relationship of cutaneous chemical sensitivity to the somesthetic senses.

Given that chemesthesis derives from elements of the cutaneous senses, one would expect to observe similarities between the psychophysical characteristics of chemical irritation and those of the thermal and nociceptive senses. Interactions between chemical and other somatic stimuli should also be commonplace. That is, if chemical stimuli are detected by nerve fibers that also respond to thermal and painful stimuli, then activation of these nerve fibers by one type of stimulation ought to affect the sensitivity to other types of stimulation. As will be seen below, these expectations have been borne out by both physiological and psychophysical research. However, chemical irritation also has unique sensory characteristics that appear to be dictated by the biochemical properties (e.g., rate of permeation through the epithelium and the manner of interaction with neural membranes) of chemical stimuli.

Discussed below are two of the most salient properties of oral chemical irritation: a strong interaction with temperature and striking temporal effects. The thermal properties illustrate the close relationship between chemesthesis and the temperature senses; the temporal effects appear to derive from the combined effects of chemical permeation through the epithelium and chemical activation (and inhibition) of nociceptors.

INTERACTIONS WITH TEMPERATURE

Because the temperature of foods and beverages are rarely close to the oral norm of 37°C, temperature is a potentially important factor whenever something is consumed. It has long been known that temperature can affect the absolute sensitivity to virtually all gustatory stimuli (see Pangborn et al., 1970, for review), but recent evidence indicates that at suprathreshold levels thermal effects may be significant only for sweet and bitter substances (Green and Frankmann, 1987). Similarly, the tactile sensitivity of the oral cavity can be reduced by cooling, but not for all types of mechanical stimulation (Green, 1987).

In contrast, temperature has a strong and relatively uniform effect on the sensations produced by chemical irritants (Szolcsányi and Jancsó-Gábor, 1973; Sizer and Harris, 1985; Green, 1986a, 1990), and the perception of temperature can similarly be altered by the presence of an irritant (Green, 1986a, 1990). Such effects are not confined to perceptually "hot" chemicals, as is apparent from the strong effect menthol has on the perception of cooling. Examples of the diversity and magnitude of chemical-thermal interactions are discussed below.

Chemical "Heat" and Temperature

Common experience testifies that sipping a cool beverage after consuming a hot spice like chili pepper relieves even the most intense burning sensation. Yet, as striking as this relief can be, it is always fleeting; moments after the beverage is swallowed the burn begins to reassert itself. It is now clear that the primary source of this temporary relief is not the rinsing action of the beverage, but rather its capacity to cool. Szolcsányi and Jancsó-Gábor (1973) were the first to demonstrate in the laboratory that cooling the site of chemical irritation could significantly attenuate chemogenic irritation. More recently, Sizer and Harris (1985) reported that the threshold for detecting the presence of capsaicin in the mouth was significantly higher when the vehicle that contained capsaicin was cooled. At about the same time, a study performed in my laboratory (Green, 1986a) measured the changes in sensations of burning produced by warming or cooling a capsaicin solution. The results were in general agreement with Szolcsányi and Jancsó-Gábor's initial observations: Cooling the solution to 13°C nearly eliminated the burning sensation produced by 2-ppm capsaicin, whereas warming the solution to 45°C greatly increased the burning sensation beyond what it was at the normal oral temperature (about 37°C). Figure 3.1 contains the results for temperatures between 34 and 45°C. In an ancillary experiment performed on a confined area of the lip, we were able to confirm Szolcsányi and Jancsó-Gábor's finding that cooling the skin to a sufficiently low temperature could extinguish the chemogenic sensation completely (Green, 1986a). Failure to eliminate the burn using whole-mouth rinses was likely due to an inability to cool uniformly all the areas of the mouth exposed to capsaicin.

The modulation of the capsaicin burn by temperature is consistent with the well-supported hypothesis that the perception of capsaicin is mediated by polymodal nociceptors, which are pain fibers that normally respond both to extreme pressure and to intense thermal stimulation (hence the term "polymodal") (Szolcsányi, 1977; Konietzny and Hensel, 1983; Simone et al., 1987; Szolcsányi, 1990). Essentially, the activation of polymodal nociceptors by capsaicin may be thought of as a sensitization process: capsaicin induces the fibers to discharge at innocuous temperatures to which they would otherwise be indifferent (i.e., it lowers their thermal threshold). Raising temperature in the innocuous range further excites the nociceptors, with the result that the heat pain threshold is sometimes reached at an abnormally low temperature (e.g., as low as 35–40°C

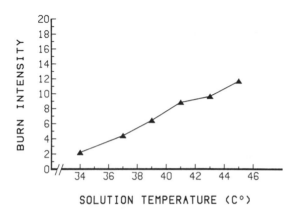

FIG. 3.1 The relationship between the perceived intensity of burning sensations in the oral cavity and stimulus temperature for temperatures near and above normal oral temperature. The stimulus was a solution of 2 ppm capsaicin in a water and ethanol vehicle (EtOH conc. < 0.25%) *(Adapted from Green, 1986a.)*

rather than 45°C or above). On the other hand, cooling offsets the sensitizing effect of the chemical until, when skin temperature is sufficiently low (ca. 20–25°C), capsaicin becomes an ineffective stimulus (Green, 1986a).

The effect of temperature on sensations of irritation is not limited to capsaicin. It was found in a subsequent experiment that the irritations evoked by piperine (black pepper), NaCl (in high concentrations), and ethanol could all be modulated by sipping warm or cool water shortly after the compounds had been applied to the tongue tip (Green, 1990). The thermal effects were not equally strong for each compound, however, suggesting either that the chemicals stimulated somewhat different populations of afferent fibers (e.g., including some that are less sensitive to temperature) or that differences in the rate at which the chemicals were washed from the mucosa interacted with the thermal effects.

Just as common as the experience of obtaining relief from chemical irritation by sipping a cool drink is the experience of perceiving the mouth to be "hot" when eating chili pepper. So common is this experience that peppers are routinely reffered to in terms of their "heat." It is important to consider, however, that a sensation of "burning" need not carry with it the perception of "heat." Burning is often used to describe the sensation of intense cold, and the sensation produced, for example, by an antiseptic (e.g., alcohol) applied to a skin abrasion may be described as "burning" but not "hot."

Surprisingly, until very recently there had been no study of the relationship between pepper irritation (or concentration) and perceived warmth or heat. In the same study that measured the effect of temperature on the magnitude of the capsaicin sensation, ratings were obtained of the perceived warmth (or coolness) of the capsaicin solution (Green, 1986). The result for a series of warm solutions can be seen in Fig. 3.2. As one would expect, the presence of capsaicin significantly increased the perceived warmth of the solution. However, the effect on warmth was not uniform over the range of temperatures tested. The perception of mild temperatures was not significantly affected; solutions were judged to be warmer in the test condition only at the higher stimulus temperatures (> 40°C).

This result carries with it two notable implications. First, the absence of an enhancement of warmth at the lower temperatures demonstrates that subjects were able to attend selectively to sensations of irritation and temperature. Burning sensations from the capsaicin were present at 37°C (Fig. 3.1), yet they did not significantly alter the subjects' ratings of perceived temperature. Second, the lack of effect at lower temperatures coupled with the increasing magnitude of the effect at higher temperatures is consistent with capsaicin stimulating or sensitizing heat-sensitive nociceptors rather than warm fibers. Because warm

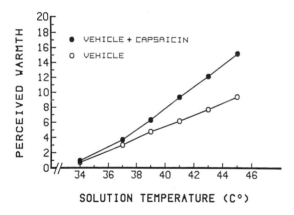

FIG. 3.2 The effect of capsaicin on the perceived warmth of the oral cavity for temperatures near and above normal oral temperature. The magnitude of thermal enhancement grows with temperature, indicating that sensations of warmth and "heat" were affected more strongly at temperatures where nociceptors would be expected to become active. *(Reprinted with permission from Green, 1986a.)*

fibers normally respond to innocuous temperatures that evoke sensations of mild warmth rather than sensations of burning or pain, the activation of these fibers by capsaicin would be expected to elevate ratings of the warmth of solutions having temperatures of 34–38°C. To date, no convincing physiological evidence has been found that capsaicin affects cutaneous warm fibers.

Why, then, do chili peppers generate sensations of "heat?" At least two possibilities exist. One is that capsaicin excites nociceptors that normally respond to high temperatures (heat-sensitive nociceptors), and that these fibers convey a quality code for heat as well as for irritation. The other possibility is that although capsaicin does not itself stimulate warm fibers, warm fibers are nevertheless active whenever the solution or food matrix that contains capsaicin is near or above oral temperature. The majority of spicy "hot" foods are consumed under just such conditions. Even when a food is cool or cold, warm fibers become active as soon as the mucosa begins to warm again, which because of its ample superficial blood flow, occurs almost immediately after a bolus is swallowed. Thus it is difficult to produce lasting conditions in the mouth that do not result in simultaneous input from warm fibers and nociceptors, a pattern of neural activity that would seem to provide the prototypic quality code for "hot."

Chemical "Cooling" and Temperature

Chemical cooling agents, although probably less commonly consumed than hot spices, provide an even clearer example of the somatosensory basis of oral chemoreception. Menthol, and hundreds of related (mostly synthetic) compounds (see Watson et al., 1978), are capable of generating strong sensations of coolness on virtually any area of skin. It has been known since 1951 that this effect is due primarily to an excitatory action of such compounds on fibers that normally respond to cooling (Hensel and Zotterman, 1951). However, because menthol is volatile, its sensory effect is amplified under conditions that allow significant amounts of evaporative cooling (e.g., when one inhales through the mouth while consuming a mentholated candy). Thus when studying menthol as a chemical stimulus, it is essential to at least minimize, and preferably eliminate, evaporation.

Until recently, the magnitude of menthol's cooling effect had not been systematically studied in relation to either stimulus concentration or stimulus temperature. During initial investigations designed to provide this missing information, it was discovered that menthol was more than simply an "artificial cooling agent" (Green, 1985).

First, when subjects held cool or cold menthol solutions in the mouth for only 5 seconds before rating perceived warmth or coolness, the cooling effect did not appear. Instead, as can be seen in Fig. 3.3, ratings during presentations of *warm*

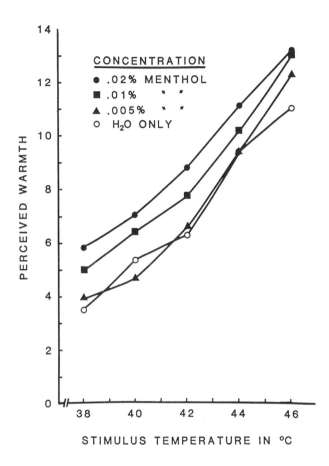

FIG. 3.3 The perceived warmth of the oral cavity is shown when warm solutions of menthol and water were sipped and held in the mouth for 5 sec. Perceived warmth was enhanced by the presence of menthol at the two highest concentrations. *(Reprinted with permission from Green, 1985.)*

menthol solutions were significantly *higher* than ratings during presentation of warm water alone. The latter effect became significant (over a temperature range of 38–46°C) when menthol was dissolved in deionized water in concentrations of 0.01 and 0.02%, but not 0.005%. Under those conditions, menthol was an "artificial warming agent."

Second, when subjects prerinsed with solutions of 0.02% menthol (37°C) for 5 minutes before receiving the test solutions, the typical cooling effect appeared.

In contrast, ratings of the warmth of the mouth were no longer enhanced when menthol was present; rather, the mouth was perceived as *less* warm. Apparently, at least a few seconds is required for the menthol molecule to penetrate to and sensitize cold receptors. Less time appears necessary to reach and affect warm receptors, and the effect is at first excitatory and later inhibitory.

An additional experiment revealed still another aspect of menthol's chemosensory effects. When the period of menthol preexposure was extended to 10 minutes (from the original 5), the cooling effect remained but the inhibitory effect on warmth disappeared. Thus menthol's influence on perceived warmth during whole-mouth stimulation depended entirely upon the temporal pattern of exposure, changing from enhancement to inhibition to no-effect.

It was decided to investigate menthol's surprising inhibitory effect on perceived warmth in another paradigm that allowed better control of skin temperature (Green, 1986b). Moving to the vermilion border of the lower lip and using a peltier thermoelectric module to vary skin temperature, it was found that warmth inhibition (shown in Fig. 3.4) varied directly with menthol concentration. The graph in Fig. 3.4D illustrates just how strong the inhibition was: although the effect diminished as temperature increased, at 43°C perceived warmth was still reduced by more than 50% when menthol was present. The results of this experiment also indicated that inhibition could last for at least 15 minutes. Why inhibition persisted longer on the lip than in the whole mouth is not clear. Differences in skin type (mucosal vs. keratinized) and stimulus vehicle (mineral oil was used on the lip so that higher concentrations could be tested) must both be considered as possible factors. Unfortunately, the method of stimulus application and thermal testing in the second study did not allow assessment of possible warmth enhancement during the first few seconds of exposure.

An important characteristic of menthol's inhibitory effect on warmth is that it is limited to innocuous temperatures: the heat pain threshold is unaltered by the presence of menthol. The convergence of the functions for menthol versus vehicle alone in Fig. 3.4 hints at this fact, and a subsequent experiment confirmed it (Green, 1986b). Restriction of the inhibitory effect to innocuous temperatures provides indirect evidence that it involves some sort of direct action of the menthol molecule on warm-sensitive neurons.

It should also be noted that in addition to these complex effects of menthol on the thermal senses, the chemical is also used as a "counterirritant" in topical analgesic ointments and balms (e.g., Ben-Gay). Menthol readily generates sensations of burning and stinging on hairy skin, and we have observed the same to be true at high concentrations in the oral cavity (unpublished observations). Its ability to produce sensations of irritation implies that menthol also excites or sensitizes some type of cutaneous nociceptor.

In summary, studies with menthol have illustrated that the cutaneous thermal

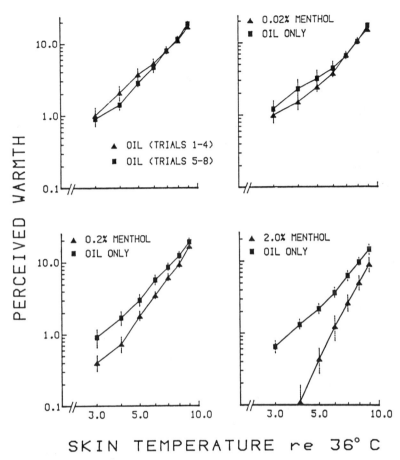

FIG. 3.4 The capacity of menthol to suppress sensations of warmth when exposure is for minutes rather than seconds is illustrated here for three concentrations. Testing was on the vermilion border of the lower lip, and heating was via peltier thermoelectric module. *(Reprinted with permission from Green, 1986b.)*

senses that innervate the oral cavity (and the rest of the integument) can be affected in both excitatory and inhibitory ways by an exogenous chemical. Although it would be misleading to designate the thermal senses as "chemosensitive" systems, it is fair to say that they are an important element of chemesthesis in the oral cavity.

CHEMICAL SENSITIZATION

Unlike gustatory sensations, which tend to adapt with repeated exposures, chemesthetic sensations usually increase in intensity with repeated exposures. Indeed, the perceptual sensitization that often occurs with chemical irritants is one of the most notable characteristics of somatosensory chemoreception. Although it had earlier been shown that repeated stimulation by volatile irritants resulted in a growth in nasal irritation intensity with time (Cain, 1976), only recently was there any indication that a similar effect could be found in the oral cavity. Stevens and Lawless (1987) reported that when two capsaicin or two piperine stimuli were presented in rapid succession, the second stimulus produced a stronger sensation than did the first. An intriguing aspect of this order-effect (which the authors termed "enhancement") was that it was amplified when the second stimulus was different from the first (i.e., when capsaicin followed piperine, or when piperine followed capsaicin). More recently, Green and Gelhard (1989) showed that this sequential enhancement was part of a cumulative sensitization process that can extend over many stimuli and for several minutes. It was discovered that when subjects sipped moderately-concentrated solutions of NaCl in water at the rate of once per minute (with no water rinses between), the irritation component of the sensation of saltiness grew steadily over time. This phenomenon is visible in Fig. 3.5, which shows that for the 0.8 M NaCl stimulus the sensation of irritation more than doubled in strength after 15 trials. Not shown in the figure are the data for salt taste per se (ignoring the sensations of irritation), which showed no significant change over the same number of trials. It was therefore concluded that the irritative component of saltiness becomes a more prominent feature as exposure to moderate or high levels of salt continues. A similar (though not as distinct) trend was also found when KCl was the stimulus.

Having observed irritation to grow over time with two salts, an obvious question was whether sensitization was a feature of all oral chemical irritation. We quickly discovered that although it happens frequently enough to be considered a common characteristic of chemesthesis, not all chemicals produce sensitization. In particular, it was learned that for most subjects tested, ethyl alcohol did not generate a progressively stronger sensation during repeated exposures (Green, 1990). However, the same study demonstrated that the irritation produced on the tongue tip by a given concentration of ethanol was greatly increased following several exposures to NaCl. That is, cross-sensitization was observed between the two compounds, albeit in only one direction. When taken together, the latter two findings have several interesting implications: First, the occurrence of a cross-sensitization from NaCl to ethanol indicates that the two chemicals stimulate many, and perhaps all, of the same irritant-sensitive nerve

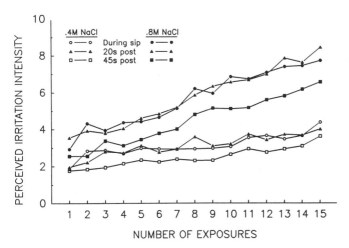

FIG. 3.5 Ratings of sensations of irritation produced by exposure to two concentrations of NaCl are plotted as a function of time. Solutions were held in the mouth for 15 sec and then expectorated; ratings of salt irritation were made 5 sec after the solution was sipped, and at 15 and 40 sec after it was expectorated. A new solution was presented every 60 sec, and no rinsing was allowed between stimuli. *(Reprinted with permission from Green and Gelhard, 1989.)*

fibers. Second, failure to obtain sensitization with ethanol demonstrates that mere stimulation of such endings is insufficient to produce sensitization. Third, if the latter conclusion is correct, sensitization results primarily or entirely from a peripheral, stimulus-dependent process; i.e., chemical sensitization in the mouth probably does not depend upon some sort of central integration or temporal summation.

The likelihood that perceptual sensitization is a peripheral process is again consistent with the hypothesis that oral irritants stimulate nociceptors. C-polymodal nociceptors have been shown to become sensitized after exposure to intense thermal or chemical stimuli (Lynn, 1977; Konietzny and Hensel, 1983; Simone et al., 1987; Lang et al., 1990).

CHEMICAL DESENSITIZATION

Probably the most striking feature of chemethesis is the ability of some chemical irritants to induce desensitization. The opposite of sensitization, desensitization is a loss of sensitivity brought about by repeated, usually intense chemical stimulation. Capsaicin, the compound responsible for the irritation ("heat") produced by chili pepper, is the most widely studied desensitizing agent. Since its first demonstration by Janscó in 1960, capsaicin desensitization has been studied extensively under a multitude of conditions and in many different neural systems (see Buck and Burks, 1986; Szolcsányi, 1990). The oral cavity was not, however, among the earliest sites chosen for study. The first investigation of desensitization in the mouth was reported by Szolcsányi and Jancsó-Gábor in 1973, and although Lawless (1984) obtained results consistent with desensitiza-tion, it was not until 16 years later that the phenomenon was systematically studied a second time (Green, 1989). No doubt the chief reason for the interval between these two oral studies is the dearth of researchers working on chemical irritation in the oral cavity. But it is also likely that the necessity of submitting one's tongue to a punishing series of capsaicin stimuli persuaded many in-vestigators to study oral chemoreception in other ways. One of the notable findings of the more recent study was, however, that stimulation need not reach painful levels to induce desensitization. Whereas Szolcsányi and Jancsó-Gábor had used a capsaicin concentration of 1.0% (10,000 ppm), Green (1989) obtained virtually complete desensitization using only 3 ppm.

The key to obtaining desensitization with the lower concentration—and per-haps with any concentration—is the insertion of a hiatus in stimulation between the desensitizing ("conditioning") stimuli and the test stimuli. If stimulation occurs too rapidly, not only does desensitization fail to occur, but *sensitization* develops.

Figure 3.6 illustrates how sensitization and desensitization were related to one another in the recent study. Following as few as five 30-sec exposures to capsaicin over a 5-min period, desensitization was significant after a 15-min hiatus in nominal stimulation. On the other hand, stimulation could be con-tinued for as long as 25 min without the occurrence of desensitization; only after a 15-min delay was the perceptual response to capsaicin depressed. A subsequent experiment in the same study provided evidence that the delay had to be longer than 2.5 min but not longer than 5 min for capsaicin to be rendered an ineffective stimulus (Green, 1989). Furthermore, the dual phenomena of sensitization and desensitization are not limited to capsaicin; piperine, the pun-gent compound in black pepper, produces a similar pattern of results (unpub-lished data).

In view of these data it is obvious that for at least some compounds the effect of the temporal pattern of stimulation on oral chemesthetic sensations is virtually the mirror image of what occurs in taste. For taste as for virtually all sensory modalities, frequent stimulation results in a progressive loss in sensitivity (adaptation), and a break in stimulation enables sensitivity to recover to pre-adapted levels. For chemesthesis frequent stimulation produces a heightened

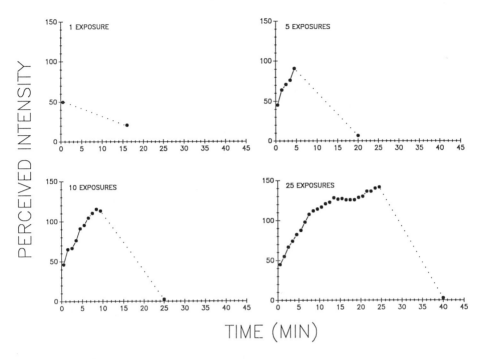

FIG. 3.6 The perceived intensity of sensations of irritation produced by repeated applications of 3 ppm capsaicin (on filter paper discs) to the tip of the tongue. Sensitization is indicated by the growth in irritation when stimuli are applied at 60-sec intervals; desensitization is indicated by the reduced level of the final stimulus delivered in each case 15 min after the last "sensitizing" stimulus. *(Reprinted with permission from Green, 1989.)*

perceptual response, whereas a respite in stimulation results in a loss of sensitivity. The only sensory system that shares at least some of these characteristics with chemesthesis is the pain sense, which under certain conditions exhibits a rise in excitability during and after repeated heat stimuli (i.e., hyperalgesia) (LaMotte, 1979). Once again, the similarity between the characteristics of pain and chemesthesis reflects the likely fact that both modalities stimulate at least some of the same sensory fibers—polymodal nociceptors.

SUMMARY AND IMPLICATIONS

As is abundantly clear from the above examples, oral chemesthetic sensations can be strongly affected by both temperature and the temporal pattern of exposure. Furthermore, chemicals that are mediated by oral somatosensory fibers can dramatically alter the perception of temperature in the mouth. Together these characteristics make oral chemesthesis a highly interactive and unique sensory modality about which a great deal remains to be learned. However, what we already know about chemesthesis suggests that greater attention should be paid to its role in normal oral perception.

The interactions between chemical irritants and temperature mean that the presence of an irritant in a food will likely change the perceived temperature of the food, and the temperature at which the food is served will in turn influence the intensity of the chemesthetic sensation. The occurrence of sensitization makes is likely that the chemesthetic component of a flavor will increase during the course of a meal relative to the gustatory and olfactory components, a fact that was evident in the study of the irritative component of saltiness (Green and Gelhard, 1989). Cross-sensitization raises the possibility that the irritation or burn produced by one food item will raise the intensity of a chemesthetic sensation produced by another food item (e.g., the irritation produced by a salty tortilla chip might accentuate the burn of a chili dish). Finally, because of desensitization and the necessity of a hiatus in stimulation for inducing it, consuming one spicy "hot" food a few minutes before eating another could result in the second food being perceived as less "hot." Although this effect can be readily demonstrated, it is probably rarely encountered under normal circumstances simply because too little time elapses between successive mouthfuls of spicy foods, or even between successive courses. It is probably for just this reason that until it was demonstrated in the laboratory, neither the occurrence of short-term desensitization nor the need for a hiatus in stimulation to induce it had been appreciated.

It should be noted that it remains unclear what the relationship is between the short-term desensitization observed in the laboratory and the long-term desensitization that seems to occur in individuals who regularly eat and spicy foods. Rozin and Schiller (1980) and Rozin et al. (1981) have reported that the threshold for detecting capsaicin is higher in individuals who frequently consume chili pepper than in those who do not, and Lawless et al. (1985) and Cowart (1987) found parallel but even larger differences between the two groups in ratings of the "burn" produced by suprathreshold concentrations of capsacin. The question is, do long-term effects occur simply because frequent consumption prevents full recovery from short-term desensitization, or because chronic consumption produces lasting changes in neurosensory function? A detailed discussion of the neurophysiological bases of desensitization is beyond the scope of this chapter; however, the interested reader can find a succinct review of what is currently known about the mechanisms that underlie capsaicin desensitization in a recent article by Maggi et al. (1990).

Whether or not we fully understand the physiological bases of these phenomena, it is clear that their perceptual consequences must be taken into account whenever sensory assessments are made of foods and beverages that contain chemesthetic stimuli. Thus, evaluations of spicy foods at temperatures other than those at which the foods are typically consumed, or at temporal intervals different from those encountered in normal eating, are likely to lead to false conclusions about questions as basic as what the optimal concentration of a spice in food should be.

ACKNOWLEDGMENTS

Preparation of this chapter, and most of the author's work reported in it, was supported by a grant from the National Institutes of Health (DC00249).

REFERENCES

Buck, S. H. and Burks, T. F. 1986. The neuropharmacology of capsaicin—review of some recent observations. *Pharmacol Rev.* 38: 179.

Cain, W. S. 1976. Olfaction and the common chemical sense: Some psychophysical contrasts. *Sensory Processes* 1: 57.

Cowart, B. J. 1987. Oral chemical irritation: Does it reduce perceived taste intensity? *Chem. Senses* 12: 467.

Green, B. G. 1985. Menthol modulates sensations of warmth and cold. *Physiol Behav.* 35: 427.

Green, B. G. 1986a. Sensory interactions between capsaicin and temperature in the oral cavity. *Chem. Senses* 11: 371.

Green, B. G. 1986b. Menthol inhibits the perception of warmth. *Physiol. Behav.* 38: 833.

Green, B. G. 1987. The effect of cooling on the vibrotactile sensitivity of the tongue. *Percept. Psychophys.* 42: 423.

Green, B. G. 1989. Capsaicin sensitization and desensitization on the tongue produced by brief exposures to a low concentration. *Neurosci. Lett.* 107: 173.

Green, B. G. 1990. Effects of thermal, mechanical, and chemical stimulation on the perception of oral irritation. In *Chemical Senses,* vol. 2: *Irritation* (B. G. Green, J. R. Mason, and M. R. Kare, Ed.). Marcel Dekker, Inc., New York.

Green, B. G. and Frankmann, S. P. 1987. The effect of cooling the tongue on the perceived intensity of taste. *Chem. Senses* 12: 609.

Green, B. G. and Gelhard B. 1989. Salt as an oral irritant. *Chem. Senses* 14: 259.

Green, B. G. and Lawless, H. (in press). The psychophysics of somatosensory chemoreception in the nose and mouth. In *Smell and Taste in Health and Disease* (T. V. Getchell, R. L. Doty, L. M. Bartoshuk, and J. B. Snow, Eds.). Raven Press, New York.

Green, B. G., Mason, J. R., and Kare, M. R. (Ed.) 1990. *Chemical Senses,* vol. 2: *Irritation* Marcel Dekker, Inc., New York, pp. v–vi.

Harada, S., Maeda, S., and Kasahara, Y. 1987. Trigeminal nerve fiber responses to the temperature, tactile, and chemical stimulation applied on the rat tongue. In *Proceedings of the 21st Japanese symposium on taste and smell* (T. Sato, Ed.).

Hensel, H. and Zotterman, Y. 1951. The effect of menthol on thermoreceptors. *Acta Physiol. Scand.* 4: 27.

Jancsó, N. 1960. Role of the nerve terminals in the mechanism of inflammatory reactions. *Bull. Millard Fillmore Hosp,* 7: 53.

Konietzny, F. and Hensel, H. 1983. The effect of capsaicin on the response characteristic of human c-polymodal nociceptors. *J. Therm. Biol.,* 8: 213.

LaMotte, R. H. 1979. Intensive and temporal determinants of thermal pain. In *Sensory Functions of the Skin of Humans* (D. R. Kenshalo, Ed.). Plenum Press, New York.

Lang, E., Novak, A., Reeh, P. W., and Handwerker, H. O. 1990. Chemosensitivity of fine afferents from rat skin in vitro. *J. Neurophysiol.* 63: 887.

Lawless, H. 1984. Oral chemical irritation: psychophysical properties. *Chemical Senses* 9: 143.

Lawless, H., Rozin, P., and Shenker, J. 1985. Effects of oral capsaicin on gustatory, olfactory and irritant sensations and flavor identification in humans who regularly or rarely consume chili pepper. *Chem. Senses* 10: 579.

Lynn, B. 1977. Cutaneous hyperalgesia. *Brit. Med. Bull.* 33: 103.

Maggi, C. A., Astolfi, M., Donnerer, J., and Amann, R. 1990. Which mechanisms account for the sensory neuron blocking action of capsaicin on primary afferents in the rat urinary bladder? *Neurosci. Lett.* 110: 267.

Okuni, Y. 1978. Response of chorda tympani fibers of the rat to pungent spices and irritants in pungent spices. *Shika Gakuho* 77: 73.

Pangborn, R. M., Chrisp, R. B., and Bertolero, L. L. 1970. Gustatory, salivary and thermal responses to solutions of sodium chloride at four temperatures. *Percept. Psychophys.* 8: 69.

Parker, G. H. 1912. The relation of smell, taste and the common chemical sense in vertebrates. *J. Acad. Natural Sci. ser. 2,* 15: 221.

Rozin, P. and Schiller, D. 1980. The nature and acquisition of a preference for chili pepper by humans. *Motiva. Emotion* 4: 77.

Rozin, P., Mark, M., and Schiller, D. 1981. The role of desensitization to capsaicin in chili pepper ingestion and preference. *Chem. Senses* 6: 23.

Silver, W. L. 1987. The common chemical sense. In *Neurobiology of Taste and Smell* (T. E. Finger and W. L. Silver, Ed.). John Wiley and Sons, New York.

Simone, D. A., Ngeow, J. Y. F., Putterman, G. J., and LaMotte, R. H. 1987. Hyperalgesia to heat after intradermal injection of capsaicin. *Brain Res.* 418: 201.

Sizer, F. and Harris, N. 1985. The influence of common food additives and temperature on threshold perception of capsaicin. *Chem. Senses* 10: 279.

Stevens, D. A. and Lawless, H. T. 1987. Enhancement of responses to sequential presentation of oral chemical irritants. *Physiol. Behav.* 39: 63.

Szolcsányi, J. 1977. A pharmacological approach to elucidation of the role of different nerve fibers and receptor endings in mediation of pain. *J. Physiol.* 73: 251.

Szolcsányi, J. 1990. Capsaicin, irritation and desensitization. In *Chemical Senses* vol. 2: *Irritation* (B. G. Green, J. R. Mason, and M. R. Kare, Ed.). Marcel Dekker, Inc., New York.

Szolcsányi, J. and Jancsó-Gábor, A. 1973. Capsaicin and other pungent agents as pharmacological tools in studies on thermoregulation. In *The Pharmacology of Thermoregulation* (E. Schonbaum and P. Lomax, Ed.). Basel, Karger.

Watson, H. R., Hems, R., Roswell, D. G., and Spring, D. J. 1978. New compounds with the menthol cooling effect. *J. Soc. Cosmet. Chem.* 29: 185.

4

Specific Anosmias: Implications for the Physiological Mechanisms of Quality Discrimination

Robert J. O'Connell

The Worcester Foundation for Experimental Biology
Shrewsbury, Massachusetts

INTRODUCTION

Sensory receptor neurons each provide a route for the initial interaction between an organism and the various environmental energies that impinge upon it and thus set the stage for, and are intimate determinants of all of the subsequent neural processing that ultimately gives rise to particular perceptual experiences. The sense organs act as the initial filters which determine the absolute range of environmental energies that may be detected, define the relative limits of sensitivity to each of the selected environmental stimuli, and delineate by subsequent neural processing both the modality of the sense experience to be perceived and its discriminitive content (Kuffler et al., 1984). The central nervous system does not create real sensory experiences out of whole cloth. It only processes and interprets those views of the environment provided by sensory receptors.

Although conventional wisdom provides a fairly detailed description of the range of human sensory experience within any one of the several available sensory domains, there are individuals who fail to experience some subset of these normal sensory perceptions. The range and magnitude of these specific deficits, within and across any one sensory domain, vary widely from individual to individual. The various forms of human color blindness and the pitch defects known as tonal islands are two common examples of specific alterations in the range of visual and auditory sensory experience. It is common, then, and often correct to assume that these narrow or notch defects in the range of our sensory experiences within any one perceptual domain are the result of specific deletions or alterations in the response properties of the particular sensory receptor neurons actually involved with the initial interaction with environmental stimuli. For example, an individual with a tonal island may fail to perceive a particular pitch because the hair cells in the inner ear responsible for the detection of that frequency have been damaged.

Specific Anosmias

Individuals with specific anosmias are generally defined as those who seem to have a good sense of smell for most odors, but lack the ability to perceive the aroma characteristic of a selected odor compound (Amoore, 1967, 1969; Amoore et al., 1968). This odor blindness (specific anosmia) is always expressed by an elevated threshold for the particular odor and is often accompanied by a pronounced shift, when compared to normal (osmic) subjects, in the perceived odor quality of just liminal concentrations of the target compound. The range of sensory experiences which exists between the total absence of the sense of smell (anosmia) and the specific alterations in the perception of particular odorous compounds (specific anosmias) discussed here is quite broad. Unfortunately, we do not have a ready array of descriptive terms to describe the large number of disparate perceptual differences which occur across this span. For convenience, we will continue to refer to specific anosmias and the mechanisms that give rise to them as though they arose through defects or disorders even though we know that they may arise as easily from normal differences in neural processing as from particular lesions or defects.

The diastereoisomeric ketone, cis-4-(4'-t-butylcyclo-hexyl)-4-methyl-2-pentanone (pemenone), elicits an intense, urine-sweaty type odor (Ohloff et al., 1983) in about 20% of the individuals we have tested (O'Connell et al., 1989). The remaining 80% appear to have a specific anosmia for this material, as they usually find it to be inodorous or to be a rather weak and generally inoffensive smell that elicits a range of different nonurinous odor reports. This compound is remarkable in that the sensory experiences of individuals may differ on each of three descriptive dimensions of olfactory sensory experience, that is, the intensity of an odor, its quality, and the hedonic responses that it elicits.

There are many specific anosmias distributed across the whole range of olfactory experience (Amoore, 1969). Within any one odor quality type it is possible to find several different compounds for which specific anosmias exist. For example, a specific anosmia for 5α-androst-16-en-3-one (androstenone) has been described (Griffiths and Patterson, 1970; Wysocki and Beauchamp, 1984). It shares with pemenone a distinctive but very unique urine-sweaty type odor for those individuals that are osmic for it (O'Connell et al., 1989). In a subsequent study, which compared the thresholds for these two odors and several others, it was observed that individuals anosmic to pemenone were also anosmic to androstenone. The same was true for those osmic for pemenone. Several of the other compounds evaluated produced multimodal distributions of odor thresholds (Fig. 4.1) suggesting that specific anosmias for other compounds coexisted in this one pool of subjects (Stevens and O'Connell, 1991). Subsequent analysis failed to find simple patterns of anosmia across the range of odors tested, save for the original relationships among the compounds with urine-like odors.

In this study, pemenone thresholds were determined for several hundred human subjects, a matched set of 62 anosmic and osmic subjects were selected for further study. All of the subjects were judged to have normal olfactory capabilities based upon their ability to correctly identify a battery of 10 common odorants. Thresholds were determined for all six odorants under consideration, and quality and hedonic reports were elicited at threshold and perithreshold concentrations of each substance. Both ANOVA and principal-component analyses were performed. When the thresholds, quality, and hedonic reports of subjects were compared, a number of significant interactions were observed. A subject's threshold and quality report for pemenone was significantly correlated with the threshold and quality report elicited by androstenone. The distribution of quality reports at threshold suggests that, although these substances share the urine-sweaty note, there were several differences among them which are especially obvious for those individuals initially characterized as anosmic for pemenone. The distribution of thresholds for most of the compounds evaluated was bimodal except for isovaleric acid see (Fig. 4.1). The matrix of correlations between the threshold concentrations of each odor is shown in Table 4.1. The relationships among odor thresholds were further analyzed by a principal-component analysis. This produced a solution whose first component had three orthogonal factors. These included pemenone, androstenone, and isovaleric acid thresholds. This observation suggests that these compounds all interact, perhaps differentially, with a common pool of sensory receptor neurons or perceptual channels (see p. 139).

In order to determine if this postulated array of interactions gives rise to different quality perceptions for particular compounds, the quality labels applied to stimuli at near threshold concentrations were evaluated. Relationships between odor quality and threshold were examined by comparing the relative

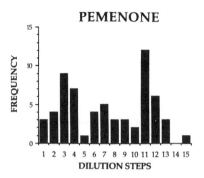

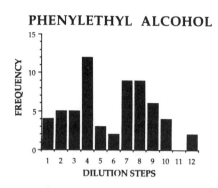

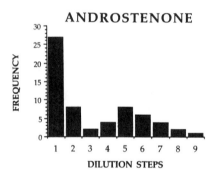

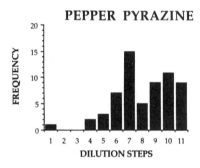

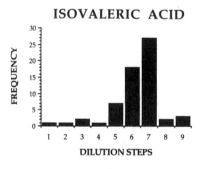

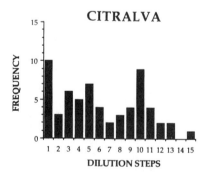

FIG. 4.1 Frequency histograms showing the distribution of subjects' thresholds for each of the indicated odorants. In each case, a relative binary dilution scale is employed such that the concentration at threshold for the least sensitive subject has been termed step 1. For comparison purposes, the concentration of step 5 for each compound is: 25.0 mM for pemenone; 42.2 μM for androstenone; 30.5 μM for isovaleric acid; 61.0 μM for phenylethyl alcohol; 0.4 μM for pepper pyrazine; and 48.8 μM for citralva. *(From Stevens and O'Connell, 1991.)*

TABLE 4.1 Correlation Matrix for Relative Threshold Concentrations

	Pemenone	Androstenone	Isovaleric acid	Phenylethyl alcohol	Pepper pyrazine
Androstenone	.813*				
Isovaleric acid	.425*	.230			
Phenylethyl alcohol	−.027	.141	−.235		
Pepper pyrazine	.347*	.292*	.192	.250	
Citralva	.040	−.081	.090	−.113	.171

*$p < 0.025$

Source: Stevens and O'Connell (1991).

frequencies of modal, nonmodal, and no-odor reports for subjects above and below the antimodes of the threshold distributions. The modal quality classifications for pepper pyrazine included vegetable, grass, and green peppers; for isovaleric acid, androstenone, and pemenone they included rancid, sweaty, and urine; for phenylethyl alcohol and citralva they included floral, fruity, and perfume. For each compound, the frequencies of modal, nonmodal, and no-odor responses were compared for subjects whose thresholds were above (relatively osmic) or below (relatively anosmic) the antimode for that compound and for each of the other compounds. These matrices were analyzed by Chi-square tests which indicated that subjects relatively anosmic for pepper pyrazine reported a nonmodal quality relatively more often than did the osmic subjects (Yate's Chi-square = 4.33, df = 1, $p < 0.05$). For both pemenone and androstenone, those subjects relatively anosmic for the compound reported a nonmodal odor or no odor more often than did relatively osmic subjects (Chi-square = 7.24, df = 1, $p < 0.05$ and 18.14, df = 1, p < 0.001, respectively). The same relationship was found for androstenone when subjects were classified by their relative sensitivity to pemenone (Chi-square = 17.01, df = 1, $p < 0.001$). Figure 4.2 shows the frequencies of odor qualities reported for threshold concentrations of pemenone among those subjects relatively osmic and relatively anosmic to pemenone. Those subjects osmic for pemenone are most likely to characterize both pemenone and androstenone as putrid, while those anosmic to pemenone report a range of other odor qualities for both substances. There were no age differences in the distributions of odor qualities. The pronounced shifts in the odor quality report of individuals for pemenone suggest that this one compound likely interacts with multiple receptor neurons and perceptual pathways that ultimately give rise to different quality reports and suggests that the difference between osmics and specific anosmics for this substance is the particular array of odor qualities that it generates (see p. 141).

Hedonic responses were related to sensitivity for isovaleric acid and pemenone. For the former, the less sensitive the subject the more unpleasant the odor

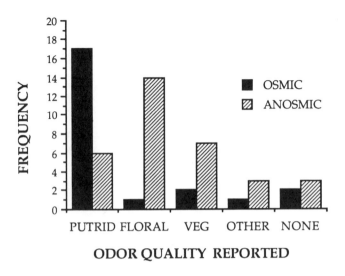

FIG. 4.2 Frequency histogram of the quality reports generated by just liminal concentrations of pemenone for subjects classified as osmic or anosmic for pemenone. Individual quality reports have been collapsed across labels into the indicated classes. VEG includes both woody and vegetable descriptors. *(From Stevens and O'Connell, 1991.)*

($r = -0.33$, df) (48, $p < 0.02$). For pemenone, the less sensitive the subject the less unpleasant the odor ($r = +0.66$, df = 52, $p < 0.001$).

Although an array of findings similar to those detailed above was suggested by our earlier study with suprathreshold odor concentrations (O'Connell et al., 1989), the individual differences in odor quality observed could have arisen from differences in the intensities of the odors evaluated. Accordingly, we then determined the relationship between individual thresholds and odor quality. The underlying assumption for this study was the notion that stimulation with just liminal concentrations of odorants equates the intensity of sensation elicited by different odorants. Thus, we are fairly confident that the relative sensitivities exhibited by individual subjects to androstenone and pemenone are highly correlated, and that the array of odor quality reports generated by these substances differed substantially between different odorants and in particular between pemenone osmic and anosmic groups. It is likely that the constellation of odor reports elicited by an odorant with a characterized anosmia is a reflection of a deficit in one or more of the shared perceptual channels. The factors produced by the analysis employed here are orthogonal. Differences in sensitivity to isovaleric acid was an important contributor to two of those factors. This finding is also consistent with a multiple profile model of odor interaction (Polak, 1973). Indeed, it is difficult to reconcile these findings with models in which individual odorants interact with only a single perceptual channel (these ideas are more fully described later).

Specific anosmias involve both shifts in intensity and, for many subjects, profound shifts in odor quality. For any one odor, anosmic subjects may be characterized by the particular quality report they produce. The differences in the threshold and odor quality reported reveal that there must be large individual differences in the perceptual pathways that give rise to different quality reports. The role of individual differences in flavor perception is explored more fully by David Stevens in Chapter 10.

Recalling the earlier cautionary note that specific anosmias could also arise from central nervous system alterations, let us now explore the variety of transduction mechanism that are currently thought to occur in olfactory receptor neurons in order to see what physiological mechanisms are available for the encoding of olfactory information and how disorders in these mechanisms may give rise to specific perceptual defects.

TRANSDUCTION MECHANISMS

It is reasonable to suppose that the existence of specific anosmias, especially those that are characterized by a complete inability to detect maximal concentrations of only one or a few odors reflects deficits or alterations in the response properties of individual olfactory receptor neurons and that these receptor changes collectively contribute to the deficit in olfactory perception. Although there are a great number of indirect transport and perireceptor mechanisms, each of which may influence the access of particular odor molecules to their sites of interaction with olfactory receptor neurons, it is generally conceded that the subsequent direct interactions of odors with receptor neurons are the primary events responsible for the detection and ensuing encoding of olfactory information (Getchell and Getchell, 1987; Carr et al., 1990). The important steps in olfactory transduction seem to depend on relatively nonspecific, that is, energetically weak interactions which, nonetheless, have to be reliable and give rise to exquisitely specific effects. If we consider for a moment the incredible range and diversity of chemical compounds present in the environment of an evolving organism the extraordinary demands made on the olfactory system readily become apparent. At the molecular level, exceedingly small changes in structure (e.g., optical isomers) must be discriminated. Therefore, the molecular diversity that must be handled by the system continues to increase as other organisms (including organic chemists) synthesize new compounds that have never existed before. The concentrations of these potential chemical signals vary widely from as little as one molecule of odor/cc of medium up to hundreds of trillions of molecules/cc. To cope with this, an animal needs a system with enormous gain so that very dilute stimuli like those emanating from potentially willing, but remote, sexual partners will be detected. The system's resolving power must be prodigious as well, since important chemical signals will be buried in a sea of other nonsignificant compounds. The olfactory system should also have a very wide dynamic range so that small differences in local concentration gradients may be used to locate the sources of important chemical signals. Finally, all of these requirements must be packaged within the space and energy requirements imposed upon a nervous system.

These diverse requirements could be satisfied by systems that include a set of broad range detectors and their central connections, which are responsible for discriminating among a very large number of potentially important odor signals and a narrow band, high sensitivity system of detectors and their central neural apparatus, which are responsible for rapidly detecting and encoding chemical signals with large amounts of survival value. These systems may operate independently or be chained together in various ways so that discriminations made in one may affect the discriminations made in another. Thus individual olfactory receptor neurons should vary in their relative sensitivity, specificity, and speed of interaction with particular chemical stimuli.

Unfortunately, the mechanisms associated with olfactory transduction and the subsequent neural and computational procedures for the coding of olfactory information in the nervous system remain an enigma largely because the physical characteristics of the environmental energies responsible for odor are still largely undefined. That is, much of the power that has historically been brought to bear on the analysis of other sensory systems by an intimate knowledge and control of the relevant stimulus variables is absent in olfaction. We continue to lack in a fundamental way insight into those aspects of chemical composition which are responsible for the unique odor signature observed for each odor molecule. In the absence of this information it is difficult to discern the common molecular features that make a particular array of stimuli effective modulators of neural activity and, thus, comparably difficult to unmask the binding, transduction, coding, and decoding mechanisms responsible for the perception of odors. In spite of these limitations, recent advances and new technologies drawn from biochemistry, biophysics, and molecular biology have begun to make it possible to consider the broad array of mechanisms that appear to be available for olfactory transduction.

There is now general, but not universal (Lerner et al., 1988; Kurihara et al., 1989), agreement that all odor molecules must interact directly with specialized odor receptor molecules (binding sites) which are expected to be found in the ciliary membranes of olfactory receptor neurons (Getchell and Getchell, 1987). These proteinaceous binding sites are thought to be present on the extracellular mucosal surface where they interact with odors and couple, through a variety of mechanisms, the presence of a stimulus molecule in the extracellular space to alterations in transmembrane current flow. These latter currents typically arise because of the presence in the membrane of a variety of membrane-spanning, single-channel proteins. Activation of these single channels provides a reversible and ready avenue for the movement of ions into or out of the cell (Hille, 1984). The transmembrane current flow that results may then modulate the action potential production mechanism in the neuron so that alterations in the train of electrical activity transmitted to the central nervous system over olfactory receptor neuron axons is altered in characteristic ways.

Odor-binding sites have not yet been isolated and identified from any olfactory receptor neuron. Accordingly we have little information about their presumed specificity, the sensitivity of their interaction with odors, or the range of mechanisms which might couple these binding events to the channel proteins responsible for transmembrane current flow. We also have little idea how many different kinds of binding sites there may be or how they are distributed across the whole population of olfactory receptor neurons. In contrast, there is substantial information about the rich diversity of binding and coupling mechanisms available in the signal transduction pathways of other tissues. Components with properties similar to those seen in these pathways are now being found in olfactory receptor neurons suggesting that some array of them are involved with the mechanisms which will satisfy all of the presently known requirements for olfactory transduction.

The following three schemes incorporate many of the concepts and principles that have been elucidated in other neural tissues which connect the interaction of a ligand molecule with the resulting transmembrane current flow (North, 1989). None of the three have been experimentally verified as the mechanisms responsible for one or more of the various aspects of olfactory transduction in any olfactory receptor neuron.

Directly Gated Channels

In the first coupling mechanism, the odor-binding site is a structural component of one or more of the single channel proteins that are collectively responsible for the transmembrane current flow that arises in response to odor stimulation. Interaction with odor results in direct activation of the binding site which is coupled directly to the mechanism responsible for modulating the rate of channel opening and closing. This direct gating mechanism provides for substantial signal amplification because a single odor-binding event may directly control and modulate the flow of thousands of ions/ms across the cell membrane. Moreover, since the binding site is an integral structural part of the channel protein, there are few intervening steps to slow the activation of transmembrane current. Thus, directly gated odor channels provide a transduction mechanism that must result in the most rapid electrical response to odor in conjunction with a chemical specificity and sensitivity that is an unmodified function of the specificity and sensitivity of the initial binding events. Thus, we should expect this coupling mechanism to be involved with the processing of biologically relevant chemical signals like sex attractant pheromones. There is substantial information available on the rich diversity of binding site subtypes that have been found in a variety of directly gated channels (Schofield et al., 1990). Molecular analysis has demonstrated that this receptor heterogeneity often arises because there exists a gene family encoding for highly related but functionally distinct receptor subtypes.

Although directly gated ligand channels are well described in a large number of signal transduction schemes in a variety of other tissues (North, 1989; Schofield et al., 1990), only a few reports of their existence in the olfactory system have appeared (Vodyanoy and Murphy, 1983; Labarca et al., 1988). Recently, proteins isolated from frog olfactory cilia were reconstituted into artificial lipid planar bilayers (Labarca et al., 1988) and their single channel currents were measured with standard patch clamp techniques which make it possible to directly measure the minute current flows that pass through individual channel proteins (Bruch and Teeter, 1989). In addition to several channels whose conductances and ionic selectivity were reminiscent of channels seen in other excitable cells, an odor-activated cation selective channel was observed. This latter channel was activated by nM concentrations of pepper pyrazine and citralva in the absence of the normal modulators associated with other indirectly gated channels. These modulators normally include adenosine triphosphate (ATP), guanosine triphosphate (GTP), cyclic adenosine monophosphate (cAMP), cyclic guanosine monophosphate (cGMP), and other soluble cytosolic components, many of which have been implicated in other transduction mechanisms involving indirectly gated second messenger systems (Labarca et al., 1988).

Although it is not yet clear what array of odor-gated channels might exist in a particular olfactory system, it seems likely that some compounds may be transduced solely by their interaction with directly gated channels (McClintock and Ache, 1989). It is also likely that selected compounds may well be detected by multiple transduction pathways. For example, citralva is know to be a robust activator of the adenylyl cyclase (AC) found in preparations of frog olfactory cilia, which has proved to be responsible for the GTP-dependent conversion of ATP to the important second messenger compound cAMP (Sklar et al., 1986). As noted above, citralva also activates a directly gated channel found in the same tissue (Labarca et al., 1988). An alternate explanation to account for these observations would require that the current flow produced by the directly gated channel should influence the activation of AC (see p. 138).

Indirectly Gated Channels

In the second type of coupling mechanism, activation of the odor-binding site by chemical stimuli modulates one or more intracellular second messenger systems, which, in turn, affects one or more single-channel macromolecules (Shirley and Persaud, 1990; Trotier, 1990). Although there are a number of well-described second messenger systems in other tissues, we will focus here on those that involve the modulation of cyclic AMP (cAMP) as this compound has been widely implicated in olfactory transduction.

Considerable evidence exists to support the contention that cAMP serves as an intracellular second messenger in a variety of hormone and transmitter activated tissues. Signal transduction in these tissues involves pathways that couple ligand binding to receptor proteins at the cell surface with the activation of specialized cytosolic, membrane associated GTP-binding proteins. These G proteins modulate the activity of adenylyl cyclase and thus regulate the rate of conversion of ATP to cAMP (Pace and Lancet, 1986; Bruch and Kalinoski, 1987). A large number of different G proteins have been identified. Some are stimulatory and result in an increase in the rate of production of cAMP, whereas others are inhibitory and result in a decrease in the rate of cAMP production (Birnbaumer et al., 1987; Gilman, 1987).

The involvement of second messenger systems in olfactory transduction was first suggested by Kurihara and Koyama (1972) as a result of the elevated levels of adenylyl cyclase that they observed in the rabbit olfactory system. Later, derivatives of cAMP were shown to alter the electro-olfactogram (EOG) in the frog (Menevse et al., 1977). Subsequent experiments demonstrated that the magnitude of the currents elicited by odorants in planar lipid bilayers containing homogenized rat olfactory epithelium is mimicked by added cAMP (Vodyanoy and Vodyanoy, 1987). In a like fashion, small patches of ciliary membrane removed from isolated toad olfactory receptor neurons contain a nonselective cation channel that is activated by μM concentrations of cGMP or cAMP (Nakamura and Gold, 1987). Olfactory ciliary membranes prepared from several different vertebrates contain an adenylyl cyclase with a high specific activity (Pace et al., 1985; O'Connell et al., 1990), which apparently arises, in part, because the enzyme is 100-fold more abundant in cilia than in other tissues (Pfeuffer et al., 1989). The activity of this enzyme is modulated by a large number of different volatile compounds (Sklar et al., 1986). Recently, a number of different GTP-binding proteins including a unique stimulatory GTP-binding protein (G_{olf}) have been isolated from rat olfactory cilia and implicated in the cascade of odor-induced events which are thought to lead to odor transduction in primary receptor neurons (Jones and Reed, 1987; 1989; Reed, 1990).

The relative contribution of directly gated and indirectly gated second messenger systems to odor transduction is not known precisely in any species.

Although the role of odor-modulated adenylyl cyclase in the transduction of a large number of different odors has been implicated in several studies, it is usually the case that a fairly large number of compounds fail to modulate cAMP production (Sklar et al., 1986; Lancet et al., 1989). This remains true even when both biochemical and electrophysiological estimates of odor potency are shown to covary in the same species (Lowe et al., 1989). This suggests that some compounds may be ineffective modulators of adenylyl cyclase because they are ineffective odors for the particular species or alternatively, because they are processed by other transduction mechanisms (Lancet et al., 1989; Lowe et al., 1989). In many cases, an independent estimate of the olfactory potency of a particular compound for the species in question is not available because the compound has never been evaluated with modern behavioral or psychophysical techniques. Explorations of odor transduction are difficult to evaluate if they inadvertently include compounds that are not actually detected and discriminated by the test species.

We recently compared the relative biochemical and electrophysiological efficacies of a number of compounds, several of which are known to be produced by the female hamster and sequestered in the odor signals she normally uses to signal behavioral receptivity (Singer et al., 1976). One of these compounds, dimethyl disulfide (DMDS), is an extremely potent activator of male investigatory behavior with a response threshold of ~1 pM (O'Connell et al., 1979). As we show in Table 4.2 DMDS is a perfectly good activator of a large

TABLE 4.2 Biochemical and Electrophysiological Responses to Odor Stimuli in the Male Hamster Olfactory System

Odorant	% of basal AC activity[a]	% of neurons responding[b]	Average responses 10 nM[c]
Amyl acetate	141 ± 6	88 (16)	13.1 (9)
Butanol	103 ± 1	53 (17)	15.7 (13)
DMDS	101 ± 2	81 (21)	14.8 (12)

[a]Adenylyl cyclase (AC) activity stimulated by each odorant presented at 100 μM.
[b]Percent of receptor neurons out of the total stimulated (in parentheses) which produced responses (impulses/5 s) more than twice the maximum prestimulus level.
[c]Mean responses (impulses/5 s) elicited in the total number (in parentheses) of receptor neurons stimulated with 10 nM of the indicated odorant.

Source: O'Connell et al. (1990).

fraction (81%) of the male hamster olfactory receptor neurons we have studied and, on average, elicits from them relatively large amounts of electrical activity. In contrast, DMDS does not activate adenylyl cyclase within the concentration range of 10nM to 100 μM because it fails to elicit a measurable alteration in the level of cAMP (O'Connell et al., 1990). A second constituent of hamster vaginal discharge, n-butanol, is also a good activator of olfactory receptor neurons yet is an ineffective stimulator of both adenylyl cyclase activity and male investigatory behavior. Amyl acetate is a robust activator of adenylyl cyclase and a potent electrophysiological stimulus.

Collectively, these facts argue strongly for the involvement of multiple transduction mechanisms in the hamster olfactory system, especially with regard to the processing of the hamster sex attractant, DMDS (O'Connell et al., 1990). The possibility of multiple mechanisms has been suggested for other species, in part to account for the observations that small shifts in cAMP production are routinely elicited by a range of different compounds which appear to be perfectly good "odorants" (Sklar et al., 1986; Anholt et al., 1987; Labarca et al., 1988; Anholt, 1989; Dionne, 1989). In this regard, we should note that fully 75% of the receptor neurons that we exposed to 10 nM concentrations of each of the three odorants (amyl acetate, n-butanol, and DMDS, $n = 12$), responded with significant amounts of electrical activity to each of the three compounds. From this we may infer that if multiple transduction mechanisms are actually responsible for the detection and processing of these three compounds, then it is reasonable to assume that they all must coexist in at least some olfactory receptor neurons.

The involvement of intervening enzymatic steps in olfactory transduction has a number of consequences: (1) Since there are additional biochemical steps interposed between the interaction with odor and the production of a transmembrane current flow, the latency of response to odors must be longer than is the case in directly gated channels. The actual increase in latency is not yet known precisely, but the fastest indirectly gated channel so far studied in frog olfactory receptor neurons (Firestein and Werblin, 1989) has response latencies in the order of 140–200 ms. In contrast, directly gated channels may be two to three orders of magnitude faster. For example, small patches of neuronal membrane removed from crayfish chemoreceptors respond to the rapid application of high concentrations (~ 1 mM) of purine nucleotides (5'-GMP) with response latencies that are measured in the range of 0.4–0.8 ms (Hatt, 1989). (2) The relative amount of signal amplification to be expected in second messenger systems is much larger than the amplification observed in directly gated channels because each step in the enzymatic cascade has the potential to amplify the effect of the initial binding event. It is likely, then, that a simple cascade, like the one described above, will amplify an individual odor-binding event so that it may ultimately effect hundreds of transmembrane current channels. (3) Finally, each of the steps in a second messenger cascade may have multiple points of both positive and negative control (Codina et al., 1987; Sternweis and Pang, 1990). In turn, the products of each step may have multiple targets for subsequent modulation of neural activity (Light et al. 1990). For example, noradrenalin activates G proteins, which then open potassium channels and close calcium channels in guinea pig submucosal neurons (North et al., 1988).

Receptor Enzyme Channels

In the third type of coupling mechanism postulated, the odor-binding site is incorporated into a membrane-bound enzyme such as tyrosine kinase, adenylyl cyclase, or guanyly cyclase. Activation of these receptor-coupled enzymes alters the rate of production of their respective intracellular products, which, in turn, modulate the gating of single channel currents (North, 1989). Another variant of this mechanism would bring the odor-binding site, the G protein and the enzyme or channel protein together to form a single transmembrane macromolecule. Recent molecular information on the structure of voltage sensitive ion channels (Catterall, 1988) and on the adenylyl cyclase isolated from bovine brain (Krupinski et al., 1989) suggests that the latter enzyme may also incorporate a transmembrane structure reminiscent of other G protein–regulated Ca^{2+} channels. This same protein includes regions which incorporate multiple N-linked glycosylation sites suggesting that this region could represent the stabilized folds that form an extracellular odor-binding site (Krupinski et al., 1989). If each of the subunits of this receptor enzyme retained its original capabilities, as described in other tissues, than one might imagine that odor activation of this hypothetical entity would result in nearly maximal response speed coupled with substantial levels of amplification. Depending on the relative specificities of the binding proteins involved, this would result in an olfactory receptor neuron system with response properties intermediate between those primarily endowed with directly gated or indirectly gated channels.

RECEPTOR NEURON MECHANISMS

At this stage in our understanding of olfactory transduction, it is not possible to decide if the various receptor neuron transduction mechanisms described in earlier sections are singly or collectively involved in odor detection. It may be that all the second messenger systems including others not mentioned previously like guanylyl cyclase (Bruch and Teeter, 1989) and guanine nucleotide-dependent inositol phosphate production (Huque and Bruch, 1986) interact in various ways with other transduction mechanisms including odor-gated channels

(Anholt et al., 1987; Labarca et al., 1988; Anholt, 1989; Anholt and Rivers, 1990) to shape the ultimate response properties of individual olfactory receptor neurons.

In some cases, the low levels of activation of adenylyl cyclase by certain odors may indicate the presence of additional stages of signal amplification involving other enzymatic (Lancet et al., 1989; Anholt and Rivers, 1990) or single-channel (Nakamura and Gold, 1987) effectors elsewhere in the cell. For example, Anholt and Rivers (1990) have suggested a model based on their recent studies which demonstrated that frog cilia contain reasonable amounts of calmodulin. This compound proves to be a potent GTP and calcium dependent stimulator of the odor-activated pathways that each result in increases in intracellular cAMP. In one, odor directly activates a channel that gates an inward calcium flux, which, in turn, forms a complex with calmodulin that strongly activates the olfactory adenylyl cyclase. In the second, high concentrations of odors are thought to directly activate the olfactory G protein, which would, in turn, activate adenylyl cyclase in the normal fashion. In either case, the resulting increase in cAMP could act directly on ion channels or it could stimulate a protein kinase whose phosphorylation events could then directly gate other single channel proteins (Anholt and Rivers, 1990). It is also likely that odors modulate second messenger levels over short time scales that are not readily detected by the standard enzymatic assay (Breer et al., 1990).

It should be obvious at this point that there are a large number of reasonable transduction mechanisms that could exist in olfactory receptor neurons. Particular combinations of the various components within individual receptor neurons will produce coupling systems that vary widely in their sensitivity, specificity, and temporal acuity. As noted earlier, the resulting response properties of the receptor neurons are the initial determinants of olfactory perception. The mechanisms responsible for the transmission and encoding of odor information in the afferent activity of the olfactory nerves and the computational capabilities of the central olfactory pathways are still largely unknown. The interested reader is referred to the recent review by Kauer (1987) for a consideration of alternative coding mechanisms which may be operational in the olfactory system.

PHYSIOLOGICAL IMPLICATIONS OF SPECIFIC ANOSMIAS

The discovery of specific anosmias for particular compounds and their subsequent psychophysical evaluation provides interesting opportunities for brief glimpses into some of the inner workings of the olfactory system. Among the many intriguing questions that recent studies of the specific anosmias raise are: where do the responsible alterations occur in the central nervous system and how do they give rise to the described perturbations in quality perception? Answers to these questions may well tell us much about the detection and encoding of odors and the overall cognitive operation of an important sensory system.

Anatomical Locus of Anosmias

To the extent that sensory perceptions arise as a result of both peripheral and central neural processing of particular environmental simuli, it remains true that specific changes in sensory experience could easily arise from alterations or defects anywhere in the sensory system. The specific anosmia for pemenone that we discussed earlier involves shifts in all three of the dimensions about which odor may vary. Therefore, we may inquire into the implications of this for the location in the nervous system of the pemenone defect. That is, does a specific anosmia involve primarily ofactory receptor neurons or does it involve more central regions like the olfactory bulb and the pyriform cortex? In order to begin to answer this question, we will have to determine what these particular areas are responsible for in the whole of olfactory experience.

In primate color vision, it is clear that color blindness is often due to the congenital absence of one or more of the visual pigments that are normally found segregated in different types of cones (Jacobs, 1986). In audition, tonal islands in pitch perception seem to be the result of damage to specific hair cells which are located in certain regions of the cochlea. The well-known fact that a very great number of different colors can be perceived in a system with only three kinds of broadly tuned primary color receptors has long supported the contention in olfaction that there are a small number of primary receptor neurons whose collective response properties code for odor quality. However, since the earliest electrophysiological studies of individual olfactory receptor neurons (Gesteland et al., 1965; O'Connell and Mozell, 1969) it has been clear that a unique class of receptor neuron, each responsible for detecting one of the discriminable odor qualities, does not exist. Olfactory receptor neurons in vertebrates regularly respond to odors which seem drawn from a large number of different quality classes.

In spite of this, it remains likely that defects in the kind or number of primary olfactory receptor neurons may account for specific sensory defects. Unfortunately, these properties of vertebrate olfactory receptor neurons are not well known and thus it is impossible to identify a specific sensory defect as arising solely from changes in primary olfactory receptor neurons. Recent evidence suggesting that a specific anosmia to androstenone may be reversed in some anosmic individuals by regular exposure to the target odor (Wysocki et al., 1989) has led to the idea that this inculcation is the result of a peripheral process that engenders new olfactory receptor neurons from primordial basal cells and imbues them with the ability to respond to androstenone. However, it is also likely that the reversal of this specific anosmia is the result of a central process involving the learning or discrimination of a new odor perception. Moreover, there are other defects later in the central nervous system, in one or another sensory system, that are known to give rise to specific types of sensory deficits (Damasio et al., 1980; Pantev et al., 1989).

In some hierarchical sensory systems, notably vision, it seems clear that many of the more central regions of the system are specialized for the processing of more abstract aspects of sensory experience (Gouras, 1981). In primate vision, for example, it is now clear that there are more than 30 different anatomical areas in the macaque cerebral cortex and more than 300 identified interconnecting neural circuits that are involved with one or another aspect of vision (Gibbons, 1990). These cortical areas and their connections form a series of interconnected, but largely parallel, pathways in the central nervous system, each associated with a different component of visual information processing including, among others: color, form, depth, movement, and texture (Livingstone and Hubel, 1987). Neural damage restricted to one or another of these areas can result in specific deficits in sensory perceptions. The relative complexity of the resulting sensory defect seems to be a function of how far central in the pathway the anatomical disruption occurs.

It is also true that, in the case of higher order cognitive processing of sensory information, the largely sensory areas of the cortex interact with or are modulated by other cortical areas comprising a series of associative systems. There is a growing appreciation of the fact that these associative neural structures may be specialized for processing more abstract information about sensory experiences including pattern recognition and various aspects of short- and long-term memory. For example, Damasio (1990) has shown that visual agnosias can arise in patients with circumscribed bilateral lesions in ventral occipital and temporal association cortices. Affected individuals can recognize familiar people based upon their voice or upon visual representations of their gait or posture, but they cannot recognize the same individuals from a visual image of their face (face agnosia). These individuals have no difficulty recognizing components or categories of familiar faces, but they cannot recognize the face as a whole as belonging to a particular person. Thus, defects in these associative, apparently nonsensory areas may also result in specific disruptions of complex sensory perceptions.

Mechanisms of Quality Coding

It is common to assume that the perception of odor quality is analogous to color and pitch perception in the same way that odor intensity is analogous to brightness and loudness. If this is true, then we may inquire into the olfactory analogs of form, movement, pitch, and other higher order visual and auditory features, and ask how they are differentially processed in the olfactory system. These questions remain inextricably enmeshed with the previous topic of anatomical localization of function. In order for the analogies with other sensory systems to have much weight one must posit that the olfactory system is also organized into hierarchical parallel pathways with different regions responsible for different percepts.

If odor quality, intensity, and hedonics are encoded in this fashion, then it should be possible to find certain individuals with peripheral defects that alter all three olfactory dimensions and others with more central defects in which only one or another of these percepts is disrupted. This is similar to the place specificity in the visual system where it is possible to find individuals with specific deficits in vision which are related to defects in primary receptors (Lennie, 1984; Nathans et al., 1986) and others in which their visual deficits arise from central disturbances. For example, there are subjects with cortical achromatopsia, a selective loss of color perception, that prove to have neural damage in one of the central visual areas anterior to cortical area 18 (Damasio et al., 1980).

We have identified individuals with the ability to detect pemenone and androstenone but with much reduced sensitivity. Many of them exhibit very large shifts in reported odor quality and hedonic ratings to presentation of these materials. We suggest that these specific anosmias arise from peripheral defects largely because they interfere with all of the perceptual attributes of the compounds. On the other hand, several of our androstenone anosmic individuals continued to apply the correct odor quality and hedonic rating suggesting that their anosmia could have been the result of changes in either peripheral or central levels of processing. That is, it could be that urinous odor is a higher order percept or alternatively that androstenone interacts with the normal receptor mechanism, but with greatly reduced efficiency. One of the great difficulties in unraveling these complexities harkens back to the simple fact that what we routinely call odor quality may be a complex central percept more analogous to face recognition, for example, than to the detection of a red object.

We know that color vision does not arise simply by detecting the wavelength composition of the light reflected from an object's surface. It also takes into account the relationship between the object and its surround. Thus, in a very real way the small differences in wavelength between an object and its background are used to enhance the contrast in the visual scene where gradients of light intensity are often small (Gouras, 1981). Could it be that odor quality is similarly a matter of contrast? Given the general ubiquity of odors in the natural world, there is every reason to believe that every stimulus situation includes a rich odor background, either as a result of biological process in the environment or in the perceiving animal. In a similar vein, uniform featureless fields of color input do not elicit color sensations presumably because of the lack of contrast. Thus the graying out of a uniform color field may be analogous to olfactory adaptation where a constant background odor in the environment gradually fades from notice.

Although there are many similarities between color vision and olfaction, there are also a number of substantial differences. A prime difference lies in the relative ability to match stimuli by combinations of other stimuli. In color vision it is possible to duplicate any one perceived color by combinations of different wavelengths (e.g., appropriate mixtures of green and yellow wavelengths can match a blue that is spectrally different). A corresponding ability to match the quality of a single odor with mixtures of others is not seen in the olfactory system. Instead, it is common to find that when odors combine they produce new odor sensations that are unique. That is, chemically distinct compounds that have odors are always discriminable. As far as one knows, there are no two unique odor compounds that have indistinguishable odors. Moreover, it is well known that there are a large number of dissimilar chemical entities that produce similar types of odor.

Perhaps the failed attempts to do odor mixture experiments analogous to color mixture experiments indicate that odors qualities are not the elemental primal attributes of individual compounds but are actually higher order percepts that arise from the computational abilities of the central olfactory system. This line of arguments leads naturally to the hypothesis that individual odor molecules are more analogous to a complete visual scene than to the visual elements that make up a scene. Thus we can return to the earlier questions concerning the level of computation required for an odor quality discrimination and where these computations may be carried out in the central nervous system. Are odor hedonics and quality partitioned into different anatomical regions in the same way that form and color are segregated into the parvocellular and magnocellular divisions of the lateral geniculate nucleus and their central connections (Livingstone and Hubel, 1987)? We may also ask, what aspects of olfactory perception are performed in the main and accessory olfactory bulbs and in the pyriform cortex? To a large extent we have no ready answers for these questions. However, in animal models there is some evidence which suggests that the accessory olfactory bulb plays an important part in the olfactory recognition and odor memory that must be involved with the strange male odor pregnancy block phenomenon in mice (Kaba et al., 1989).

In a like fashion, Haberly and coworkers have drawn attention to the similarities between the anatomical structures of the olfactory and accessory olfactory bulbs and the pattern of connections required by certain associative, content-addressable memory models (Haberly and Bower, 1989). Others have pointed to the fact that odors elicit 40–60 Hz oscillations in both the olfactory bulb and cortex which may be entrained to the rhythms of brain activity (θ-rhythm) that are normally associated with the motor patterns involved in sniffing. Similar patterns of 40 Hz synchronous activity have been reported in other sensory systems and are thought to be involved with coherence detection (Granger et al., 1989; Barinaga, 1990; Bressler, 1990). This latter process is thought to arise as a direct consequence of the parallel architecture of certain portions of the olfactory nervous system and seems to be involved with detecting particular patterns of stimulus input and comparing them with patterns of activity previously stored in odor memory. A mechanism which provides for coherence detection seems required in the olfactory system where the detection of odor differences seems a large part of odor quality discrimination.

CONCLUSIONS

In summary then, specific notch defects in sensory experience are observed in a variety of sensory systems and can involve alterations in both the quantitative and qualitative domains of sensory experience. In each case, these defects provide important clues to the organization of the system and provide the opportunity to decipher some of the important mechanisms in normal sensory processing. The evaluation of specific anosmias provides similar opportunities in the olfactory system.

The relative sensitivities exhibited by individual subjects to androstenone and pemenone are highly correlated. The array of odor quality reports generated by these substances differed substantially between pemenone osmic and anosmic groups and, as we expected, between different odorants. These systematic relationships between sensitivity and the quality reports elicited by different compounds are consistent with a multiple profile model of odor quality encoding (Polak, 1973; Beets, 1974). This model suggests that most individual odor compounds interact with more than one receptor-binding process and that these processes are differently distributed across the whole neural array found in the olfactory epithelium. The odor quality elicited by any one compound then is thought of as a computational composite of all of the interactions that are possible. Thus, any individual alteration in the relative specificity or number of individual binding sites or, for that matter, in any of the other normal receptor neuron processes in any one neural channel should alter both the overall pattern of interaction for any one odor and thus should give rise to alterations in both intensity and quality of the odor perceived. Because of the shared nature of these interactions, they might well alter the quality reports elicited by any of the other compounds that normally share some of the same array of receptor mechanisms.

Any model of olfactory perception which postulates that odors produce their sensations by interacting uniquely with a single perceptual channel can easily account for the shift in sensitivity which characterizes the anosmia to a particular odorant by a simple reduction in the number of receptor neurons uniquely responsive to the odor in question (Gennings et al., 1977; Persaud et al., 1988). However, the distribution of different quality reports routinely observed for pemenone anosmics cannot be explained in this way. It seems more likely that the constellation of different odor reports elicited by an odorant like pemenone, in individuals with a characterized urine-sweaty anosmia, is a reflection of a deficit in one or more of the shared perceptual channels that normally contribute to the generation of the particular odor quality. The observation that there appears to be several different kinds of pemenone anosmics, as distinguished by the odor qualities that they report, then provides an estimate of the number of different perceptual channels normally involved with the perception of pemenone. In a similar vein, the fact that isovaleric acid is an important contributor (see Table 1) to two of the three orthogonal factors produced by the principal component analysis could suggest that normal perception of this compound also involves molecular interactions with more than one perceptual channel. Indeed, it is difficult to reconcile this array of findings with models of olfactory perception in which individual odorants interact uniquely with separate individual perceptual channels.

ACKNOWLEDGMENTS

We thank Dr. G. Ohloff, Director of Laboratory Research, FIRMENICH SA, Geneva, Switzerland, for providing the sample of pemenone, Dr. R. Patrick Akers for help preparing the figures, and A. Houser, E. McCabe, J. Mini, and C. St. Remy for their assistance with subject testing. Portions of these data were presented at the 12th annual meeting of the Association for Chemoreception Sciences, April 1990. Supported by NIH grants DC00131 and DC00371 from NIDCD to RJO.

REFERENCES

Amoore, J. E. 1967. Specific anosmia: A clue to the olfactory code. *Nature* **214:** 1095–1098.

Amoore, J. E. 1969. A plan to identify most of the primary odors. In *Olfaction and Taste III*. C. Pfaffmann (Ed.) pp. 158–171. The Rockefeller University Press, New York.

Amoore, J. E., Venstrom, D., and Davis, A. R. 1968. Measurement of specific anosmia. *Percept. Motor Skills* **26:** 143–164.

Anholt, R. R. H., Mumby, S. M., Stoffers, D. A., Girard, P. R., Kuo, J. F., and Snyder, S. H. 1987. Transduction proteins of olfactory receptor cells: identification of guanine nucleotide binding proteins and protein kinase *C*. *Biochem.* **26:** 788–795.

Anholt, R. R. H. and Rivers, A. M. 1990. Olfactory transduction: Cross-talk between second-messenger systems. *Biochem.* **29:** 4049–4054.

Anholt, R. R. H. 1989. Molecular physiology of olfaction. *Am. J. Physiol.* **257:** C1043–C1054.

Baringa, M. 1990. The mind revealed? *Science* **249:** 856–858.

Beets, M. G. J. 1974. Odor, taste, and molecular structure. In *The Contribution of Chemistry to Food Supplies.* I. Morton and D. N. Rhodes (Ed.) pp. 99–152. Butterworth & Co., London.

Birnbaumer, L., Codina, J., Mattera, R., Yatani, A., Scherer, N., Toro, M. J., and Brown, A. M. 1987. Signal transduction by G. proteins. *Kidney International* **32:** S14–S37.

Breer, H., Boekhoff, I. and Tareilus, E. 1990. Rapid kinetics of second messenger formation in olfactory transduction. *Nature* **345:** 65–68.

Bressler, S. L. 1990. The gamma wave: a cortical information carrier? *Trends in Neuroscience* **13:** 161–162.

Bruch, R. C. and Kalinoski, D. L. 1987. Interaction of GTP-binding regulatory proteins with chemosensory receptors, *J. Biol. Chem.* **262:** 2401–2404.

Bruch, R. C. and Teeter, J. H. 1989. Second-messenger signalling mechanisms in olfaction. In *Chemical Senses, Volume 1, Receptor Events and Transduction in Taste and Olfaction,* J. Brand, J. Teeter, R. Cagan, and M. Kare (Ed.) pp. 283–298. Marcel Dekker, Inc. New York.

Carr, W., Gleeson, R., and Trapido-Rosenthal, H. 1990. The role of perireceptor events in chemosensory processes. *Trends in Neuroscience* **13:** 212–215.

Catteral, W. 1988. Structure and function of voltage-sensitive ion channels. *Science* **242:** 50–61.

Codina, J., Yatani, A., Grenet, D., Brown, A., and Birnbaumer, L. 1987. The α subunit of the GTP binding protein G_k opens atrial potassium channels. *Science* **239:** 442–445.

Damasio, A. R. 1990. Category-related recognition defects as a clue to the neural substrates of knowledge. *Trends in Neuroscience* **13:** 95–98.

Damasio, A., Yamada, T., Damasio, H., Corbett, J., and McKee, J. 1980. Central achromatopsia: behavioral, anatomic, and physiologic aspects. *Neurology* **30:** 1064–1071.

Dionne, V. E. 1989. Odor detection and discrimination. Can isolated olfactory receptor neurons smell. In *Chemical Senses, Volume 1, Receptor Events and Transduction in Taste and Olfaction*, J. Brand, J. Teeter, R. Cagan, and M. Kare (Ed.), pp. 415–426. Marcel Dekker, Inc., New York.

Firestein, S. and Werblin, F. 1989. Odor-induced membrane currents in vertebrate-olfactory receptor neurons. *Science* **244:** 79–82.

Gennings, J. N., Gower, D. B., and Bannister, L. H. 1977. Studies on the receptors to 5α-androst-16-en-3-one and 5α-androst-16-en-3α-ol in sow nasal mucosa. *Biochem. et Biophys. Acta* **496:** 547–556.

Gesteland, R. C., Lettvin, J. Y., and Pitts, W. H. 1965. Chemical transmission in the nose of the frog. *J. Physiol.* **181:** 525–559.

Getchell, T. V. and Getchell, M. L. 1987. Peripheral mechanisms of olfaction: Biochemistry and neurophysiology. In *Neurobiology of Taste and Smell*, T. E. Finger, and W. L. Silver (Ed.), pp. 91–123. John Wiley & Sons, New York.

Gibbons, A. 1990. New maps of the human brain. *Science* **249:** 122–123.

Gilman, A. G. 1987. G proteins: transducers of receptor-generated signals. *Ann. Rev. Biochem.* **56:** 615–649.

Gouras, P. 1981. Visual system IV: Color vision. In *Principles of Neural Science*, E. R. Kandel and J. H. Schwartz (Ed.). pp. 249–257. Elsevier/North-Holland, New York.

Granger, R., Ambros-Ingerson, J., and Lynch, G. 1989. Derivation of encoding characteristics of layer II cerebral cortex. *J. Cognitive Neuroscience* **1:** 61–87.

Griffiths, N. M. and Patterson, R. L. S. 1970. Human olfactory responses to 5α-androst-16-en-3-one—principal component of boar taint. *J. Sci. Fd. Agric.* **21:** 4–6.

Haberly, L. B. and Bower, J. M. 1989. Olfactory cortex: Model circuit for study of associative memory? *Trends in Neuroscience* **12:** 258–264.

Hatt, H. 1989. Stimulus-driven chemosensory membrane channels on crayfish sensory cells. In *Chemical Senses, Volume 1, Receptor Events and Transduction in Taste and Olfaction*, J. Brand, J. Teeter, R. Cagan, and M. Kare (Ed.), pp. 415–426. Marcel Dekker, Inc., New York.

Hille, B. 1984. *Ionic Channels of Excitable Membranes* pp. 138–147. Sinauer Associates, Inc. Sunderland, MA.

Huque, T. and Bruch, R. C. 1986. Odorant and guanine nucleotide-stimulated phosphoinositide turnover in olfactory cilia. *Biochem. Biophys. Res. Commun.* **137:** 36–42.

Jacobs, G. H. 1986. Color vision variations in non-human primates. *Trends in Neuroscience* **9:** 320–323.

Jones, D. T. and Reed, R. R. 1987. Molecular cloning of five GTP-binding protein cDNA species from rat olfactory neuroepithelium. *J. Biol. Chem.* **262:** 14241–14249.

Jones, D. T. and Reed, R. R. 1989. G_{olf}: An olfactory neuron specific-G protein involved in odorant signal transduction. *Science* **244:** 790–795.

Kaba, H., Rosser, A., and Keverne, B. 1989. Neural basis of olfactory memory in the context of pregnancy block. *Neuroscience* **32:** 657–662.

Kauer, J. S. 1987. Coding in the olfactory system. In *Neurobiology of Taste and Smell* T. E. Finger and W. L. Silver (Ed.), pp. 205–231. John Wiley & Sons, New York.

Krupinski, J., Coussen, F., Bakalyar, H., Tang, W., Feinstein, P., Orth, K., Slaughter, C., Reed, R., and Gilman, A. 1989. Adenylyl cyclase amino acid sequence: possible channel- or transporter-like structure. *Science* **244:** 1558–1564.

Kuffler, S. W., Nicholls, J. G., and Martin, A. R. 1984. *From Neuron to Brain* p. 380. Sinauer Associates Inc., Sunderland, MA.

Kurihara, K., Kashiwayanagi, M., Nomura, T., Yoshii, K., and Kumazawa, T. 1989. Chemical stimulus discrimination by specific and nonspecific receptor mechanisms and their transduction sequences. In *Chemical Senses, Volume 1, Receptor Events and Transduction in Taste and Olfaction,* J. Brand, J. Teeter, R. Cagan, and M. Kare (Ed.), pp. 283–298. Marcel Dekker, Inc., New York.

Kurihara, K. and Koyama, N. 1972. High activity of adenyl cyclase in olfactory and gustatory organs. *Biochem. Biophys. Res. Commun.* **48:** 30–34.

Labarca, P., Simon, S. A., and Anholt, R. R. H. 1988. Activation by odorants of a multistate cation channel from olfactory cilia. *Proc. Natl. Acad. Sci. USA* **85:** 944–947.

Lancet, D., Shafir, I., Pace, U., and Lazard, D. 1989. Receptor-activated chemosensory enzyme cascades. In *Chemical Senses, Volume 1, Receptor Events and Transduction in Taste and Olfaction* J. Brand, J. Teeter, R. Cagan, and M. Kare (Ed.), pp. 263–281. Marcel Dekker, Inc., New York.

Lennie, P. 1984. Recent developments in the physiology of color vision. *Trends in Neuroscience* **6:** 243–248.

Lerner, M. R., Reagan, J., Gyorgyi, T., and Roby, A. 1988. Olfaction by melanophores: What does it mean? *Proc. Natl. Acad. Sci. USA* **85:** 261–264.

Light, D. B., Corbin, J. D., and Stanton, B. A. 1990. Dual ion-channel regulation by cyclic GMP and cyclic GMP-dependent protein kinase. *Science* **344:** 336–339.

Livingstone, M. S. and Hubel, D. H. 1987. Psychophysical evidence for separate channels for the perception of form, color, movement and depth. *J. Neurosci.* **7:** 3416–3468.

Lowe, G., Nakamura, T., and Gold, G. H. 1989. Adenylate cyclase mediates olfactory transduction for a wide variety of odorants. *Proc. Natl. Acad. Sci. USA* **86**: 5641–5645.

McClintock, T. S. and Ache, B. W. 1989. Histamine directly gates a chloride channel in lobster olfactory receptor neurons. *Proc. Natl Acad. Sci. USA* **86**: 8137–8141.

Menevse, A., Dodd, G., and Poynder, T. M. 1977. Evidence for the specific involvement of cyclic AMP in the olfactory transduction mechanism. *Biochem. Biophys. Res. Commun.* **77**: 671–677.

Nakamura, T. and Gold, G. H. 1987. A cyclic nucleotide-gated conductance in olfactory receptor cilia. *Nature* **325**: 442–447.

Nathans, J., Piantanida, T., Eddy, R., Shows, T., and Hogness, D. 1986. Molecular genetics of inherited variation of human color vision. *Science* **232**: 203–210.

North, R. A. 1989. Neurotransmitters and their receptors: From the clone to the clinic. *Seminars in the Neurosciences* **1**: 81–90.

North, R. A., Surprenant, A., and Tatsumi, H. 1988. Potassium conductance increase and calcium conductance decrease both evoked by α_2-adrenaline and δ-opioid receptor agonists in the same submucous plexus neurones isolated from the guinea-pig. *J. Physiol.* **406**: 179P.

O'Connell, R. J., Costanzo, R. M., and Hildebrandt, J. D. 1990. Adenylyl cyclase activation and electrophysiological responses elicited in male hamster olfactory receptor neurons by components of female pheromones. *Chemical Senses* **15**: In press.

O'Connell, R. J., and Mozell, M. M. 1969. Quantitative stimulation of frog olfactory receptors. *J. Neurophysiol.* **32**: 51–63.

O'Connell, R. J., Singer, A. G., Pfaffmann, C., and Agosta, W. C. 1979. Pheromones of hamster vaginal discharge: Attraction of femtogram amounts of dimethyl disulfide and to mixtures of volatile components. *J. Chemical Ecology* **5**: 575–585.

O'Connell, R. J., Stevens, D. A., Akers, R. P., Coppola, D. M., and Grant, A. J. 1989. Individual differences in the quantitative and qualitative responses of human subjects to various odors. *Chemical Senses* **14**: 293–302.

Ohloff, G., Giersch, W., Thommen, W., and Willhalm, B. 1983. 127. Conformationally controlled odor perception in 'steroid-type' scent molecules. *Helvetica Chimica Acta* **66**: 1343–1354.

Pace, U., Hanski, E., Salomon, Y., and Lancet, D. 1985. Odorant-sensitive adenylate cyclase may mediate olfactory reception. *Nature* **316**: 255–258.

Pace, U. and Lancet, D. 1986. Olfactory GTP-binding protein: Signal-transducing polypeptide of vertebrate chemosensory neurons. *Proc. Natl. Acad. Sci. USA* **83**: 4947–4951.

Pantev, C., Hoke, M., Lütenhöner, B., and Lehnertz, K. 1989. Tonotopic organization of the auditory cortex: pitch versus frequency representation. *Science* **246:** 486–488.

Persaud, K. C., Pelosi, P., and Dodd, G. H. 1988. Binding and metabolism of the urinous odorant 5α-androstan-3-one in sheep olfactory mucosa. *Chem. Senses* **13:** 231–245.

Pfeuffer, E., Mollner, S., Lancet, D., and Pfeuffer, T. 1989. Olfactory adenylyl cyclase, identification and purification of a novel enzyme form. *J. Biol. Chem.* **264:** 18803–18807.

Polak, E. H. 1973. Multiple profile-multiple receptor site model for vertebrate olfaction. *J. Theor. Biol.* **40:** 469–484.

Reed, R. R. 1990. How does the nose know? *Cell* **60:** 1–2.

Schofield, P. R., Shivers, B. D., and Seeburg, P. H. 1990. The role of receptor subtype diversity in the CNS. *Trends in Neuroscience* **13:** 8–11.

Shirley, S. G. and Persaud, K. C. 1990. The biochemistry of vertebrate olfaction and taste. *Seminars in the Neurosciences* **2:** 59–68.

Singer, A., Agosta, W., Pfaffmann, C., Bowen, D., Field, F., and O'Connell, R. J. 1976. Dimethyl disulfide: an attractant pheromone in hamster vaginal secretion. *Science* **191:** 948–950.

Sklar, P. B., Anholt, R. R. H., and Snyder, S. H. 1986. The odorant-sensitive adenylate cyclse of olfactory receptor cells. *J. Biol. Chem.* **261:** 15538–15543.

Sternweis, P. C. and Pang, I. H. 1990. The G protein-channel connection. *Trends in Neuroscience* **13:** 122–126.

Stevens, D. A. and O'Connell, R. J. 1991. Individual differences in thresholds and quality reports of human subjects to various odors. *Chemical Senses* **16:** In press.

Trotier, D. 1990. Physiology of transduction in olfaction and taste. *Seminars in the Neurosciences* **2:** 69–76.

Vodyanoy, V. and Murphy, R. 1983. Single-channel fluctuations in bimolecular lipid membranes induced by rat olfactory epithelial homogenates. *Science* **220:** 717–719.

Vodyanoy, V. and Vodyanoy, I. 1987. ATP and GTP are essential for olfactory response, *Neurosci. Letters* **73:** 253–258.

Wysocki, C. J. and Beauchamp, G. K. 1984. Ability to smell androstenone is genetically determined. *Proc. Natl. Acad. Sci. USA* **81:** 4899–4902.

Wysocki, C. J., Dorries, K. M., and Beauchamp, G. K. 1989. Ability to perceive androstenone can be acquired by ostensibly anosmic people. *Proc. Natl. Acad. Sci. USA* **86:** 7976–7978.

5

Toward the Unification of the Laws of Sensation: Some Food for Thought

Kenneth H. Norwich

University of Toronto
Toronto, Ontario, Canada

INTRODUCTION

Like the patter of so many raindrops on a windowpane, the molecules of an odorant gas impinge randomly upon their receptor. Now three molecules strike; now five; now only one. But despite the chaotic drumming of the particles of the odorant gas, the sensation of smell is relatively steady. The sensation does not reflect the randomness of the molecular motion. A kind of order emerges out of the chaos. The receptor *informs* the organism of the *prevailing* concentration of the odorant in air (or solute in solution) despite the incessant, random fluctuation of particles in the quantum world. It *informs* and, therefore, it provides *information* about the prevailing or *mean* concentration or density of the stimulus gas. But in what language is the receptor's output expressed? How are means encoded? This is the fundamental issue addressed by the *informational* or *entropic* view of sensation. However, in order to understand this theory of perception, we should review the concept of "information."

You may have leafed through this chapter, seen a number of equations, and be tempted to skip past it to something less mathematical. Please don't! If you are familiar with statistical analysis to the level of means and variances, and you know high school algebra and what a logarithm is, you can make your way through the chapter. For readers who wish to delve a little deeper into the mathematical analysis, a number of boxes have been inserted. You don't have to study the boxes to get the gist of the chapter. A few mathematical rules will get you through:

(a) Remember that $log_b a$ means *the logarithm of a to the base b.*

(b) Logarithms can be changed to a different base by multiplying by a conversion factor.

(c) $log\ (1) = 0$ and $log\ (0) \rightarrow -\infty$.

(d) $log\ ab = log\ a + log\ b;\ log\ a/b = log\ a - log\ b;$ and $log\ a^n = n\ log\ a.$

(e) Recall (to understand Eq. (10)) that $\int_{x=a}^{x=b} d(log\ x) = log\ b - log\ a.$

(f) For Eq. (18) it is helpful to know that $X(I_1) - X(I_2) \simeq (\partial X/\partial I)\ \Delta I$ if the difference on the left-hand side is small.

(g) Equation (1) will be differentiated to obtain Eq. (7). Differentiation is just a technique in the calculus for obtaining *the rate of change of* one variable with respect to another. So if you know x in terms of y, you can differentiate to obtain dy/dx, which is the rate of change of y with respect to x.

And now on to *information.*

Information, as we commonly understand the term, represents *loss of uncertainty* or loss of randomness. When the chemoreceptor is first exposed to its stimulus, it begins in a state of maximal incertitude about the *mean* stimulus concentration. Due to random motions of the molecules, the first sample has concentration C_1, the second sample has concentration C_2, the third C_3. There is not yet any compelling mean or consensus. If s is the standard deviation of m *samples* of the stimulus gas, then $s/\sqrt{m}$ is an estimator of the standard error of the mean which is, initially, quite large. As the receptor continues to sample its stimulus, m, the number of samples, increases, and $s/\sqrt{m}$ decreases. *Uncertainty,* as encoded by the standard error or the mean, has decreased and, therefore, *information* has been received. If we represent uncertainty by the symbol H and approximate the sample variance s^2 by the population variance, σ^2, then

$$H = f\ [\sqrt{\sigma^2/m}]$$

This is, uncertainty, H, might be represented as some function, f, of $\sqrt{\sigma^2/m}$. Recalling that information is a change in uncertainty, it is given, therefore, by a change in $H = \Delta H = \Delta f$, which is the change in f as m, the number of samples, increases.

The variance σ^2/m gives a measure of the range of possibilities of values for the mean, and the reader may remember that information is always calculated from the *logarithm* of the number of possibilities.[1] For example, the information obtainable from the throw of a fair die is given by log 6, because there are 6 sides to the die. The base of the logarithm is arbitrary, but if 2 is used as base (that is, $\log_2 6$), then the result is reported in *bits* of information, and if $e = 2.718 \ldots$ is used as base, then the result is expressed as *natural units*. Conversion from bits to natural units is by a simple multiplicative constant (see Box 6). Since the square root of the variance $\sqrt{\sigma^2/m}$ is a measure of the range of possible values of the mean stimulus concentration, analogous to the 6 faces on which a die may fall, we are led to suspect that the function, f, is a log function, so that

$$H = log \ (\sigma^2/m)^{1/2}$$

This is not a bad guess, but we note that asymptotically, as m becomes very large, the uncertainty, H, will approach *log* 0, which is infinite, so the H-function must be fixed up a little. An effective change is

$$H = log \ (1 + \sigma^2/m)^{1/2} = (1/2) \ log \ (1 + \sigma^2/m)$$

so that H approaches *log* 1 = 0 as m becomes large. This makes sense because uncertainty about the mean concentration of the stimulus solution *should* approach zero as the number of samples of that solution becomes large.

The H-function will be much more useful if we can relate the quantities σ^2 and m to quantities measurable in the laboratory. For the present, I suggest that we regard the chemoreceptor as sampling its stimulus at a constant rate; that is, I recommend taking m proportional to t, the duration of the stimulus.[2] Time, t, of course, is directly measurable. The variance, σ^2, may be handled by *restricting all stimuli under discussion to be constant stimuli*. That is, we shall consider as

---

[1]The concept of the logarithm as a measure of information may be traced to the physicist Ludwig Boltzmann toward the end of the nineteenth century, but explicit use of the term *information* is due to the works of Wiener and Shannon in 1948. Norbert Wiener in his seminal book *Cybernetics* wrote: "We know *a priori* that a variable lies between 0 and 1, and *a posteriori* that it lies on the interval (a,b) inside (0,1). Then the amount of information we have from our *a posteriori* knowledge is $-log_2$ [measure of (a,b)/measure of (0,1)]."

[2]It is possible, although by no means certain, that for audition (L. Ward, personal communication), the rate of sampling diminishes immediately following imposition of a stimulus, and then very quickly approaches a constant.

stimuli only solutions and gases whose macroscopic (laboratory measured) concentrations are unchanging with time. We shall then relate variance to mean using the relation

$$\sigma^2 \propto I^n$$

this is, variance is proportional to *mean intensity, I,* raised to the power *n.* The justification for this equation is simply that larger quantities tend to manifest larger fluctuations, and that many physical quantities such as density of a gas do, indeed, comply with such a mean-variance rule. At this stage we are still concerned solely with the physics of the stimulus signal; specifically with the manner in which molecular movements produce fluctuations in the density of a gas. The psychophysics is yet to come. For dilute solutions of substances such as sodium chloride, the exponent, *n,* might be expected to be approximately equal to one (Norwich, 1984), but this is not certain, and we must leave the mean-variance relation as an assumption of the theory, with *n* still undetermined from theoretical considerations.

Making the above substitutions, and selecting the natural log function for mathematical simplicity (remember, the base of the logarithms is arbitrary) the H-function takes on the form

$$H = (1/2) \, ln \, (1 + \beta I^n / t) \qquad (1)$$

where β is a quotient of the proportionality constants, I is the mean (laboratory measured) stimulus concentration, and t, the stimulus duration, may not be less than the time required for the receptor to draw its first sample of the stimulus. Equation (1) has been derived somewhat more rigorously elsewhere (e.g., Norwich, 1981, 1987), but we capture the flavor of it here. It is the only equation we shall need in this chapter in order to derive many of the equations governing the chemical senses that have been discovered empirically (by measurement) over the years.

Let's review the significance of Eq. (1). H is the uncertainty (or officially, the *entropy*) of a chemoreceptor as a function of the stimulus intensity, I, and the stimulus duration, t. The stimuli considered here are concentrations of odorants or of solutions tasted. β and n are constant parameters, greater than zero, that will vary from stimulus to stimulus. For constant stimulus duration, t, entropy H increases with I, since more concentrated solutions possess greater absolute entropy—fluctuations, disorder—than dilute solutions. For constant concentration, I, entropy or uncertainty H decreases with time, since uncertainty about the mean concentration (in a fluctuating environment) decreases with the number of samples of solution taken. In ordinary language, the greater the number of samples taken from a population, the more certain we are that the mean of these samples is a good measure of the mean of the whole population. Since information received is associated with a decrease in entropy, we can say that a receptor gains information during the period of sampling. Notice that we now have a general "currency" (namely, information) with which to assess the sensory transactions of the various modalities of sensation. This informational currency applies equally well to gustation and to vision. The manner in which we are using information is related to the quantum structure of sensory signals, and is very different from the manner in which information theory was used by psychologists in the 1950s and 1960s.

Hitherto, we have concerned ourselves with the quantized or discrete inputs to sensory receptors.[3] Let us now turn our attention back to the matter of the code or language used as the receptor's *output* (for example, the impulse rate in a sensory afferent neuron). By conjecture, let us represent the receptor output function by F, where

$$F = kH \tag{2}$$

where k is a constant greater than zero (Box 1). Combining Eqs. (1) and (2),

$$F = (1/2) \, k \, ln \, (1 + \beta I^n/t), \tag{3}$$

where all quantities are now measurable in the laboratory. Equation (3) can, therefore, be tested in the laboratory in ways which we shall presently explore.

MECHANISTIC LAWS VS. CONSERVATION LAWS

We shall use Eqs. (1) to (3) to derive some of the laws of psychophysics and sensory physiology, and it is important to understand, philosophically, what these derivations signify. As implied above, Eq. (2) is a kind of input-output relationship at the level of a sensory receptor, which we can comprehend best by anthropomorphizing the receptor. We can regard the input as the uncertainty, H, of a chemoreceptor about the mean or average concentration of a solution. The output may then be the impulse frequency, F, which encodes numerically the value of this uncertainty in a sensory afferent neuron. As receptor input un-

---

[3]That is, until now we have concerned ourselves with the nature of input to a sensory receptor, and with the state of certitude in which a receptor might find itself at each stage in the process of sampling a stimulus solution or gas. However, we have made no biological commitment; we have not committed the receptor to act upon its state of certitude. This step is taken by the primary assumption of the theory: $F = kH$ [Eq. (2)].

Box 1

The equation $F = kH$ represents a rather audacious transformation of H, which is best understood as the human attribute "uncertainty," into a quantity F (an impulse transmission rate, say). F can be measured in the laboratory. The magnitude of the constant, k, depends on the physiological system being measured. This process is not without precedent. Ludwig Boltzmann has written $S = kH$ (or $S = -kH$, depending on how H is defined) to relate the *thermodynamic entropy*, S, to the quantity H. The constant, k, in the latter equation is the famous Boltzmann constant. S, or at least the change in S during a physical or chemical reaction, can be measured in the laboratory. I suspect that these two equations are parallel, and may even find unification in some higher level of theory capable of embracing both perception and statistical mechanics (a branch of physics).

---

certainty declines (H decreases with sampling and hence information is gained), receptor output, F, declines proportionately. The difference between any two F-values, $F(t_1) - F(t_2)$, encodes the gain of information by the sensory receptor in the language of the neuron. In other words,

> Number of units of information received by the receptor
> = number relayed to the brain

We deal here with a kind of conservation law: If information is not lost by the receptor, then the total quantity of information "obtained from the stimulus" is relayed to the brain. Note though, that it is not the stimulus quality itself (e.g., very sweet, not so sweet, etc.) that is transmitted to the brain, but the *state of certitude of the receptor*. The brain, we shall see, is capable, to a limited extent, of decoding stimulus quality from state of certitude. It is important at this point to realize that laws of the type we shall be deriving in this chapter are "information balance" laws (akin to conservation laws), and not laws describing the mechanisms of sensory receptors.

Recall how classical conservation laws complement laws of mechanisms. Suppose a chemical reaction is written:

$$\text{reagents} \longrightarrow \text{products}$$

We can say using only the law of conservation of mass that the total mass of reagents is equal to the total mass of products (in the absence of nuclear reactions). If at a later time we learn the mechanism of the reaction,

$$A + B \longrightarrow C + D$$

we have in no way violated the conservation law. On the contrary, we have complemented it. Every law of conservation or balance can and *must* have an accompanying law of mechanism. So it is with the information balance law, $F = kH$. While it may provide useful derivations of sensory laws, it, too, must eventually be complemented by laws of sensory mechanism. The two types of law work in unison.

ADAPTATION

We can now follow the sequence of events as a chemoreceptor is exposed to a stimulus gas or solution.[4] The receptor output encodes the uncertainty of the receptor about the concentration of the stimulus gas at any instant of time. For small t, uncertainty is high and so, too, its "mirror," F. As t increases, the uncertainty drops, H drops, and so, too, F. The process of sensory adaptation, then, may find expression as a process of gain in receptor certitude. For a receptor capable of recording a sufficiently large number of samples, the values of m, and hence of t, will become very large, and, therefore, F will approach zero or total adaptation [see Eq. (3)]. *Whence: One cannot perceive a certainty.* When the nature of a stimulus is absolutely certain to a receptor, the receptor is silent. It has no output. In order to adapt Eq. (3) to describe an experiment on gustatory or olfactory adaptation, we fix stimulus intensity, I, to be constant, so the equation simplifies to the form

$$F = (1/2) \; k \; ln \; (1 + A/t). \qquad (4)$$

If one is willing to set subjective magnitude proportional to neuronal impulse rate, as suggested by the work of Borg et al. (1967), we can then replace F by a general *perceptual variable, P,* which can refer to *either* impulses per second *or* subjective magnitude. Thus,

$$P = kH \qquad (2a)$$

Adaptation curves can now be looked upon in quite a new way. The total *excursion* of the perceptual variable from P_{max} at small t to P_{min} at larger t defines a drop in uncertainty (entropy) equal to the gain in information by the receptor. If we are speaking about subjective magnitude, then it is necessary to assume that multiple sensory receptors are working in parallel, so that the report of the many to the brain is equivalent to the report of a single receptor. This is not totally satisfactory, but it is the best we can do at the moment. Since $H = P/k$ from Eq. (2a), we can calculate values for H and write for the excursion of an adaptation curve,

---

[4]Olfactory molecules, it seems, must pass through a mucus layer before reaching their receptor. Perhaps, therefore, we should always speak about the *stimulus solution*.

$$Information = H_{max} - H_{min} \; [natural \; units].$$

We may refer to the quantity $H_{max} - H_{min}$ as the *information content of the stimulus*. The extraordinary result of calculations performed on adaptation data from various stimuli (such as solutions of sodium chloride or of sucrose of various molarity) is that the information content is usually in the range of 1.8 to 2.5 bits per stimulus (Box 2).

In this chapter we shall be analyzing many different types of sensory experiment using the information law, Eq. (1). You may often be tempted to say "I'll bet I could find a mathematical function that fits the data set better than that!" And you probably could. The idea here, however, is to fit *all* sensory data relating a stimulus of constant intensity *I*, held for a time, *t*, to some perceptual variable, *P*, using the *same* function [Eqs. (1) and (2a)], not just for adaptation data, but for discrimination measurements, reaction times, etc. *Unification* of the laws of sensation is the goal.

Equation (1) can be modified (and has been modified) to fit the known data more precisely (for example, by refining the sampling rate from the default

---

Box 2

Illustrating the method of calculating the information content of a gustatory stimulus from adaptation data. The following data, digitized from a graph of Gent and McBurney (1978), describe the decrease in subjective magnitude, *P*, following application of a 0.32 M solution of sodium chloride to the tongue.

Time(s)	Median magnitude est.
5	8.57
15	6.00
30	3.85
45	2.86
60	1.86
75	1.43
90	1.00

The data have been curve-fitted to Eq. (4) using a simplex algorithm, but a method of curve-fitting using a programmable calculator is given by Norwich (1984). In this way one can obtain numerical values for the constants *k* and *A*.

$$P = (1/2) \, k \, ln \, (1 + A/t) = (7.63/2) \, ln \, (1 + 46.0/t)$$

From the data we have $P_{max} = 8.57$ and $P_{min} = 1.00$.
Since $P = kH$, therefore $P_{max} = kH_{max}$ or $H_{max} = P_{max}/k = 8.57/7.63 = 1.12$
Similarly, $H_{min} = P_{min}/k = 1.00/7.63 = 0.131$
Therefore, the information content of the stimulus = 1.12 − 0.13 = 0.99 natural units. Dividing by ln 2, we can convert this value to 1.26 bits of information per stimulus (see Box 6). If we extrapolate and say that *P* would eventually have reached zero, then the total information per stimulus would be at least $(8.57 - 0)/(7.63 \, ln \, 2) = 1.62$ bits per stimulus.

function $t \propto m$), but the resulting equation becomes much more complicated. At this introductory level we are striving for an overview of the simple manner in which apparently different sensory phenomena cohere and interrelate. For example: What may adaptation times and reaction times have in common? Does Stevens' power law give the Weber fraction its characteristic shape? The principles now; the details later.

THE PSYCHOPHYSICAL LAW

In our exploration of adaptation phenomena above, we fixed stimulus intensity, I, in Eq. (3), and expressed F as a function of stimulus duration, t, alone. To derive the "psychophysical law" or the law of sensation, we proceed in analogous fashion by fixing stimulus duration at some constant value and expressing F as a function of I alone. Condensing β/t into the constant, γ, Eq. (3) becomes

$$F = (1/2) \, k \, ln \, (1 + \gamma \, I^n) \tag{5}$$

Now, we associate the psychophysical law with Fechner's simple logarithmic law, or with the power law popularized by Stevens, but Eq. (5) does not look like either. In fact, though, the informational law given by Eq. (5) *is both* Fechner's and Stevens' laws, and will transform naturally into one or the other form with slight changes in the magnitude of the quantity $\gamma \, I^n$. Over a large range of value of concentration, I, Eq. (5) is an acceptable approximation of both Fechner's and Stevens' laws, which accounts for the effective use of both laws by some investigators to describe the same set of data (Stevens, 1970) (Box 3).

Fechner's Law

When $\gamma \, I^n \gg 1$, we may drop the "1" from Eq. (5) with impunity, leaving

$$F = (1/2) \, k \, ln \, (\gamma \, I^n) = (1/2)kn \, ln \, I + (1/2) \, k \, ln \, \gamma$$

That is (recalling that we can change from natural logs, ln, to logs of any base by just multiplying through by a constant (see Box 6),

$$F = a \, log \, I + b \tag{6}$$

where a and b are constants. We recognize Eq. (6) as Fechner's law. It has been derived here solely from informational principles. The relationship between the logarithmic nature of Fechner's law and the logarithmic character of the information function was recognized by Moles (1966).

Box 3

The Psychophysical Law

The informational law (5), $F = (1/2) k \, ln \, (1 + \gamma I^n)$, is readily seen to become Fechner's law when $\gamma I^n \gg 1$, and can be seen to give rise to Stevens' power law when $\gamma I^n \ll 1$. The basis for the latter is the Taylor series, valid for $x \leq 1$:

$$ln \, (1 + x) = x - x^2/2 + x^3/3 - \ldots$$

Therefore, when x is very much less than unity, all terms in the series except the first become vanishingly small (e.g., when $x = 0.01$, then $x^2/2 = 0.00005$, etc.) Under these conditions, $ln \, (1 + x) = x$ effectively, for most scientific purposes. Applying this rule to Eq. (5) above, we see that for $\gamma I^n \ll 1$, $F = (1/2) k\gamma \, I^n = \lambda \, I^n$, which is Stevens' power law.

Because Fechner's law and Stevens' law are approximations to the more general information law, it is expected that when I assumes values that violate the conditions of approximations, Fechner's and Stevens' laws would fail to account adequately for the data. This has been observed many times. Stevens' power law is usually demonstrated by plotting the logarithm of F (or subjective magnitude, P), against the logarithm of stimulus intensity, I. If the law is valid, the data should fall on a straight line. We expect, though, that when I becomes sufficiently large we will require not just the first term of the Taylor series for $ln \, (1 + x)$, but at least the first *two* terms:

$$F = (1/2) \, \gamma I^n - (1/4) \, \gamma^2 I^{2n}$$

That is, on a graph of *log F* against *log I*, the data will fall *below* the straight line for larger values of I, because of the second term in the series. In fact, this is exactly what was observed by Atkinson (1982) when he examined Stevens' law for many sensory modalities. Similarly, Fechner's law is demonstrated by plotting F (not *log F*) against *log I*. If the law is valid, the data will fall on a straight line. The reader should now have no difficulty in showing that for stimulus intensities that are sufficiently *small*, the condition for approximation of the information law by Fechner's law are violated, and data should rise *above* the straight line. This is not observed invariably, but see the data of McBride (1983a). In Fig. 5.1, the data of Stevens (1969) (subjective magnitude vs per cent composition of sodium chloride solution) have been plotted in two ways. In Fig. 5.1a a full logarithmic plot has been made, showing the near-validity of the power law of sensation. In Fig. 5.1b a semilog plot has been made, showing the near-validity of Fechner's law. Notice that both laws can hold concurrently (nearly). Both laws are embraced by the single information equation (Fig. 5.1c).

If the reader is still skeptical about how two seemingly different laws emerge

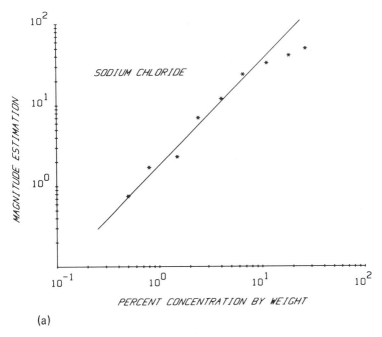

(a)

FIG. 5.1 Subjective magnitude of taste relative to the concentration of a sodium chloride solution (Stevens, 1969, Figure 8). (a) A full logarithmic plot and a straight line plotted by eye through all but the upper few data points, demonstrating the near-validity of the power law of sensation. (b) A semilogarithmic plot of the same data, and a straight line by eye through all but the lowest few data points, demonstrating the near-validity of Fechner's logarithmic law of sensation. (c) The information function (Eq. (5)) fitted to the data using a simplex algorithm. Stevens' and Fechner's laws may be seen to be approximations to the information law.

Box 3 continued

from the single information equation, the following may convince him/her. Let us represent the information equation in its "generic" form: $y = ln(1 + x)$, and construct the following table:

x	y	$ln\ x$	$ln\ y$
0.5	0.41	−0.69	−0.89
0.6	0.47	−0.51	−0.76
0.7			
.			
.			
2.0	1.10	0.69	0.10

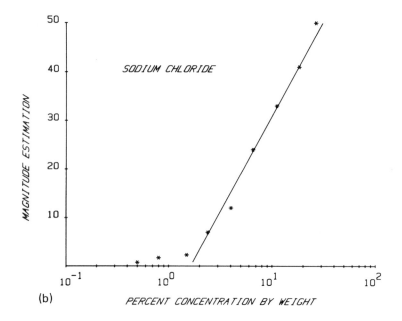

(b)

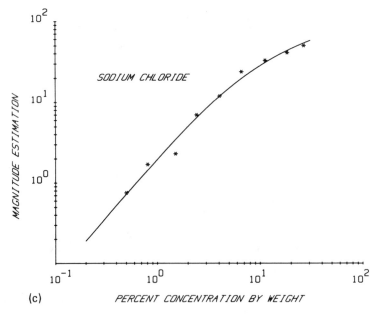

(c)

FIG. 5.1 *(cont.)*

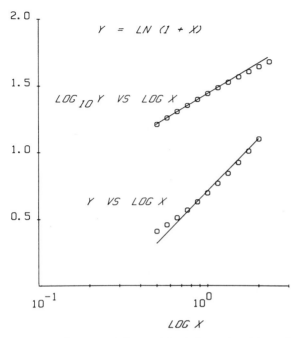

FIG. 5.2 The general form of the informational law of sensation (Eq. (5)) is $y = ln(1 + x)$. The data points in this graph are obtained by plotting $y = ln(1 + x)$ in two ways. The lower graph shows y plotted against $log x$ ("Fechner plot"), and the upper graph shows $log_{10}y$ (+ 1.6 to shift the graph upwards) plotted against $log x$ ("Stevens plot"). It may be seen that for a specified range of x, the linear portions of the two graphs correspond to those seen with experimental data, as shown in Fig. 5.1a and b. We argue here again that $ln(1 + x)$, or, more fully, $(1/2)k \, ln(1 + \gamma I^n)$, is the correct form for the simple law of sensation.

Box 3 continued
We shall now plot $ln \, y$ vs. $ln \, x$ (Stevens plot), and y vs. $ln \, x$ (Fechner plot) and compare the results (Fig. 5.2). We see that two quite respectable straight lines have emerged (the full log plot is a little better than the semi-log plot). If we allow for the fact that measured data will invariably contain some noise, so the points will scatter somewhat about their respective straight lines, one can see *graphically illustrated* how the two empirical versions of the psychophysical law emerge from the single informational equation.

The Power Law of Plateau, Brentano, and Stevens

When $\gamma I^n < 1$, the right-hand side of Eq. (5) can be expanded in a Taylor series (Box 3), which is just a sum of power functions. When γI^n is sufficiently small, one is justified in neglecting all terms in the series except the first, leaving

$$F = (1/2)\, k\gamma\, I^n$$

or

$$F = \lambda I^n$$

where λ is constant.

Thus we can see that the single informational law gives rise to both logarithmic and power laws of sensation. There is often a degree of incredulity expressed about the apparent synonymy of the logarithmic and power laws. The two laws are, however, just different approximations of the single informational equation, and the domains of the approximations overlap considerably. (See Box 3 for more intensive study.)

The informational law provides meaning to the two totally empirical renditions of the psychophysical law. It states, essentially, that fluctuations in the magnitude of larger signals are greater than fluctuations in the magnitude of smaller signals (H is greater for large I than for small I). For example, a constant, small volume of *concentrated* solution, sampled on two occasions, may differ in concentration by a few thousand molecules. However, the same constant, small volume of *very dilute* solution, sampled on two occasions, may differ by only a few dozen molecules. The *absolute magnitude* of fluctuations of large signals is greater. Translated into informational language, that is the psychophysical law.

THE WEBER FRACTION, I

In our exploration of the information equation we have held I constant and let t vary to analyze adaptation data; and we have held t constant and let I vary to derive the psychophysical law. We continue now by holding t constant again (all stimuli applied for the same duration), and we differentiate H [see rule (g)] as given by Eq. (1) with respect to I. We then replace the differentials dI and dH by their respective finite differences, ΔI and ΔH, and rearrange the equation algebraically to give

$$\frac{\Delta I}{I} = \frac{2\Delta H}{n\,\gamma}\left[\gamma + \frac{1}{I^n}\right] \tag{7}$$

where the constant $\gamma = \beta/t$. ΔI is interpreted as the change in intensity that corresponds to one just noticeable difference (jnd) in stimulus concentration, measured by any desired experimental criterion, and ΔH is the corresponding change in entropy. A simpler but not completely accurate way of expressing the above is that ΔH is the gain in information as the stimulus concentration is increased by molarity ΔI. The fraction $\Delta I/I$ is known as the *Weber fraction*. Note the ubiquity of the Stevens exponent, n, as it appears in almost every equation governing sensory function.

Throughout this section we shall adopt Fechner's well-known conjecture that the subjective magnitude of the jnd is constant. From Eq. (2), using the perceptual variable, P, in place of the neuronal frequency, F, we have $\Delta H = \Delta P/k$, where ΔP is the subjective magnitude of the jnd. Therefore, if ΔP is constant (Fechner), ΔH must also be constant. This permits us to sketch a graph of the Weber fraction, $\Delta I/I$ vs I (Fig. 5.3). The dependency of the Weber fraction on I^{-n} shows that the fraction becomes very large for small[5] I, and approaches a constant *(Weber's constant)* for larger values of I. This will be referred to as the *plateau region* of the curve.

$$\text{Weber's constant} = 2\Delta H/n \tag{8}$$

The theoretical curve matches that observed experimentally for taste (e.g., Lemberger, 1908) with reasonable accuracy. There is, however, a tendency for Weber fraction curves in many sensory modalities to bend upward for large values of I (Fig. 5.3). This terminal bend is not captured by Eq. (7), and we shall return to it in the following section (Weber Fraction II). Nevertheless, there is a great deal that can be learned from the present model of the Weber fraction.

Just a word here on the jnd. The jnd is not, of course, an absolute quantity; it changes in magnitude according the method and criteria used to measure it. Therefore, a quantity equal to the "number of jnd's" must not be used in an equation representing a physical or biological process unless it is "counterbalanced" by multiplicative factors that render the product independent of the manner in which the jnd is measured. The resulting mathematical expression is called an *invariant*. This will become much less abstract as we proceed.

The plateau region of the Weber fraction curve conceals a secret [Eq. (16) below] that we may learn if we just recall a lession taught by Riesz (1928). Riesz demonstrated how to calculate the total number of jnd's, δN, between any two intensities, I_1 and I_2, using the Weber fraction. His equation, developed in Box 4, may be written as follows:

---

[5]When I is small, ΔI becomes quite large, so that the approximation of dI by ΔI is invalid. This approximation was, however, used in deriving Eq. (7). Therefore Eq. (7) for the Weber fraction weakens, for example, for very dilute solutions.

Box 4

Riesz reasoned as follows: Let dI be a small interval of intensity (e.g., small range of concentration of solution), and let ΔI be the magnitude of the jnd. Then $dI/\Delta I$ is the fraction of the jnd that "fits into" the element dI and

$$\int_{I_1}^{I_2} \frac{d\,I}{\Delta I}$$

is the sum of all jnd's between two intensity (concentration) levels I_1 and I_2. With a little algebraic prestidigitation, the integrand can be transformed:

$$\frac{dI}{\Delta I} = \frac{dI/I}{\Delta I/I} = \frac{d(\ln I)}{\Delta I/I}$$

Therefore

$$\text{Sum of all the jnd's between } I_1 \text{ and } I_2 = \int_{I_1}^{I_2} \frac{d(\ln I)}{\Delta I/I}$$

We recognize in the denominator of the integrand the Weber fraction, $\Delta I/I$.

$$\delta N = \int_{I_1}^{I_2} \frac{d(\ln I)}{\Delta I/I} \tag{9}$$

Let us take I_1 as the lower bound of the plateau and I_2 as the upper bound. In the plateau region, $\Delta I/I$ is equal to the Weber constant, and, since it is constant, may be removed from under the integral sign. Therefore (mathematical rules e and d),

$$\delta N = \frac{1}{\Delta I/I} \int_{I_1}^{I_2} d(\ln I) = \frac{\ln (I_2/I_1)}{\text{Weber constant}} \tag{10}$$

from which we notice that

$$(\delta N)\ (\text{Weber constant}) = \ln (I_2/I_1). \tag{10a}$$

That is, the product of the number of jnd's spanned by the plateau multiplied by the value of the Weber constant is equal to a constant determined only by the boundaries of the plateau. So (δN) *(Weber constant) is an invariant that is not dependent on the manner in which the Weber fraction is measured.* As δN, the number of jnd's beneath the plateau becomes larger (as a consequence of the criteria for a jnd adopted by the experimenter), the Weber constant became proportionally smaller

Introducing Eq. (8) into Eq. (10),

$$\delta N = \frac{ln \ (I_2/I_1)}{2\Delta H/n} \tag{11}$$

Equation (11) is interesting in its own right, but it becomes even more illuminating when we operate on it somewhat further. Since ΔH, the information content of the jnd, has been taken to be constant, the total number of jnd's spanning any entropy band δH (or, since $P = kH$, spanning any subjective magnitude band, δP) is just equal to $\delta H/\Delta H$. That is,

$$\delta N = \delta H/\Delta H \tag{12}$$

Hence, replacing $1/\Delta H$ by $\delta N/\delta H$ in Eq. (11) and cancelling δN from both sides of the equation,

$$\delta H = (1/2n) \ ln \ (I_2/I_1). \tag{13}$$

Now, δN, in Eq. (11), is the number of jnd's between I_1 and I_2, the lower and upper bounds respectively of the plateau in the Weber fraction curve. I_2 is the maximum physiological intensity of the stimulus; that is, it is the maximum (tolerable?) concentration of a solution or odorant. δH, which now replaces δN [in Eq. (13)], is the entropy change between I_1 and the upper bound of concentration.[6]

Equations (11) and (13) are parallel. δN is proportional to δH. We can speak equally well about the number of jnd's in some stimulus intensity interval, or about the entropy change in the same interval.

The question we ask now is: How does δN, the number of jnd's spanned by the plateau, compare in magnitude to $(\delta N)_{max}$, the total number of jnd's between threshold and maximum concentration? Since δN does not contain those jnd's spanned by the early, falling portion of the Weber fraction curve, and the jnd becomes progressively smaller as one proceeds toward smaller concentrations, one might think that $(\delta N)_{max}$ is much greater than δN. In fact, when the calculations are made, δN is always $\geq$ 70% of $(\delta N)_{max}$ and often $\geq$85% of $(\delta N)_{max}$ (Box 5). That is, most jnd's lie beneath the plateau of the Weber fraction curve. The significance of this finding is that *the number of jnd's spanned by the plateau of the Weber fraction curve is an approximation to (but, of course, less than) the total number of jnd's from threshold to maximum stimulus intensity.*

---

[6]I know. The concept of entropy change tends to be rather abstract. Onc can think of it this way. The entropy change between the two boundaries of the plateau of the Weber fraction curve, I_1 and I_2, is equal to the difference in the *information content* of the two stimuli of concentrations I_1 and I_2. To review *information content*, see the section on Adaptation.

Box 5

Analysis of the data of Lemberger (1908).

In order to compute the number of jnd's between threshold and maximum stimulus concentration from the measured Weber fraction using Eq. (9), we require a table or graph of *the reciprocal of the Weber fraction, $(\Delta I/I)^{-1}$* vs. *ln I*. The integral in Eq. (9) represents the area under the curve produced by plotting the reciprocal of the Weber fraction against *ln I*. This area can be measured quite well by hand, or by using a computer to evaluate the integral numerically by quadrature. Let's do it manually. From Lemberger's data for sucrose, we plot the graphs in Figs. 5.3 and 5.4. Adding rectangles under the curve in Fig. 5.4, we find that the area under the entire curve and, therefore, the maximum number of jnd's for sucrose is 21, while the area under the plateau from approximately ln $I = 0.62$ to *ln I* $= 2.93$ equals 14.6 jnd's. Therefore, the number of jnd's spanned by the plateau of the Weber fraction curve is 70.5% of the total number. (For simplicity, we have not considered the final, rising portion of the Weber fraction curve.)

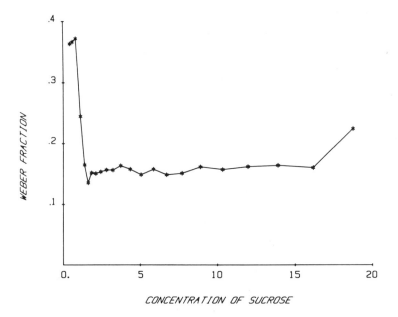

FIG. 5.3 Weber fraction, $\Delta I/I$, is plotted against I (concentration of sucrose). We observe the plateau region where Weber's law, $\Delta I/I =$ *constant*, holds closely, flanked on both extremes by a rising component of the Weber fraction. *(From Lemberger, 1908.)*

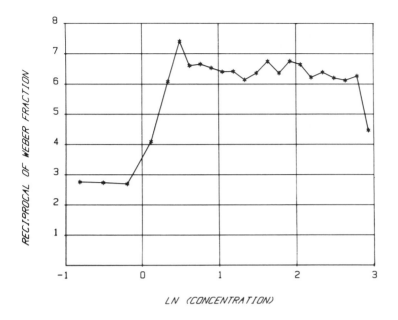

FIG. 5.4 Using the same data as in Fig. 5.3, the *reciprocal* of the Weber fraction is plotted against the natural logarithm of concentration. By counting, one can easily see that about 21 rectangles (that is, 21 jnd's) are contained beneath the entire curve, and about 15 rectangles (15 jnd's) are contained beneath the plateau region [see Box 4 and Eq. (9)].

This approximation extends to other sensory modalities such as audition and vision.

Recalling the proportionality between δH and δN, we can also say *the entropy change spanned by the plateau of the Weber fraction curve is an approximation to (but, of course, less than) the total entropy change from threshold to maximum stimulus intensity.* But this entropy change, threshold to maximum concentration, is equal to the information received by a chemoreceptor as it adapts from F_{max} (or P_{max}) to zero (complete adaptation). That is, the entropy change is equal to the information content of the stimulus, and we have seen above (section on Adaptation) that the information content of the gustatory stimulus is about 2 bits of information or $2\ ln\ 2$ natural units of information (Box 6).

In other sensory modalities, δH is, perhaps, more closely approximated by 2.5 bits of information per stimulus, so we might replace δH by $2.5\ ln\ 2$ natural units. Then Eq. (12) can be rewritten:

Box 6
Converting from natural units of information to bits of information.
We use the formula for converting logarithms from one base to another: The
logarithm of x to the base y is written $log_y x$. Then $ln\ x = log_e x = (log_e 2)(log_2 x)$. (You can use the mnemonic $x/e = (2/e)(x/2)$)
Let $[\delta H]_2 = log_2 q$ bits of information
and
$[\delta H]_e = log_e q$ natural units of information.
Then, using the above rule,
$[\delta H]_e = ln\ 2\ [\delta H]_2$, or

natural units = bits multiplied by the natural log of 2

or

natural units = (0.69)(bits)

$$\delta N = 2.5\ ln\ 2/\Delta H \tag{14}$$

The quantity δN may be eliminated between Eqs. (11) and (14):

$$\frac{2.5\ ln\ 2}{\Delta H} = ln\ (I_2/I_1)\ \frac{n}{2\Delta H}$$

ΔH, the constant information content of the jnd (Fechner's conjecture), cancels
from both sides of this equation, leaving

$$n\ ln\ (I_2/I_1) = (2)(2.5)\ ln\ 2 \tag{15}$$

We can convert from natural logs to common logs (base 10) by multiplying both
sides of Eq. (15) by a constant:

$$n\ log_{10}(I_2/I_1) = 5\ log_{10}2 = 1.51 \tag{16}$$

Expressing this equation in words,

(Stevens' exponent) (common log of "range" of plateau) $\simeq 1.51$

This is the "secret" mentioned above, which is concealed by the plateau region of
the Weber fraction curve. The reader may recognize this rule as one discovered
empirically by Teghtsoonian (1971) using data by Stevens and Poulton (1967).

We have now seen a derivation of the rule within an informational model of the process of sensation, the same model which provided a derivation of Fechner's and Stevens' laws. Teghtsoonian obtained for the constant on the right-hand side of Eq. (16) the value 1.56, purely empirically, by a process of curve fitting. We now gain a good deal of understanding of this number, as we see that it is equal to the product of $log_{10}2$ with twice the maximum information content of a stimulus. The number will be constant across the modalities of sensation only if the maximum information content of all stimuli is about 2.5 bits, corresponding to the absolute recognition of $2^{2.5} \simeq 6$ categories. The derivation is also seen to rest on the veracity of Fechner's conjecture that the subjective magnitude of the jnd (that is, of $\Delta H = \Delta P/k$) is constant. We observe also (Eq. (10a)) that the factor $log_{10}(I_2/I_1)$ can be replaced by the invariant product 0.43 $(\delta N)(Weber$ $constant)$, where 0.43 converts natural to common logs. It is, of course, tacitly assumed in the derivation that a region in the Weber fraction curve can be identified as a "plateau region" and that upper and lower bounds to this region can be assigned unambiguously. Using the data of Lemberger (1908) (see Figs. 5.3 and 5.4), the common log of the range of the plateau is approximately equal to log_{10} $18.75/1.4 = 1.13$. The power function exponent for sucrose is 1.3 (suggested by McBride, 1983b). Therefore, the product of exponent with range is about 1.46 (cf. the predicted value of 1.51 from Eq. (16)).

THE WEBER FRACTION, II

In our initial exploration of the Weber fraction, we simply differentiated the H-function [Eq. (1)] to give dH/dI, which we approximated by $\Delta H/\Delta I$. The ΔI was taken as the change in stimulus concentration corresponding to an information change, ΔH (solutions with greater concentrations have greater absolute fluctuations in solute concentration, so that changing concentrations, ΔI, produces a change in uncertainty, and hence a change in information received, ΔH, when they are tasted). Let us now approach the Weber fraction again, but now with closer attention to the manner in which it might be measured experimentally. The subject is presented with a solution to taste. A short time later, perhaps after rinsing with distilled water, he or she is presented with a second solution to taste, and is given a two-alternative, forced choice: the solutions are the same concentration or they are different concentrations.

In order to transcribe the experiment into mathematical terms, we retain Eq. (1) in full generality; that is, we let entropy, H, remain a function of the *two* variables, I and t. By convention, therefore, we shall have to use the symbol "∂" to indicate a derivative; that is, $\partial H/\partial t$ is the rate of change of entropy with time. Then $\partial H/\partial t$ Δt is a measure of the total change in H during the time interval Δt

(rule f). For example, if a stone is flying through the air at an instantaneous speed of 10 meters per second (dx/dt), then in one-half of a second (Δt), the stone will travel about dx/dt Δt, or 5 meters. There are problems with this *first-order approximation*, because the derivative must be taken at the correct time within the time interval Δt, but for the sake of simplicity, we shall ignore them in this discussion. Then $\Delta H_1 = \partial H_1/\partial t$ Δt is the amount of information received by the subject when he or she tastes the first solution (concentration I_1) for time Δt, and $\Delta H_2 = \partial H_2/\partial t$ Δt is the amount of information gained when he or she tastes the second solution (concentration I_2) for time Δt. Let us define the difference between these two quantities of information by the symbol $\Delta^2 H$. That is,[7]

$$\Delta^2 H = \Delta H_1 - \Delta H_2 \tag{17}$$

When $\Delta^2 H$ is too small, the subject will not be able to distinguish the difference in concentration between the two solutions. The smallest value of $\Delta^2 H$ that permits discrimination of the stimuli is the *informational difference limen* for taste for this solution. $\Delta^2 H$ now plays the same role that ΔH did in the previous section: it defines the magnitude of the jnd for taste. Expanding Eq. (17),

$$\Delta^2 H = \frac{\partial H_1}{\partial t} \Delta t - \frac{\partial H_2}{\partial t} \Delta t$$

We observe that $\Delta^2 H$ is a difference, just like ΔH_1 and ΔH_2, so we can apply the same first-order approximation, this time using intensity, I, as the independent variable. Thus (see mathematical rule f),

$$\Delta^2 H = \frac{\partial}{\partial I} \left[\frac{\partial H}{\partial t} \Delta t \right] \Delta I$$

---

[7]The superscript "2" indicates a difference between two differences. It is not a squared quantity. Read "Δ^2" as "delta-2" rather than "delta-squared." Since from Eq. (2a) subjective magnitude, $P = kH$, therefore, multiplying Eq. (17) by k,

$$\Delta^2(kH) = \Delta(kH_1) - \Delta(kH_2)$$

That is, $\Delta^2(kH)$ (or equivalently $\Delta^2 H$) approaches the difference limen or jnd for taste.

That is, the informational difference limen (*not* quite the same as the informational difference limen in the previous section) is given by a second partial derivative[8]

$$\Delta^2 H = \frac{\partial^2 H}{\partial I \, \partial t} \Delta t \, \Delta I \tag{18}$$

Solving this equation for the Weber fraction,

$$\frac{\Delta I}{I} = \frac{1}{I} \frac{\Delta^2 H}{\Delta t} \left[\frac{\partial^2 H}{\partial I \, \partial t} \right]^{-1} \tag{19}$$

While there are more accurate ways of proceeding than the above, by avoiding the first-order approximations (Valter, 1988), the derivation provided here has the advantage of showing clearly which operations have been carried out.

---

[8]The meaning of $(\partial^2 H / \partial I \, \partial t) \Delta t \, \Delta I$:

Let us continue the example of the stone flying through the air. Suppose now that the distance, x, traveled by the stone depends on two variables: time, t, and the stone's mass, m. That is, $x = x(t,m)$; x is a function of t and m. The stone's velocity is then represented by $\partial x / \partial t$, the *first partial derivative* of x with respect to t, rather than dx/dt, as before. However, this quantity now depends on the mass of the stone (perhaps larger stones travel faster). The rate of change of distance traveled with mass is then $\partial / \partial m \, [\partial x / \partial t \, \Delta t]$, so that for a small change in mass, Δm, the difference in distance traveled is $\partial / \partial m \, [\partial x / \partial t \, \Delta t] \, \Delta m$ $= \partial / \partial m \, [\partial x / \partial t] \, \Delta t \, \Delta m$. The rate of change of the stone's velocity with mass, $\partial / \partial m \, [\partial x / \partial t]$, is written $\partial^2 x / \partial m \, \partial t$, which is the *second partial derivative* of x with respect to m and t.

It must be cautioned, however, that the above is not a general approach to the study of approximate values of functions of two independent variables such as $x(t,m)$, for which Taylor's formula must be used. $\partial^2 x / \partial m \, \partial t \, \Delta t \, \Delta m$ is only one term in this formula.

With the above in mind, we return to the Weber fraction. $-\partial H / \partial t$ is the rate at which information is gained with time. $\partial H / \partial t \, \Delta t$ is the amount of information (neglecting algebraic sign) gained in time, Δt, as the subject perceives a stimulus. $\partial / \partial I [\partial H / \partial t \, \Delta t]$ is, then, the change in amount of information gained with a change in stimulus intensity (more intense stimuli transmit information more rapidly); and $\partial / \partial I [\partial H / \partial t \, \Delta t] \, \Delta I$ is the difference in the information transmitted when stimulus intensity changes by ΔI. This quantity has been written $\Delta^2 H$ and, therefore,

$$\Delta^2 H = (\partial^2 H / \partial I \, \partial t) \Delta t \, \Delta I$$

Recall that Fechner postulated a constant change in subjective magnitude, ΔP, for every jnd. With this constraint, recalling Eq. (2a),

$$\Delta P = \Delta^2 (kH) = k \Delta^2 H = \text{constant}$$

Thus, $\Delta^2 H$ has been used as a constant parameter in Eq. (20).

Since H is available from Eq. (1), we can find the second derivative, $\partial^2 H / \partial I \, \partial t$, explicitly and substitute its value into Eq. (19). We can approximate Δt by t for simplicity, although this results in some loss of accuracy. The approximation is, of course, better for smaller t, and t must be greater than zero. We obtain, finally,

$$\frac{\Delta I}{I} = \frac{-2\Delta^2 H}{n\gamma} [I^{-n/2} + \gamma \, I^{n/2}]^2 \tag{20}$$

where $\gamma = \beta/t$ as in Eqs. (5) and (7). In evaluating the above expression we regard t, the duration of exposure to the stimulus, as constant. As an extension of Fechner's conjecture, we regard the difference limen, $\Delta^2 H$, also as constant. That is, the information change per jnd is constant. Therefore, there are three constant parameters: n, γ, and $-\Delta^2 H$. The latter parameter, as we have defined it, will be greater than zero.

Equation (20) should be compared with Eq. (7), our first rendition of the Weber fraction. If we regard γ as a small quantity, so that the term $\gamma I^{n/2}$ becomes prominent only for large I, then we see that the early part of the curves (lower values of I) described by functions (7) and (20) are identical. The curves differ primarily in the shape of the latter part of the curves (higher values of I). While Eq. (7) predicts a plateau, Eq. (20) predicts an upward turn in the curve for larger I (see Fig. 5.3 or Holway and Hurvich, 1937). This is in accord with more usual experimental observation. For more details see Valter (1988).

THE INFORMATIONAL CONCEPT OF THRESHOLD

The concept of threshold finds very natural expression in the language of information. We have seen how the adaptation process is perceived within this theory as a process of progressive acquisition of information. *A threshold might then be defined as the smallest quantity of information (smallest decrease in entropy), ΔH, that enables a motor response by the subject.* The response could be verbal or nonverbal such as pressing a button. Thresholds to sensory stimulation will be accompanied by conscious awareness of the stimulus, but the concept of "consciousness" is not needed for the definition of threshold.

Consider the events that follow the application of a steady taste stimulus to the tongue. Entropy begins to drop at the taste receptor immediately after the stimulus is applied. Information is accumulated progressively with time, but the subject does not yet respond. Finally, the entropy reduction reaches ΔH bits, and the subject responds to the stimulus, for example, by pressing a button.

Threshold has been reached. The subject may or may not be consciously aware of the presence of the stimulus at the instant he or she has reacted (Libet, 1985).

Simple Reaction Time

The time interval following the imposition of the stimulus required for the subject to receive information, ΔH, at the receptor level is designated t_r. Now, the shortest time interval following the imposition of the stimulus at which a subject executes some motor task such as pressing a button is called the *simple reaction time*, t_R. The critical question is, of course, how are the times t_r and t_R related? The most obvious answer is that reaction time, t_R, exceeds the time of information receipt, t_r, by some positive quantity, t_{lag}, which is the time required for conduction of neural impulses from sensory receptor (to cerebral cortex?) to muscle and for muscle to contract. However, in the study of auditory and visual reaction times (Norwich et al., 1989), it was not apparent that t_{lag} was significantly greater than zero, which is strange to say the least. So let us proceed here with a study of gustatory reaction times on the assumption that $t_R = t_r$. However, I do encourage the reader to explore the matter further.[9]

Let us begin, as we have each time, with Eq. (1), but constraining I to be constant. That is, we are concerned here with a stimulus of constant magnitude applied for a varying period of time. We evaluate H at two different times, t_o and t_r, where $t_o < t_r$. Define t_o as the time at which H is maximum. Then, from the physical interpretation of entropy, H,

$$\Delta H = H(t_o) - H(t_r)$$

is the threshold information.

$$\Delta H = (1/2) \ln (1 + \beta \, I^n/t_o) - (1/2) \ln (1 + \beta I^n/t_r) \tag{21}$$

---

[9]The "conserved quantity" (the "currency"?) for achievement of threshold is taken here as information, in contradistinction to energy, as is usually done in the study of visual or auditory threshold. The term "energy summation" is often used in the study of vision and audition. Since stimulus intensity, in these modalities, has the dimensions of power [Joule/second], the product of intensity with time, $I.t$, is energy. It has been observed that when this energy reaches a constant value, visual threshold is reached (Bloch's law). There are a number of problems with this formulation. One problem is that Bloch's law and, therefore, rigorous energy summation, is valid only for very brief stimuli; for stimuli of longer duration Bloch's law is violated. Another problem is that the law does not generalize to the chemical senses, where the product of I with t is not an energy. The use of information as the conserved currency does permit the generalization of the threshold concept to all sensory modalities. Also, the use of information conservation for both brief and extended stimuli is possible, but there are, admittedly, difficulties here.

Solving for t_r, the reaction time,

$$t_r = \left[\frac{e^{-2\Delta H}}{t_o} - \frac{1 - e^{-2\Delta H}}{\beta I^n}\right]^{-1} \tag{22}$$

If t_r is plotted against I it is seen (Fig. 5.5) that t_r decreases with increasing I, as determined experimentally for gustatory reaction times by Bujas (1935).

There are two limiting cases that are of interest.

(a) The maximum value of t_r occurs for the minimum value of I, which we can call I_{thresh}. We obtain the maximum value of t_r by setting the quantity in square parentheses in Eq. (22) equal to zero. We then solve for $I = I_{thresh}$.

$$I_{thresh} = \left[\frac{t_o(e^{2\Delta H} - 1)}{\beta}\right]^{1/n} \tag{23}$$

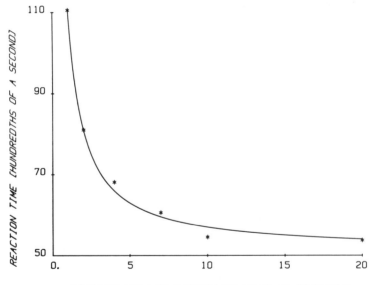

FIG. 5.5 Gustatory reaction time decreases as a function of stimulus intensity. The same trend is observed in simple reaction times to auditory and visual stimuli. The smooth curve was drawn by curve fitting the data of Bujas (1935) for sucrose to Eq. (26), the informational equation for simple reaction time.

In this model $t_r \to \infty$ for $I \to I_{thresh}$, and this is at variance with experimental observation, where t_r is never really observed to be greater than a finite value, usually less than one second. It will cause us problems when we come to analyze data.

(b) The minimum value of $t_r = t_{rmin}$ occurs for the largest value of I. Letting $I \to \infty$ in Eq. (22), we obtain

$$t_{rmin} = t_o e^{2\Delta H} \tag{24}$$

With a little algebra we can now rewrite Eq. (22) in the simpler form (Norwich et al., 1989)

$$t_r = \frac{t_{rmin}}{1 - \left[\dfrac{I_{thresh}}{I}\right]^n} \tag{25}$$

If we now introduce into Eq. (25) the approximation $[I_{thresh}/I]^n \ll 1$, Eq. (25) simplifies to the form

$$t_r - t_{rmin} = A I^{-n} \tag{26}$$

where A is constant (Box 7). This may be recognized as Piéron's empirical law for reaction time for pressure and auditory signals (Piéron, 1914, 1952). It suggests that the exponent in Eq. (26) is, in fact, the Stevens exponent, n, which appeared in the law of sensation.

Gustatory reaction times do not seem to have been investigated extensively, but they were measured by Piéron (1914, 1920) and by Bujas (1935), who placed droplets of various solutions on the tongues of subjects and measured the time for response. For solutions of citric acid, sodium chloride, saccharose, and saccharine, Bujas found that reaction time was related to concentration of solution, I, by means of the equation

$$t_r - t_{rmin} = A/I \tag{26b}$$

That is, his findings were in accord with Eq. (26) with the exponent, n, taken equal to 1, which is approximately the expected value for these substances (Norwich, 1987) (Boxes 7 and 8).

Threshold Law for Odorants

As a final offering, I put foward the following conjecture. Taking logs of both sides of Eq. (23), we obtain

Box 7
Deriving Piéron's law for simple reaction time.
Here is how we proceed from Eq. (24) to Eq. (25).
When $|x| < 1$, using the binomial theorem, we can expand $(1 - x)^{-1} = 1 + x + x^2 + \ldots$
When x is sufficiently small, we can neglect powers of x higher than the first. Therefore,

$$[1 - (I_{thresh}/I)^n]^{-1} \simeq 1 + (I_{thresh}/I)^n$$

Therefore, from Eq. (25),

$$t_r = t_{rmin}[1 + I^n_{thresh} \cdot I^{-n}]$$

or

$$t_r - t_{rmin} = (t_{rmin} \cdot I^n_{thresh})I^{-n} \qquad (26a)$$

$$t_r - t_{rmin} = A I^{-n} \qquad (26)$$

The constant, $A = t_{rmin} \cdot I^n_{thresh}$, is evaluated in Eq. (26a). Equation (26) is then an approximation, for larger stimulus concentrations, to Eq. (25), which itself gives reaction time, t_r, only to an approximation. Nonetheless, Eq. (26) does describe simple reaction time data quite well, which is why Piéron adopted it.

$$n \, log_{10}I_{thresh} = log_{10}[t_o(e^{2\Delta H} - 1)/\beta]. \qquad (27)$$

Suppose that the quantity on the right-hand side of Eq. (27) is constant for all odorants (and I know of no evidence to support this assumption). That is, the constants t_o, β, and ΔH are constant for all odorants, or, their combination as in Eq. (27) is constant. Then

$$n \, log_{10} \, I_{thresh} \simeq \kappa$$

Define

$$p_{ol} = -log_{10}I_{thresh} \qquad (29)$$

Then

$$n \cdot p_{ol} = \kappa \qquad (30)$$

Box 8

The following are simple reaction time data of Bujas (1935) to the taste stimulus sucrose. Threshold concentration is designated as one; other concentrations of the stimulus solution are given in multiples of threshold (column 1). Reaction times are in one hundredths of a second (column 2). Reaction times fitted to Eq. (25) are shown in column 3. Reaction times fitted to Eq. (25) with the value of n fixed at 1.0 are given in column 4. Reaction times fitted by Bujas to Eq. (26) are given in column 5. Recall that within the informational theory, for $I \to I_{thresh}$, $t_r \to \infty$.

$Conc^n$	Reaction time	Fitted (25) 3-parameter	Fitted (25) $n = 1$	Fitted (26)
1	110.5	110.5	112.0	110.9
2	81.0	81.2	75.0	81.0
4	68.1	67.0	64.4	66.0
7	60.6	60.2	60.7	59.6
10	54.5	57.0	59.3	57.0
20	53.7	52.4	57.8	54.0

The data are plotted in Fig. 5.5.

Equation (25), with its three parameters evaluated by a least squares procedure, is

$$t_r = \frac{t_{rmin}}{1 - (I_{thresh}/I)^n} = \frac{40.0}{1 - (0.26/I)^{0.33}}$$

The value of the exponent is smaller than would be expected for sucrose. The fitted value for I_{thresh} is necessarily smaller than the designated value of 1.0, because the value of 1.0 would produce a singularity: t_r would become infinite for $I = 1$. Equation (25), with its n fixed at 1.0 and the remaining two parameters evaluated by a least squares procedure, is

$$t_r = \frac{t_{rmin}}{1 - I_{thresh}/I} = \frac{56.4}{1 - 0.50/I}$$

Equation (26), with its parameters evaluated by least squares, is

$$t_r = A/I + t_{rmin} = 60.0\,I + 51.0$$

The value of A predicted from Eq. (26a) is $t_{rmin} \cdot I^n_{thresh} = (51.0)(1^1) = 51.0$. The value found is 60.0.

That is, the product of the power function exponent, n, with the negative logarithm of threshold concentration is expected to be constant. Lower thresholds mean lower exponents. This is the law discovered by Laffort et al. (1974) and graphed clearly by Wright (1982). Since the derivation depends on three parameters remaining constant for all odorant gases, the result is not expected to hold strictly.

SUMMARY

Unification of the laws of sensation is the goal of the entropic or informational theory of sensation. All of the derivations made in this chapter have issued from Eq. (2), $F = kH$, with entropy, H, modeled by Eq. (1). These equations state only that information about the mean value of a fluctuating stimulus (gas density or solute concentration) is assimilated gradually by the sensory receptor; that fluctuations are greater in absolute value for greater stimulus concentrations; and that information relayed neuronally to the brain will (at best) equal information received by the receptor. That is, information is, at best, conserved. No statement about how the receptor functions (chemically, electronically) has been made.

H is a function of time and of stimulus intensity. Holding intensity constant and letting H vary only with time, we derived an equation for adaptation phenomena and were able to associate the process of adaptation with the receipt of a quantity of information that was not in excess of 2.5 bits. Holding time constant, and letting H vary only with intensity, we were able to derive the law of sensation, and to show that Fechner's and Stevens' laws both emerged as approximations to the general informational law. Differentiating H with respect to intensity gave us our first representation of the Weber fraction. Introducing the concept of a fixed quantity of information received per jnd, and recalling that the maximum information receipt per stimulus is about 2.5 bits, we derived the well-known rule relating Stevens' exponent to the stimulus range. Using a second partial derivative of H, we were able to discover why the Weber function increases for higher stimulus concentrations. Introducing the concept that a threshold is the smallest amount of sensory information required for motor response, we derived an equation relating stimulus concentration to simple reaction time for gustatory stimuli, and derived Piéron's empirical law for simple reaction time. A law relating exponent to threshold for odorants (very approximately) followed easily.

Unity. Many laws derived from one. The derived laws did not always fit our data sets precisely (although some did), or with exactly the parameter values expected (again, some did), but the theory fared reasonably well. The same type of analysis extends to other modalities of sensation.

ACKNOWLEDGMENT

This work has been supported during the past 13 years by grants from the Natural Sciences and Engineering Research Council of Canada.

SYMBOLS

β Constant whose value must be established from the data. Appears first in Eq. (1).

γ $= \beta/t$ is a constant when t is constant.

Δ Represents a change or difference in the value of some quantity.

ΔH Change in entropy = gain in information.

   ΔH assumes a constant value for discrimination (differential threshold, difference limen).

   ΔH assumes a constant value for simple reaction time.

$\Delta I/I$ Weber fraction.

F A perceptual variable: Frequency of impulses in a primary sensory afferent neuron.

H Information theoretical entropy of a sensory stimulus. In general, a function of I and t.

I Intensity of a stimulus; e.g., concentration of a solution or density of an odorant.

I_{thresh} Threshold stimulus intensity.

k Scaling constant relating entropy to a perceptual variable; first appears in Eq. (2).

κ Constant defined by Eq. (30).

m Number of samplings made by a sensory receptor of its stimulus signal.

n Power function exponent.

δN Number of jnd's (just noticeable differences) between any two defined stimulus intensities; e.g. between threshold and maximum intensity.

p_{ol} $= -\log_{10} I_{thresh}$.

P A perceptual variable: Subjective magnitude.

t Time: duration of a stimulus.

t_o Duration of a stimulus for which H is maximum; a very short interval of time.

t_r Duration of stimulus required to receive information, ΔH, at the receptor.

t_R Simple reaction time.

t_{lag} $= t_R - t_r$.

REFERENCES

Atkinson, W. H. 1982. A general equation for sensory magnitude. *Percept. Psychophys.* 31: 26–40.

Borg, G., Diamant, H., Ström, L., and Zotterman, Y. 1967. The relationship between neural and perceptual intensity: a comparative study on the neural and psychophysical response to taste stimuli. *J. Physiol.* (Lond.) 192: 13–20.

Bujas, Z. 1935. Le temps de réaction aux excitations gustatives d'intensité différente. *Societé de Biologie, Comptes Rendus* 119: 1360–1364.

Gent, J. F. and McBurney, D. H. 1978. Time course of gustatory adaptation. *Percept. Psychophys.* 23: 171–175.

Holway, A. H. and Hurvich, L. M. 1937. Differential gustatory sensitivity to salt. *Am. J. Psychol.* 49: 37–48.

Laffort, P., Patte, F., and Etcheto, N. 1974. Olfactory coding on the basis of physicochemical properties. *Ann. N.Y. Acad. Sci.* 237: 193–209.

Lemberger, F. 1908. Psychophysische Untersuchungen über den Geschmack von Zucker and Saccharin (Saccharose and Krystallose). *Pflügers Archiv für die gesammte Physiologie des Menschen und der Tiere* 123: 293–311.

Libet, B. 1985. Unconscious cerebral initiative and the role of conscious will in voluntary action. *Behav. Brain. Sci.* 8: 529–556.

McBride, R. L. 1983a. Category scales of sweetness are consistent with sweetness-matching data. *Percept. Psychophys.* 34: 175–179.

McBride, R. L. 1983b. Taste intensity and the case of exponents greater than 1. *Aust. J. Psychol.* 35: 175–184.

Moles, A. 1966. *Information Theory and Esthetic Perception.* (J. E. Cohen, trans.), University of Illinois Press, Urbana. (Original work published 1958).

Norwich, K. H. 1981. The magical number seven: Making a 'bit' of 'sense'. *Percept. Psychophys.* 29: 409–422.

Norwich, K. H. 1984. The psychophysics of taste from the entropy of the stimulus. *Percept. Psychophys.* 35: 269–278.

Norwich, K. H. 1987. On the theory of Weber fractions. *Percept. Psychophys.* 42: 286–298.

Norwich, K. H., Seburn, C. N. L., and Axelrad, E. 1989. An informational approach to reaction times, *Bull. Math. Biol.* 51: 347–358.

Piéron, H. 1914. Recherches sur les lois de variation des temps de latence sensorielle en fonction des intensités excitatrices. *Année Psychologique* 20: 17–96.

Piéron, H. 1920. Nouvelles recherches sur l'analyse du temps de latence sensorielle et sur la loi qui relie ce temps à l'intensité de l'excitations, *Année Psychologique* 22: 58–142.

Piéron, H. 1952. *The Sensations*. Yale University Press, New Haven.

Poulton, E. C. 1967. Population norms of top sensory magnitudes and S. S. Stevens' exponents. *Percept. Psychophys*. 2: 312–316.

Riesz, R. R. 1928. Differential sensitivity of the ear for pure tones. *Phys. Rev. Series 2*, 31: 867–875.

Stevens, S. S. 1969. Sensory scales of taste intensity. *Percept. Psychophys*. 6: 302–308.

Stevens, S. S. 1970. Neural events and the psychophysical law. *Science* 170: 1043–1050.

Teghtsoonian, R. 1971. On the exponents in Stevens' law and the constant in Ekman's law. *Psych. Rev*. 78: 71–80.

Valter, K. M. 1988. *The entropy theory of perception as applied to intensity discrimination of auditory pure tones*. M.Sc. thesis, University of Toronto.

Wiener, N. 1948. *Cybernetics*. Wiley, New York.

Wright, R. H. 1982. *The Sense of Smell*. CRC Press, Boca Raton, FL.

6

Metaphor and the Unity of the Senses

Lawrence E. Marks

John B. Pierce Laboratory and
Yale University
New Haven, Connecticut

INTRODUCTION

In his novel *A Rebours,* J.-K. Huysmans's (1884) antihero, Jean des Esseintes, recounts the flavors of liqueurs in terms of the tonal qualities of various musical instruments:

> Each liquor corresponded, individually, by its taste, to an instrument's sound. Dry curaçao, for example, to a clarinet whose song was bitter and mellow; kummel to an oboe with sonorous and nasal tones; mint and anisette to a flute, at once sugary and peppery, piquant and sweet; and to complete the orchestra, kirsch rang furiously on the trumpet. (p. 62)

In describing liqueurs in acoustic terms, Huysmans uses an extended literary trope, an exaggerated form of what may be called "cross-modal metaphor" or "synesthetic metaphor": the transfer or translation of meaning from one sense modality to another. By implication, Huysmans is postulating the existence of a one-to-one, qualitative correspondence between the flavors of liqueurs and corresponding musical timbres.

Synesthetic metaphors are not the exclusive province of imaginative literature; they also appear in everyday language, as when vivid, highly saturated colors are called "loud," or when tastes are called "sharp," or extreme cold sensations as "bitter." Language is chock full of cross-modal expressions, expressions that interrelate the experiences of different sense modalities. And such metaphors, whether literary tropes or expressions of everyday speech, are not limited to the English language. It is notable, for example, that the German adjective "hell"—which denotes the visually bright—originally meant high-pitched (see Hartshorne, 1934).

In many instances, commonplace extensions of sensory meanings have linguistic roots that are centuries old. Williams (1976) has noted that some synesthetic transfers of meaning in English took place in the time of Chaucer—or even earlier, in the era of *Beowulf*. In fact, Williams argues that there may be a systematic sequence to the cross-modal transfer of meanings of sensory words: In general, according to his findings, as languages change over time there is a tendency for words describing attributes of somesthetic sensations—touch and temperature—to come later to apply also to taste, to color, and to sound—as in the application of "warm" and "cold" or "dry" and "sharp" to tastes, of "dull" and "heavy" to colors, of "hard" and "soft" to sounds. One of the incidental findings is that the adjective "bitter" is only secondarily a taste adjective, having originally referred to tactile experiences. Williams goes so far as to suggest that evolutionary changes in the referents of sensory words—the development of cross-modal metaphor—follow a rule of sensory evolution, with phylogenetically more primitive modalities, such as touch, serving as the sources of adjectives that later apply to phylogenetically more advanced modalities, such as vision and hearing.

The existence of a rich linguistic history in synesthetic metaphors suggests strong links among the several sense modalities, at the very least, strong links among the words we use to label our perceptual experiences. Indeed, Williams's (1976) findings fit nicely with certain commonly held notions, for example, the notion that the words we use to describe smells tend to derive from two sources: the names of objects that generate the odor experiences (e.g., "lemon," "rose") and labels borrowed, metaphorically, from other senses (e.g., "sweet," "acrid").

Williams's (1976) hypothesis speaks to the main topic of concern here: What do synesthetic metaphors tell us about the encoding and perception of sensory stimuli? Is the use of expressions like "bitter cold" or "hard sound" largely the result of abstract cognitive processes, of mental flights of fancy, of Coleridge's "secondary imagination"? Or do these expressions stand firmly on a bedrock of sensory/perceptual processes?

It is my contention that synesthetic metaphors often tell us a great deal about sensory processing, and in particular about important connections or com-munalities across different senses. To enquire in detail about these communali-ties, however, it is helpful first to ask more generally about the ties between language and perception.

TWO ACCOUNTS OF THE RELATION BETWEEN LANGUAGE AND PERCEPTION

What is the connection between language and perception? In particular, what is the connection between the multitude of dimensions, attributes, features of our perceptual experiences, and the words that we use to describe or summarize those experiences? Two largely opposing viewpoints—oversimplified, but con-venient for the present purposes—provide divergent answers to these questions: the developmental view and the Whorfian view.

The Developmental View

On the one hand, it may seem reasonable to assume that one of the purposes of language is to "represent" what is salient or significant in the world—and in particular to represent those significants that appear in perception. According to this semiotic view, the lexicon of perception—the words and phrases we use to describe our perceptual experiences—emerges from perception itself, and runs parallel to it. Accordingly, once a child is capable of making appropriate perceptual discriminations and identifications, the child then is in the position to learn words that she or he can associate with the discriminable or identifiable percepts. Subsequent to any process of learning or maturation that may be necessary to perceptual development, a child learns—for instance, from looking at picture books or television, from visits to a zoo, from eating and drinking—to associate words like "large" and "small" with horses and mice, "green" and "blue" with grass and sky, "sweet" and "salty" with candy and pretzels—in other words, to associate words with perceptual objects and attributes that are per-ceptually salient and ecologically significant. Notice, moreover, that all of this associative learning is made possible by the existence within the language itself of a lexicon of perceptual words; that is, the learning is made possible by a linguistic history through which English (or whatever language) itself developed words like "large," "small," "green," "blue," "sweet," and "salty," whose referents were highly discriminable, identifiable, and useful attributes of per-ceptual experience. For convenience, I shall call this view the developmental hypothesis of language and perception: As reasonable as it may seem, it is after all only an hypothesis, and it states that perception precedes language, both phylogenetically and ontogenetically, so that language comes to represent perception both in its linguistic history and in the learning of that language by its individual speakers.

The Whorfian View

There is an alternate account of the relation between language and perception, an account propounded by the anthropologist Benjamin Lee Whorf (1956), which flips the developmental explanation on its head. According to Whorf, it is language that largely shapes perception (and conception), not (just) the other way round. In particular, states the Whorfian hypothesis, stimuli in the environment are most readily encoded perceptually when we have words to describe them. According to this view, "salty" and "sweet," "blue" and "red" are not primarily innate categories of perceptual experience, natural categories that language subsequently comes to encapsulate in verbal labels, but instead are to a large extent artificial categories imposed on us by the ways that language forces us to chop up the otherwise continuous worlds of taste and color perception.

To ask, therefore, about the connection between cross-modal perception and the use of synesthetic metaphor is to invoke the propaedeutic question about the connection between perception and language more generally. If the developmental hypothesis is correct, then it may well be that synesthetic metaphors reflect similarities among the different senses in their underlying perceptual mechanisms. But if the Whorfian hypothesis is correct, then it is conceivable that cross-modal similarities originate in language itself, and thus constitute intrinsically metaphorical—linguistically metaphorical—entities, even when these similarities show themselves in perception.

Perhaps the best place to start is with the developmental hypothesis, for this hypothesis has guided a substantial amount of research into synesthetic metaphor and its perceptual underpinnings. In particular, the developmental hypothesis is directly linked to a theory of synesthetic metaphor through the doctrine of the "unity of the senses" (Hornbostel, 1925; Werner, 1934; Marks, 1978), this doctrine stating that vision, hearing, taste, smell, and touch intrinsically share important characteristics. Essentially, the doctrine of the unity of the senses argues for the existence of fundamental similarities in sensory perception, similarities that are captured by language in cross-modal metaphors.

THE DOCTRINE OF THE "UNITY OF THE SENSES"

We often think of the senses as largely independent modalities for transmitting perceptual information, each sense being specialized in the kind of information it transmits, whether it be the perception of objects at a distance, sensed through sight, the perception of objects close by, sensed through touch, or the perception of events at a distance, sensed by hearing. But this perspective is a narrow one, for the senses also display a goodly number of communalities, which can be characterized as either informational or as psychophysical.

By informational communalities I mean that the various sense departments can provide common or converging information about objects and events in the environment. We see a texture and we feel it; we hear a voice and we see the speaker's mouth moving. Perceptual systems are geared, either through learning or by nature, to integrate multiple sources of perceptual information. It is interesting to note that many of the strongest intersensory interactions take place under precisely those conditions where the informational content of the stimuli comes under conflict. A well-known example is the so-called "ventriloquism" effect—a change in the judged location of a sound source (e.g., a ventriloquist's voice) toward the spatial location of a visual stimulus (the dummy's mouth) when the auditory and visual stimuli are temporally entrained.

By psychophysical communalities I mean that the various sense departments share common dimensions and attributes, even when the communalities do not reflect ecologically common sources of information. Perhaps the most obvious of these is sensation intensity: The magnitudes of our sensory experience vary from weak to strong—from dim to bright in vision, from soft to loud in hearing, from mild to strong in taste, smell, and pain. Yet there need be no communality in information about visual, tactile, and gustatory stimulation: Potato chips may be either crisp or soggy, but taste salty nonetheless.

As qualitatively different as these experiences may be—and, of course, it is the qualitative differences among red and C# and sour and aching that constitute the very notion of sensory modality—nevertheless the *amount* of sensation is a universal characteristic of experience.

Synesthesia

Besides merely acknowledging that sensations all seem to have components that we may call intensive or qualitative or whatever, how else do the communalities reveal themselves? One way is in the perceptual phenomenon of synesthesia. To a small fraction of the population, well under 1%, the unity of the senses comes as part and parcel of everyday experience. To synesthetic perceivers, sounds are characterized not only by the auditory attributes of loud or soft, low or high pitch, note C# or vowel /a/, but also by attributes of another modality, perhaps bright or dark, red or blue, globular or angular. In those people who are strongly synesthetic, the experiences on the secondary modality are reliable and compelling.

What does synesthesia tell us about cross-modal communalities? Although synesthesia is marked by considerable diversity—every synesthetic individual seems to have her or his particular profile of cross-modal associations—there is also a common core, perhaps a universal thread, that runs through it; for, granting that indeed they are idiosyncratic, synesthesias nevertheless do share important characteristics (Marks, 1975, 1978). Consider, for example, a common type of synesthesia—visual hearing—where people report that sounds induce visual images. Visual-auditory synesthesia reveals five widespread, virtually universal, rules of cross-modal correspondence: (1) pitch-brightness; (2) loudness-brightness; (3) pitch-size; (4) pitch-shape; (5) pitch/duration-tempo.

First, the most salient of all synesthetic correspondences is that between auditory pitch and visual brightness: To synesthetic perceivers, the higher the pitch of the sound, the greater the brightness of the visual image or the more likely it is to be called light or bright rather than dim or dark (Riggs and Karwoski, 1934; Marks, 1975, 1978). Figure 6.1 shows an example, summarizing data reported by several dozen investigators over a period of three-quarters of a century (Marks, 1975). Plotted are scores derived from synesthetic perceivers who reported the brightness of visual images produced by five primary vowel sounds. As these vowels order themselves from low to high in pitch (/u/, /o/, /a/,

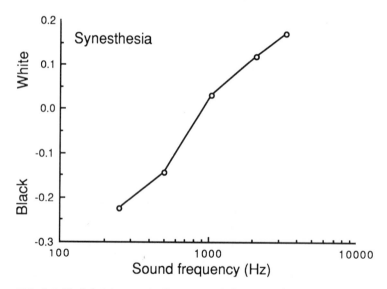

FIG. 6.1 Pitch-brightness similarity revealed in synesthesia. Plotted are measures of the average lightness or brightness of visual sensations reported by synesthetic perceivers to accompany the presentation of five primary vowels (in increasing order of sound frequency, /u/, /o/, /a/, /E/, and /i/). *(Data from Marks, 1975.)*

/E/, and /i/, corresponding to increasing second-formant frequency of the vowels, which determines their pitch), so do the vowels also order themselves from low to high in lightness or brightness, from black to white. In other words, synesthesia reveals a strong and direct cross-modal relation between pitch and lightness. One of the noteworthy features of the data that underlie Fig. 6.1 is that they were obtained from native speakers of many different languages. Native tongue per se seems to have little to do with pitch-brightness correspondence in synesthesia.

Synesthesia perceivers share other cross-modal communalities. With regard to intensity, for example, loud sounds tend to induce bright visual images. Hence loudness and brightness are linked directly, this being a second widespread cross-modal relation. Third and fourth, synesthetic perceivers also agree that high-pitched as opposed to low-pitched sounds invoke angular shapes, and small-sized images, as opposed to rounded shapes, and large-sized images. Hence pitch is linked directly with visual angularity but is linked inversely with visual size. Finally, high-pitched sounds and sounds with rapid tempos, as opposed to low-pitched sounds and sounds with slow tempos, tend to produce visual images that rapidly change their shapes, sizes, colors, and positions in the visual field.

Several of these intermodal relations were subsumed by Hornbostel (1925, 1931) in his theory of sensory brightness: According to Hornbostel, brightness is a universal dimension of sensory experience—a dimension that pertains not only to sight and sound but also to touch, taste, and smell. Moreover, brightness presumably incorporates qualitative and temporal as well as intensive character-istics. Pungent odors, high-pitched sounds, white colors—all of these are high on the scale of brightness. Furthermore, according to Hornbostel's theory of the unity of the senses, this property of brightness constitutes the same dimension in all modalities.

Cross-Modal Matching

Although synesthetic perception was one of the important perceptual phe-nomena that underlay Hornbostel's (1925) hypothesis, his main line of support came from an experiment conducted with nonsynesthetic individuals, whose task was to perform a series of cross-modality matches. In particular, Hornbostel asked subjects to match sound and color, color and odor, and odor and sound, in a round-robin. Hornbostel claimed a high degree of transitivity in these cross-modality matches, which he presumed were based on a communality in bright-ness across the modalities of vision, hearing, and smell. Karwoski et al. (1942) subsequently noted that nonsynesthetic college students often agreed with syn-esthetic perceivers as to the way that attributes of perceptions correspond across different modalities.

An analog in nonsynesthetic perceivers to the synesthetic correspondence between pitch and brightness appears in Fig. 6.2 (Marks, 1974). The figure shows the average settings of sound frequency (pitch) of a pure tone to "match" gray papers viewed against a dark background, where the surface brightness of each gray is specified by its Munsell value. Just as synesthetics say that low-pitched sounds are dark and high-pitched sounds are bright, so, analogously, when placed in an open-ended matching paradigm, do nonsynesthetics tend to match low-pitched tones to dark colors and high-pitched tones to white colors. "Free" cross-modality matching, therefore, in which subjects are not directed to match in an particular direction, can serve to quantify and substantiate the communalities in dimensions of experience on different sense modalities.

Presumably, the communalities in perceptual dimensions reveal themselves readily in undirected cross-modality matching because these communalities enjoy a kind of psychological privilege. And presumably, this psychological privilege applies alike to subjects in laboratory investigations and to panelists in applied settings. People will make cross-modal matches most easily and most reliably when their matches tap common, underlying dimensions of perceptual experience. Although vision and hearing have provided focus for most investigations of cross-modal correspondences, such correspondences doubtless exist across all of the senses. As already mentioned, intensity and perhaps brightness seem to be universal dimensions, applicable to touch, taste, and smell

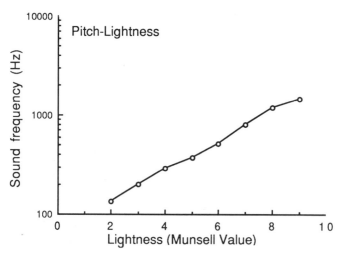

FIG. 6.2 Pitch-brightness similarity revealed in cross-modal matching by nonsynesthetic subjects. Plotted are the mean settings of sound frequency (pitch) to match each of eight levels of Munsell value (lightness) of gray surfaces. *(Data from Marks, 1974.)*

as well as to vision and hearing. Other communalities involving the senses of touch, taste, and smell, however, still seek full investigation.

The general *Weltanschauung* encapsulated by the phrase "unity of the senses" views sensory systems as fundamentally intertwined, interconnected, interrelated (Hornbostel, 1925; Werner, 1934; Marks, 1978). According to this view, sensory systems by their very nature—which is to say by dint of their neural and perhaps chemical bases—process sensory information in similar ways, encode perceptual information in a fashion that permits both common, underlying sensorineural codes (of intensity, brightness, duration, etc.) and intermodal interactions. Marks (1978) and Marks and Bornstein (1987) have suggested some possible neural codes; for example, the auditory system seems to use temporal patterning (neural response frequency) in the coding of both pitch and loudness, which may help explain the communalities between both of these dimensions and visual brightness.

Developmental Characteristics

To view the unity of the senses as a fixed, primitive property of perceptual information processing is to imply that sensory communalities should be demonstrable in children as well as adults. In many cases—though, significantly, not in all cases—this seems to be so. For example, children as young as 4 years of age will match the higher pitched of two sounds to the brighter of two lights (that is, 4-year-olds show an appreciation of pitch-brightness similarity) and will match the louder of two sounds to the brighter of two lights (show an appreciation of loudness-brightness similarity) (Marks et al., 1987), two experimental outcomes that agree well with this viewpoint. But significantly, very young children do not match the higher pitched of two sounds to the smaller of two visual shapes. Indeed, it is not until children approach adolescence that they will regularly match pitch inversely with size.

The difference in developmental timetables among different pairs of cross-modal dimensions is shown in Fig. 6.3. Virtually all 4-year-olds match pitch and brightness the way adults do (low pitch with dim and high pitch with bright); in fact, pitch-brightness matching is more reliable than is loudness-brightness matching at age 4 (88% vs 76%). With increasing age, there are modest increases in the rates of normative (adult-like) pitch-brightness and loudness-brightness matching. By contrast, the matching of pitch and size is essentially random (near 50%) at age 9, but increases markedly thereafter. The difference in timetables leads to a revised hypothesis: Whereas certain cross-modal connections may reflect inborn characteristics of perceptual experience and thus appear early in childhood, other connections may have a different origin. In the case of pitch-size similarity, there is a simple explanation in perceptual learning. Other factors

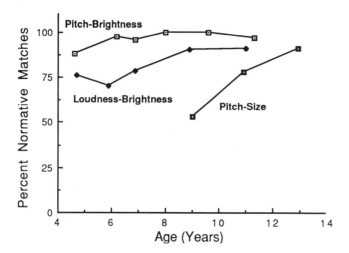

FIG. 6.3 Percentages of children at various ages who made cross-modal matches in the normative fashion of adults: pitch-brightness, matching a low pitched tone to a dim light and a high pitched tone to a bright light; loudness-brightness, matching a soft tone to a dim light and a loud tone to a bright light; and pitch-size, matching a low-pitched tone to a large circle and a high-pitched tone to a small circle. *(Data from Marks et al., 1987.)*

remaining constant, the smaller an object the higher its resonance frequency. Large objects thump, but small ones ring. Conceivably, then, children come to appreciate the similarity between pitch and size only after they have considerable experience with objects and their resonance properties.

The revised interpretation of cross-modal matching enlists perceptual learning as the source of at least this one cross-modal association. A second possibility is that the association between high pitch and small size is somehow learned not directly through perception but indirectly through language. Perhaps in language we use a common coding system for high pitch and small size.

CROSS-MODAL SIMILARITIES IN LANGUAGE: SYNESTHETIC METAPHOR

To repeat and summarize, the developmental hypothesis proposes that language reflects or represents perception. Accordingly, the hypothesis leads to the view that cross-modal or synesthetic metaphors should reflect or represent cross-modal perception. Where cross-modal similarities exist in perception, so then should these similarities be available as the underpinnings to synesthetic metaphors. Two possible schemes suggest themselves. One is simply that as long as a person can recognize a similarity perceptually, so too should he or she be able to recognize the same similarity when it is expressed verbally, through metaphor. That is, performance in a linguistic task, say, interpreting cross-modal metaphors, should precisely parallel performance in a corresponding perceptual task, say, cross-modality matching. A second possibility is that some learning—and hence some time—is required before similarities that are appreciated in perception can also be processed linguistically. That is, there might be a small developmental lag between performance on perceptual tasks and linguistic tasks.

In a wide-ranging study of cross-modal similarity, my colleagues and I have examined how adults and children interpret various kinds of synesthetic or cross-modal metaphors (Marks, 1982; Marks et al., 1987). The metaphors were constructed so as to tap the same dimensional similarities—pitch-brightness, loudness-brightness, and pitch-size—that we studied at the level of perception. The verbal materials consisted of either words or brief word combinations whose meanings the subjects rated on scales representing one of four perceptual dimensions: pitch, brightness, loudness, and size. The judgments themselves could be literal, as when subjects judged such words as "soft," "loud," "thunder," and "whisper" and such word combinations such as "soft thunder" and "loud whisper" on a scale of loudness. Or the judgments could be metaphorical, as when the subjects judged those same words and phrases on a scale of brightness, or when the subjects judged such metaphors as "dim thunder" and "bright whisper" on a scale of brightness or on a scale of loudness.

The two main results of this study can be summarized as follows: First, the ways that people interpret cross-modal metaphors follow directly the ways that people perceive cross-modal similarities. Just as, in a perceptual task, loud and high-pitched tones are matched to bright lights, and just as soft and low-pitched tones are matched to dim lights, so, in a verbal task, words that denote relatively bright as opposed to dim colors are judged to be loud and high pitched rather than soft and low pitched, and words that denote relatively loud or high-pitched as opposed to soft and low-pitched sounds are judged bright rather than dim. "Sunlight" is brighter than "moonlight"; "sunlight" is also louder and higher pitched. "Shout" is louder than "whisper"; and "shout" is also brighter.

Figures 6.4 and 6.5 give examples of the ways that subjects judged various visual-auditory metaphors on scales of loudness and brightness (Marks, 1982). "Bright" is not only brighter than "dim" (Fig. 6.5), but also louder (Fig. 6.4). And "trumpet note" is not only louder than "piano note" (Fig. 6.4), but also brighter (Fig. 6.5). Clearly, there is a direct parallel between perceptual matching and metaphorical meaning. Furthermore, the metaphorical transfers of meanings are generally symmetrical: Visual words transfer meanings substantially to sound, and acoustic words transfer their meanings substantially to sight.

One of the clearest examples of cross-modal similarity in metaphor appears in

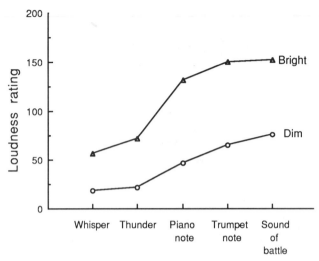

FIG. 6.4 Mean ratings on a graphic scale of loudness (0 = very very soft; 200 = very very loud) of metaphorical combinations of the visual adjectives "dim" and "bright" with each five auditory nounds, "whisper," "thunder," "piano note," "trumpet note," and "sound of battle." *(Data from Marks, 1982.)*

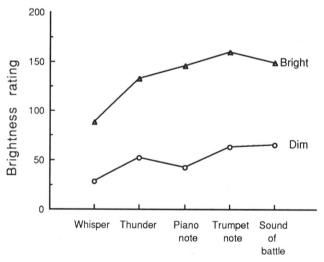

FIG. 6.5 Mean ratings on a graphic scale of brightness (0 = very very dim; 200 = very very bright) of metaphorical combinations of the visual adjectives "dim" and "bright" with each of five auditory nouns, "whisper," "thunder," "piano note," "trumpet note," and "sound of battle." *(Data from Marks, 1982.)*

196

the domain of color. Every color—or more properly every color name—is characterized not only by a prototypical brightness, but also by a corresponding prototypical pitch. White and yellow are bright colors, and correspondingly are conceived as high pitched, whereas brown and black are dark colors, and correspondingly are conceived as low pitched. Blue, red, and green are intermediate in both brightness and pitch (Fig. 6.6).

The second main result concerns the behavior of children. In general, the linguistic comprehension of cross-modal metaphors follows the same developmental timetable as does the perception of cross-modal similarity (Marks et al., 1987). Just as children as young as 4 years match pitch with brightness and loudness with brightness, so do 4-year-old children also appreciate at least some of the metaphorical correspondences between pitch and brightness and between loudness and brightness. And just as children do not appreciate the perceptual similarity between pitch and size until they approach adolescence, so do children not comprehend pitch-size metaphors until the same age. In order to compare directly linguistic and perceptual appreciation of cross-modal similarity, children's performance on tasks of metaphor comprehension is summarized in Fig. 6.7. The figure gives scores analogous to those derived from perceptual matches. To obtain a score for each child we determined, for many judgments of cross-modal metaphors, whether the average difference between corresponding pairs of expressions went in the normative fashion—for example, whether "bright thunder" was rated louder than "dim thunder." Thus average ratings of meaning

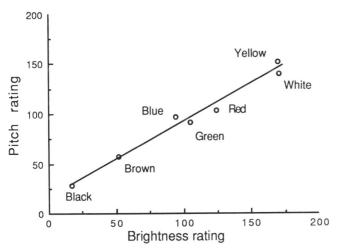

FIG. 6.6 Mean ratings of brightness versus mean ratings of pitch for the same seven color names: "black," "brown," "blue," "green," "red," "yellow," and "white." Pitch and brightness ratings are virtually proportional to each other. (*Data from Marks, 1982.*)

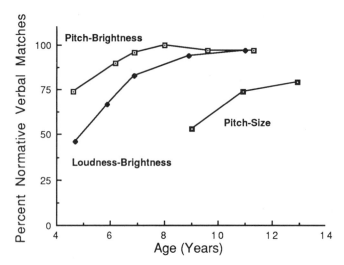

FIG. 6.7 Percentages of children at various ages who made verbal cross-modal matches (as derived from ratings of metaphorical words and expressions) in the normative fashion of adults: pitch-brightness, interpreting low pitch = dim and high pitch = bright; loudness-brightness, interpreting soft = fim and loud = bright; pitch-size, interpreting low pitch = large and high pitch = small. *(Data from Marks et al., 1987.)*

were converted into single scores of verbal cross-modal "matches" or "mismatches."

Note the strong similarity between Fig. 6.7 and the corresponding Fig. 6.3 for perceptual matching. If anything, there is a tendency for a slight developmental lag between performance on the perceptual and the verbal tasks; that is, the children's ability to appreciate cross-modal metaphors seems to lag slightly behind their ability to make "normative" cross-modality matches (matches like those made by adults). This is precisely a result one would expect from the developmental hypothesis that perception is primary and language is secondary in development—that synesthetic metaphors originate in and derive from cross-modal perception.

PERCEPTION, LANGUAGE, AND CROSS-MODAL INFORMATION PROCESSING

At this point, the evidence from cross-modal perceptions and synesthetic metaphor seems to support strongly the straightforward developmental hypothesis, to wit, that perception precedes language and that meanings derive from discriminable perceptual characteristics of objects and events; at the same time, the evidence would seem to contravene the Whorfian hypothesis that language itself shapes perception. But this is perhaps to dismiss the Whorfian view too quickly. For to hold a Whorfian view is surely not to claim that language is temporally prior to perception. Rather, it is to claim that as language is learned its significance waxes; and once language has been fully learned and incorporated into one's cognitive system, then language will have the power to shape, indeed language may even come to dominate, perception. To test the Whorfian hypothesis more fully, perhaps it is necessary to look at behaviors in which the linguistic encoding of perceptual stimuli occurs rapidly and automatically.

One of the most sensitive measures of human information processing is speed of response. For one, it is a common assumption in modern cognitive psychology that, everything else remaining constant, complex psychological processes will require more time to complete than will simpler psychological processes. Moreover, differences in response times may reflect differences in "access time" across stimuli.

One of the best-known phenomena in cognitive psychology—the Stroop (1935) effect—takes such differences in response time as evidence of a psychological priority of language over perception. Stroop found, and others have repeatedly confirmed, how language can interfere with perceptual identifications: If a person is asked to name the color in which a word is printed, but the word itself is the name of a different color, the competing color name substantially interferes with the identification of the color percept. In a nutshell, it is hard to name the color of green ink if it spells the word "red." The effect seems to be asymmetric, in that color percept has little effect on naming words. Although the interpretation of Stroop effects is complicated by the need to consider the form of the response and the compatibility of the perceptual and response codes, Stroop interference has been demonstrated even with "neutral" responses such as keypresses (e.g., Melara and Marks, 1990). One possible implication is that processes tapping well-learned verbal codes, such as the names of colors, can come to such cognitive prominence as to overrule, overpower, or simply outrun supposedly lower-level perceptual processes.

Recent experiments have applied reaction-time procedures to the study of cross-modal processing of perceptual and verbal stimuli, and the results of these experiments cast an intriguing new light on the relation between perception and language. Consider first the very possibility of cross-modal interactions in response speed during tasks of perceptual discrimination. In particular, consider a relatively simple task, such as the discrimination between a dim light and a bright light: To each stimulus the subject presses key A if the light is dim and key B if it is bright. Now add an auxiliary stimulus in a second modality, say a tone. On each trial, independent of whether the light is dim or bright, and thus informationally irrelevant to the response, the concomitant tone can be either soft or loud. How do the tone and light interact? Note that there are four types of trial: dim + soft, dim + loud, bright + soft, and bright + loud. Now, according to the rule of cross-modal similarity, dim lights resemble soft sounds and bright lights resemble loud sounds (leaving in abeyance for now the question of whether this similarity is primarily perceptual or verbal). Thus the four types of trial group themselves into two categories: congruent stimulus pairs (dim + soft and bright + loud) and incongruent pairs (dim + loud and bright + soft).

When subjects are required to respond as quickly as possible to the light, even though the sounds are informationally irrelevant, the sounds substantially affect speed and accuracy of processing (Bernstein and Edelstein, 1971; Marks, 1987; Melara and O'Brien, 1987; Melara, 1989). In particular, congruent pairs of stimuli are processed significantly more effectively than are incongruent pairs of stimuli. Figures 6.8 and 6.9 give an example. When the light was dim, the response was faster and errors were fewer when the sound was soft rather than loud; but when the light was bright the reverse was true, in that the response was faster and errors were fewer when the sound was loud rather than soft (Marks, 1987). In other words, congruent pairs of stimuli (loud + bright or dim + soft) yielded more efficient processing than did noncongruent pairs (loud + dim or soft + bright). Exactly the same pattern has emerged for interactions between pitch and brightness (Marks, 1987), between pitch and shape (Marks, 1987), and

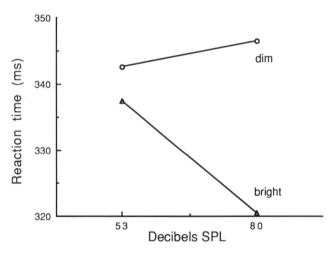

FIG. 6.8 Average reaction times, in a task of speeded discrimination, to soft (53 dB) and loud (80 dB) tones accompanied by informationally irrelevant dim (160 cd/m²) and bright (320 cd/m²) lights. *(Data from Marks, 1987.)*

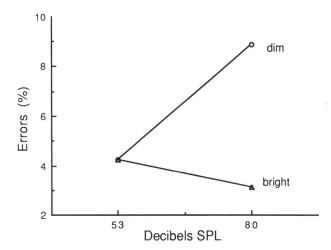

FIG. 6.9 Average error rates, in a task of speeded discrimination, to soft (53 dB) and loud (80 dB) tones accompanied by informationally irrelevant dim (160 cd/m²) and bright (320 cd/m²) lights. *(Data from Marks, 1987.)*

between pitch and spatial location (Bernstein and Edelstein, 1971). Indeed, the general rule seems to be that wherever there are cross-modal similarities in perceptual matching, so too are there analogous interactions in speeded perceptual discrimination.

Of course, these results say nothing about the locus of the cross-modal interactions. The interactions could be perceptual in nature. Or they could be verbal. The interactions could take place at some locus in perceptual processing where inputs from different sensory modalities converge. Or they could take place after the perceptual stimuli are recoded into a semantic form. That the interactions may be verbal is more than just possible—as suggested by the fact that analogous cross-modal interactions take place when one stimulus is perceptual (a tone of low or high pitch) and the other is verbal (a word such as "dim" or "bright") (Walker and Smith, 1984). In fact, cross-modal interactions involving verbal stimuli tend if anything to be greater in magnitude than interactions involving perceptual stimuli.

The results of a wide range of experiments suggest that congruity effects between different perceptual modalities may very well arise from semantic coding of stimuli rather than from direct perceptual interactions (Melara and O'Brien, 1987; Melara, 1989; Melara and Marks, 1990). Perhaps the strongest evidence in support of this contention comes from an unpublished study in which the opportunities for perceptual and verbal interactions were compared directly.

In this study, the subjects' task was rather complicated. On each trial the subject saw a word, either "WHITE" or "BLACK," and the word could be printed in either a white color or black color against the gray background. Thus there were four possible visual stimuli—"WHITE"-white, "WHITE"-black, "BLACK"-white, and "BLACK"-black. In addition, on each trial the subject heard a tone, which could be either low or high in pitch. The tone was informationally unrelated to the visual stimulus and thus was irrelevant to the response, which the subjects had to base on the visual stimulus alone. Specifically, the task was to press key A if the color and word were congruent (color white with word "WHITE" or color black with word "BLACK") and to press key B if the color and word were incongruent (white with "BLACK" or black with "WHITE").

The main question I asked was whether the pitch of the sound would interact with the color, with the word, or with both. The results were clear and un-equivocal. When the tone was low in pitch, response speed was reliably greater when the word was "BLACK" rather than "WHITE"; when the tone was high in pitch speed was greater when the word was "WHITE" rather than "BLACK." Color per se—the perceptual attribute of white or black—did not interact with sound. In other words, the results strongly support the view that certain cross-modal interactions—namely those of the present task—take place largely or exclusively at a verbal level, and not at all at a perceptual level, a finding that sits well with the Whorfian view.

To extend the argument one step further, it is conceivable that even in processing intensity, quality, and other simple attributes of sensory stimuli, which is to say nonlinguistic stimuli, people recode the attributes semantically or lexically, and it is these derivative, linguistic codes that interact. How may we account for this tyranny of metaphorical words? An in particular, how may we account for this evidence of linguistic interaction in the face of the clear developmental evidence for the primacy of cross-modal perception over cross-modal language?

To reconcile the two kinds of finding it seems necessary to synthesize the developmental and the Whorfian hypotheses. There is little reason to doubt that at least a few cross-modal similarities, similarities between attributes of experience in different modalities, are indeed fundamentally perceptual in origin, if not always in their very nature. Cross-modal similarities are evident in early childhood—indeed, some of them even appear in infancy, prior to any acquisition of language. Especially pertinent is the study of Lewkowicz and Turkewitz (1980); using deceleration in heart rate as a measure of habituation to repeated perceptual stimulation, they found evidence for a direct, auditory-visual "matching"—a kind of cross-modal generalization—between loudness and brightness in 3-week-old infants. And Wagner et al. (1981) found that one-year-olds "matched" upwardly pointing visual stimuli and tones that increased in pitch. It is difficult to conceive of a verbal basis to these kinds of cross-modal similarity, which seem to be quintessentially perceptual in nature, appearing well before language is learned.

But if perception is the origin, the starting point, of many cross-modal similarities, nevertheless language may well be the primary end station. Once we have learned a language, so that verbal labeling has become a virtually automatic response to many kinds of perceptual stimulation, it is likely that much of our processing of information—even information that initially is nonlinguistic in nature—goes on after verbal recoding. (With the possibility of intriguing exceptions: Some percepts, especially olfactory ones, may never develop good linguistic labels!) Through language, we encode one sound as "soft" and another as "loud," one color as "dim" and another as "bright." Correspondences between "soft" and "dim" and between "loud" and "bright" subsequently may depend at least partly on their semantic codes, and therefore on such semantic markers as intensity$^-$ or intensity$^+$, rather than on the perceptual values of the original stimuli.

In the long run, it is necessary to recognize that both perception and language represent cognitive systems; both are systems for understanding the world, systems that are functionally and structurally related, yet not identical (see Miller and Johnson-Laird, 1976).

What, finally, does all of this say about cross-modal perception, synesthetic metaphor, and the unity of the senses? In a nutshell, it suggests that the unity of the senses presents itself in two distinct ways: at a primary level, as an original, primitive set of perceptually given similarities among dimensions of experience in different sensory modalities, and at a secondary level, as a unity engendered by language, a unity related to and partly but perhaps not wholly derived from the temporally prior level, most significantly as a cognitive unity now substantially freed from the constraints and limitations imposed by its perceptual progenitor.

ACKNOWLEDGMENT

Preparation of this chapter was supported by NIH Grant DC 00271.

REFERENCES

Bernstein, I. H. and Edelstein, B. A. 1971. Effects of some variations in auditory input upon visual choice reaction time. *J. Exp. Psych.* 87: 241–247.

Hartshorne, C. 1934. *The Philosophy and Psychology of Sensation.* University of Chicago Press, Chicago.

Hornbostel, E. M. von 1925. Die Einheit der Sinne. *Melos, Zeitschr. Musik* 4: 290–297.

Hornbostel, E. M. von 1931. Über Geruchshelligheit. *Pflüg. Arch. Ges. Physiol.* 227: 517–538.

Huysmans, J.-K. 1884. *À rebours.* Charpentier, Paris.

Karwoski, T. F., Odbert, H. S., and Osgood, C. E. 1942. Studies in synesthetic thinking. II. The role of form in visual responses to music. *J. Gen. Psych.* 26: 199–222.

Lewkowicz, D. J. and Turkewitz, G. 1980. Cross-modal equivalence in early infancy: Auditory-visual intensity matching. *Devel. Psych.* 16: 597–607.

Marks, L. E. 1974. On associations of light and sound: The mediation of brightness, pitch, and loudness. *Am. J. Psych.* 87: 173–188.

Marks, L. E. 1975. On colored-hearing synesthesia: Cross-modal translations of sensory dimensions. *Psych. Bull.* 82: 303–331.

Marks, L. E. 1978. *The Unity of the Senses: Interrelations Among the Modalities.* Academic Press, New York.

Marks, L. E. 1982. Bright sneezes and dark coughs, loud sunlight and soft moonlight. *J. Exp. Psych.: Hum. Percept. Perf.* 8: 177–193.

Marks, L. E. 1987. On cross-modal similarity: Auditory-visual interactions in speeded discrimination. *J. Exp. Psych.: Hum. Percept. Perf.* 13: 384–394.

Marks, L. E. and Bornstein, M. H. 1987. Sensory similarities: Classes, characteristics, and cognitive consequences. *Cognition and Symbolic Structures: The Psychology of Metaphoric Transformation* (R. E. Haskell, Ed.), pp. 49–65. Ablex, Norwood, NJ.

Marks, L. E., Hammeal, R. J., and Bornstein, M. H. 1987. Perceiving similarity and comprehending metaphor. *Monogr. Soc. Res. Child Devel.* 52 (Serial No. 215).

Melara, R. D. 1989. Dimensional interaction between color and pitch. *J. Exp. Psych.: Hum. Percept. Perf.* 15: 69–79.

Melara, R. D. and Marks, L. E. 1990. Processes underlying dimensional interactions: Correspondences between linguistic and nonlinguistic dimensions. *Mem. Cogn.* 18: 477–495.

Melara, R. D. and O'Brien, T. P. 1987. Interaction between synesthetically corresponding dimensions. *J. Exp. Psych.: Gen.* 116: 323–336.

Miller, G. A. and Johnson-Laird, P. N. 1976. *Language and Perception.* Harvard University Press, Cambridge, MA.

Riggs, L. A. and Karwoski, T. 1934. Synaesthesia. *Br. J. Psych.* 25: 29–41.

Stroop, J. R. 1935. Studies of interference in serial verbal reaction. *J. Exp. Psych.* 18: 643–662.

Wagner, S., Winner, E., Ciccetti, D., and Gardner, H. 1981. "Metaphorical" mapping in human infants. *Child Devel.* 52: 728–731.

Walker, P. and Smith, S. 1984. Stroop interference based on the synaesthetic qualities of auditory pitch. *Percept.* 13: 75–81.

Werner, H. 1934. L'Unité des sens. *J. Psych. Norm. Pathol.* 31: 190–205.

Whorf, B. L. 1956. *Language, Thought, and Reality.* MIT Press, Cambridge, MA.

Williams, J. W. 1976. Synaesthetic adjectives: A possible law of semantic change. *Language* 52: 461–478.

7

Neural Networks in Sensory Perception

Peter M. Smith and L. Greg Walter

S. C. Johnson & Son, Inc.
Racine, Wisconsin

INTRODUCTION

Neural networks (NN) are a new set of computer paradigms of immense importance, interest, and potential. A paradigm is defined as a model or methodology inspiring a worldview. The NN worldview is based on the computer as a set of highly interconnected simple units with local memories. This is the reverse of traditional computers which are built from complex units working in series and accessing a common, remote memory. The single pipeline between main memory and the central processing unit is known as the von Neuman bottleneck after the father of modern computers, and for almost 50 years this has been the dominant computer architecture. Many attempts, usually classified as parallel processing, have been made to get round this bottleneck. Neural networks are often known as "massively parallel systems."

Neural networks were developed because of interest in computer models simulating the functions of learning and knowledge retention. Our brains are made up of 5–10 billion neurons, each with inputs (dendrites) and one output (the axon, which often divides into numerous endings). These form a complex network. The function of over 1 trillion synaptic junctions between axonal outputs and their respective dendritic receptions is quite well understood (Stubbs, 1988), although they are complex physiochemical sites. Needless to say, the brain is a little more complex than any artificial NN. The initial, early results from NN studies show that these systems have many of the characteristics of the brain—learning, adaptation, self-organization (Carpenter, 1989). NN can identify patterns from incomplete, noisy data such as recognizing a familiar face in a picture. So even if the best current NN have fewer capabilities by 3–4 orders of magnitude than the brain of a leech, which is not a very sophisticated mechanism, the results with rudimentary NN are enough to generate a great deal of excitement. The NN field draws from neurobiology, psychology, computer science, and mathematics to synthesize a potent problem-solving tool that has become independent of its biological roots.

Leon Cooper, Nobel Laureate, is quoted as saying "Neural Networks will be the computers of the next century." Perhaps: but we must be aware of the danger of overpromotion which can damage progress and limit funding.

HISTORICAL PERSPECTIVE

Two streams of activity converged in the mid-1980s to make NN a viable technology. One stream was the theoretical analyses of neuroscientists who were pondering biological networks ("wetware"). The second was the tremendous developments of computer chip technology which allowed powerful simulations to be readily performed on small machines—it became practical to build interconnected networks. This development was based on smaller and more dense linear processors, and as the physical limits of silicon etching became apparent, it was realized that to obtain the next order-of-magnitude improvement in power, chips would have to be developed with parallel architectures. This created a strong interest in the logic of parallel networking and brought neural networks into the limelight after 30 years of steady development. Many of the seminal papers in this field have been collected in a fine volume (Anderson and Rosenfeld, 1988). Two new journals, *Neural Computation* (MIT Press) and *Neural Networks* (Pergamon Press), as well as proceedings from frequent conferences, will continue to provide information about current developments.

Perhaps the main spark for the current explosion of interest came in 1982 with a paper by Hopfield, a noted physicist, who provided a clear and powerful link between the physics of spin-states and memories. For example a three-dimensional spin-glass (such as a magnet) has mathematical similarities to interconnected networks and biological systems (Hopfield, 1982). This paper inspired the development of physical chips to implement his networks, known as bidirectional associative memories, although their vast storage requirements have limited their realization.

Kohonen was working around this time to develop a network that was self-organizing and could successfully categorize patterns. This would become a major NN paradigm (Kohonen, 1982). In 1986 came the most-quoted paper in the NN literature (Rumelhart et al., 1986) describing the back-propagation network (BPN), although the basic algorithm had been presented by earlier workers. Hecht-Nielsen had also been working in this field and in 1986 started the Hecht-Nielsen Corporation, which developed a PC-coprocessor board for NN simulation (Hecht-Nielsen, 1988). HNC has provided both theoretical and practical consulting and done much to stimulate practial application of the technology. Since these dates there has been an exponential rise in interest in NN.

It is interesting to note that there have been few published reports of successful and practical implementation of NN systems because much of the work is proprietary and confidential. Nobody wants to reveal their successes, and of course nobody wants to discuss their failures. The subject is in its infancy. One recent paper (LeCun et al., 1989) gives the results for a large real-world application—recognition of hand-written zip codes for the U.S. Postal Service—using a five-layer BPN in which the first two middle layers were constrained to facilitate the digital processing of the image. Fast learning, good responses, and a throughput of more than 10 digits a second were obtained with this network, showing that the BPN can have excellent practical applications.

There is no real NN chip yet—the building of a true general purpose parallel architecture chip has not been easy. Such chips are 2–3 years away, although some specialized chips are available. The current interest in NN involves simulations on traditional computers. If we have come so far with these simulations, the future is indeed bright when the real tool is available. If parallel architectures show the same growth curves as traditional chips, perhaps in 10 years or so we will have NN which are as clever as the brain of a leech.

There are numerous NN paradigms (see Table 7.1), and it is not difficult to build complex networks of units with heavy mathematical support. Only a few have any real practical importance, with the BPN the clear leader. This paper will concentrate on this network, but before this, it is important to discuss the main characteristics of these fascinating systems.

CHARACTERISTICS OF NEURAL NETS

Neural Nets respond to input data, and the response is predicted by the state of the network, not by any one element of it. This state is dependent not only on the

TABLE 7.1   Main NN Paradigms[a]

Network	Examples	Characteristics	Comments
Crossbar	HOPFIELD BIDIRECT ASSOCIATIVE (BAM)	Fully interconnected networks, with large capacity requirements to store useful information	Awaits development of high-speed large density computing matrices (e.g., optical computers)
Adaptive Filters	ADELINE MADELINE	Pattern classification using a signal filtering process	Limited capacities but for simple pattern mixes, these networks are efficient and easy to use
Backprops (BPN)	STANDARD RECURSIVE	Feedforward network (input mapped to output only) via delta rule error reduction. Simple and powerful—the most popular networks in use today.	Learning times can be long. Recent variations show considerable promise for improved efficiency.
Restricted coulomb energy	NESTOR proprietary	Fast learning, real-time adaptive, good accuracy.	Applies to similar problems as the BPN. Available via the NESTOR development system only.
Self-organizing	KOHONEN COUNTER PROPAGATION	Feedforward only, with self-organized categorization of the input patterns. The Counterprop network also feeds backwards from "results" back to "input"	Used for statistical categorization of patterns. Useful functional mapping.
Adaptive resonance	ART1 ART2	Real-time learning	Highly complex networks for sophisticated applications such as robotics. Not generally applied to date.

[a]The above are the main classes of NN paradigms. There are several other architectures; however, these are the most widely used in application development.

"values" or "activity level" of the processing elements (PEs or cells) but on the values of the interconnecting links, the weights. A key element of many NN paradigms is that the output response of any one cell is determined by the weighted sum of its inputs, and this output becomes the input for other cells. The state of the network as a whole is nondeterministic because the output is the sum of numerous interactions. The dynamics and responsiveness of the network do, however, provide sensible memory and outputs when the underlying mechanics work by certain established rules.

For example, Fig. 7.1 shows a network of simple cells, with only a few of the possible links drawn. Say that the *response* you are looking for is given by the state of cell X. The *input* stimulates cell A. Now this stimulation will cause excitation of cells B, C, and any others to which A is linked, and the strength of the link, the weights W_{ab}, W_{ac} etc., will be involved. Eventually a stimulus will reach X via all its incoming links. But who could predict what that stimulus could be, or even if the mathematical form of it were tractable, what meaning would it have? It's simpler to consider the network a heuristic system and concentrate on making its responses meaningful.

If cell D became nonfunctional, would this have much effect on the response? Probably very little, because the contribution to the whole is small. This means that NN systems, unlike von Neuman computers, do not work by a sequence of precisely defined steps—they are nonalgorithmic. They tolerate faults and give answers which may not be 100% correct. They are nonlogical, they "intuit"—a capability denied to traditional computers which are rigorously programmed. The negative side of this is that NN cannot count. So while I might like a NN to

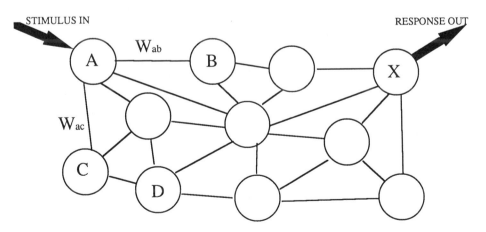

FIG. 7.1 Unstructured network-A network of cells showing possible links between stimulus (A) and response out (X). W_{ab}, W_{ac} symbolize the strength of the links between cells.

assess my loan application, I'd rather it didn't keep track of my bank balances. The point here is that NN systems are not really computers and will not replace, but rather complement, traditional computing.

NEURAL NETS AND EXPERT SYSTEMS

We should draw some comparisons between NN and expert systems since these are both branches of artifcial intelligence. They are compared in Table 7.2. Both are tools of the knowledge engineer and complement each other.

An expert system attempts to capture the knowledge and experience of an expert in a computer, usually just before the expert retires. Working with a programmer and an interviewer ("knowledge engineer") over several months, the expert describes the procedures and decision points of his work, and these are codified into a set of rules via a software "knowledge shell." When complete this system should then follow the same logic and rules that the expert follows to make decisions. Expert systems are in widespread use for medical diagnoses, mechanical testing, fault analysis, credit ratings and so on. A good and evolving expert system can be of considerable assistance to novices, and it is generally agreed that there are systems reliable in their advice some 85% of the time, usually on routine cases, leaving a comfortable 15% of more difficult cases to human judgment. Characteristic of an expert system is its ability to trace the logic followed to the answer, giving useful insights and allowing continual refinement of the rule base. Unlike "if-then-else" pathways, the power of good knowledge shells also allows some means of jumping between associations and rules so that the answers are not rigorously predicted.

Yet in many situations extracting the rules necessary to build the expert system is too difficult due to the many variables and dependencies in systems. A NN can offer a powerful alternate tool to capture knowledge. Preparing the NN side is fast and relatively easy, and it will give an instant, if heuristic, answer. The negative side to this is that a NN cannot tell you why it gave a particular

TABLE 7.2 Neural Nets vs Expert Systems

Neural net	Expert System
Computing intensive	Labor intensive
Data flexible, fault tolerant	Data inflexible
Adaptive or self-organizing	Nonadaptive; new rules must be added
No trace-back	Full trace-back
Heuristic	Logical

response and there is no way of tracing back to find the important factors. If this becomes a critical need, and the rules are accessible, then there is much to be said for using both techniques in a combined Expert Net. Here the main response to the input is given by the NN, and if reasons are required, a traceback from result to input can be done via an expert system set of rules. This is an exciting new synergistic field of study.

THE BACK-PROPAGATION NETWORK

This has been by far the most successful NN system because it is simple to apply, easy to conceptualize, and it works! So despite its limitations it remains popular and has been a significant contribution to the development of the field.

The network shown in Fig. 7.1 is too dense and intractable to be much use, and it is necessary to impose some rules and structure to make it practical. There are clear structures in the networks of the brain also (Stubbs, 1988). One rule is this: the input signal to a cell is the sum of the weighted inputs from its linked cells,

$$I = \Sigma \; wi^*xi$$

but there is only one output from this cell, although it can branch to other cells. Another rule is that the transfer function between input to output should be nonlinear, usually sigmoidal, and has a threshold value. This means that the cell only produces output if it is suitably excited and that output is not linearly dependent on the imput.

For useful networks, learning must take place, that is, they must adapt to the world. Learning can be supervised as in a classroom where the answers are known, or unsupervised as in finding our way home where we keep trying until we find the right pattern. The brain is thought to learn by Hebbian rules, namely the strength of the synapse is increased with the excitation of the neuron—the more a network pattern is exercised, the more permanent the physical implementation of the network.

The BPN is an example of supervised learning, in which the learning rule is a mathematical equation known as the delta rule, or least mean squares rule, initially developed by Widrow-Hoff in 1960 (see Caudill, 1987), which minimizes errors between the known answer and the network's response. The BPN arranges its cells in at least three layers: input, middle, and output. In a training sequence, the output of the network is compared to known values, and the errors back-propagated to the middle and input layers to adjust the weights to minimize the error (Fig. 7.2). This is repeated many times until the error between the

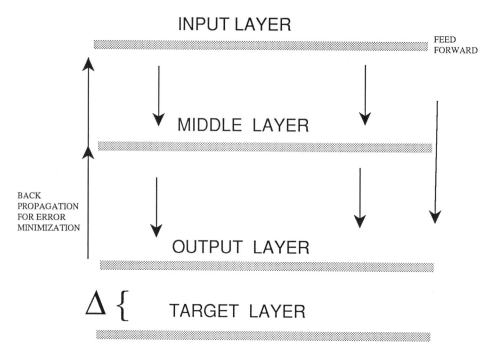

FIG. 7.2 Back-propagation structure. The error between the network output and the known target is back-propagated over repeated iterations (the learning process) to adjust the weights on the input and middle layers in order to minimize the error.

output and known values is sufficiently small. The network is then said to be trained. Now given any input pattern, its response will be a good prediction of the output. That response is immediate although it may not be 100% correct.

The BPN is essentially a mapping network, correlating one set of input vectors with another set of vectors (White, 1989, gives a full statistical analysis of mapping algorithms in connectionist models). It has powerful relevance to the real world of sensory evaluation because it provides a means of relating physical properties to human responses without knowing any rules or reasons (Fig. 7.3). For example, in food characterization, the physical characteristics of a food as defined by a chromatogram can be related to a human response—taste—and a network trained which can predict taste from chromatograms. This will be revolutionary to the theorists who are trying to elucidate the rules, although in most cases of human response analyses these rules are highly obscure. For practical purposes we can bypass these questions and provide a good predictive system with a NN and as we learn we can build the knowledge system around it.

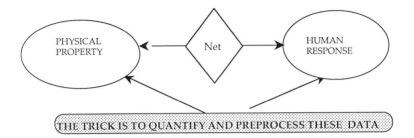

FIG. 7.3 A mapping network relates a physical characteristic, such as shape, to a human response, such as recognition.

APPLES AND BANANAS—AN EXAMPLE OF A BPN

The problem is this—how to distinguish between apples and bananas. If this sounds too easy, then we can make the problem more difficult by asking how we distinguish between Red Delicious apples and Macintosh apples. To the human brain neither of these distinctions is a problem. The fruits are distinguished by several sensory factors such as shape, color, fragrance, and taste, and by several physical factors such as texture, fat content, sugar content, and so on. Shape, color, and fragrance, in this order, are generally enough—a Red Delicious can usually be separated from a Macintosh apple because the general shape or visual pattern is distinctive.

If I now want to build a traditional computer system which will distinguish between apples and bananas, this is not so easy. Working with the shape as the main distinguishing feature, we would have to digitize the image of the fruit and then compare in sequence the digits one by one with a stored sequence, and then decide "if-then-else" enough times to pick an apple or banana. This would be slow, but for apples and bananas it might work well enough. Between apples and apples or oranges and grapefruits, it would hardly be reliable since these fruits vary individually in size and shape. This is an ideal application for a BPN. The digitized image of numerous apples would be presented to the network with the output defined "Red Delicious-Grade 1" or "MacIntosh" and this repeated as many times as necessary to minimize the error response of the network. Thereafter the network will be able to distinguish the type of apple from its digitized pattern whatever the size or shape of the sample at least as well as a human (Fig. 7.4). We would have built a real practical sorting system, and indeed such systems have been implemented. Throw in a banana and the network will respond that it cannot recognize it; include the banana in the training set and the network will tell you it's a banana.

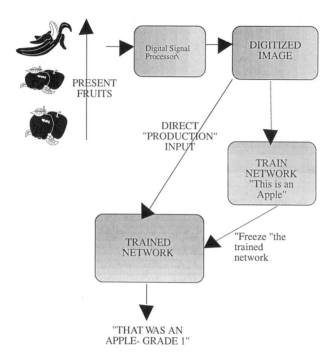

FIG. 7.4 Network schematic. This network could be trained to sort apples from bananas, or even grade apple varieties, a more difficult task.

Neural nets have been applied in the same way to the problems of image and speech recognition with remarkable results—problems far beyond the scope of sequential computers.

THE BPN ANALYZED

The ease of use of the BPN and the successful application to many elusive problems should not lead us into a wholesale acceptance of this solution. It has several drawbacks:

Slow learning rates for large input vectors
Unknown must be in the same class as the training examples
It is hard to "unlearn" input patterns and adapt to change

It is prone to "local minima" convergence, that is, the network does not properly learn from its input patterns and gets stuck with learned patterns which are inaccurate

There are several ways around these issues, and many enhancements to the basic BPN algorithm. For example, *weighted products* instead of *weighted sums* have been proposed for driving part of the input to cells (Durbin and Rumelhart, 1989). Momentum terms and learning rate factors can be included in the mathematical equations to obviate local minima. While training may be slow, response to new inputs once the network is frozen is almost real-time and can be practical with various boards added to computers and designed to accelerate the responses of networks.

The question is often raised: If the BPN is used for pattern recognition and mapping, what advantage does it have over traditional statistical analyses of these problems which are well advanced and powerful? If we are applying a least mean squares rule, what gives BPN any advantage over other algorithms (White, 1989)? In a sense there is nothing "neural" about BPN—it is simply another statistical technique for nonlinear multidimensional curve fitting. This is a good question, and there is a good answer to it. Pineda has shown that traditional techniques for N factors require at best N^2 operations to calculate the gradient descent (Pineda, 1989). The mathematics of neural networks and the BPN in particular allow the gradient descent to depend on N only. BPN works well for large collections of input data because the work involved in solving it is linearly dependent on the input vector, whereas traditional techniques of curve fitting with a quadratic dependency would be intractable. The whole explosion of interest in neural nets balances on this simple fact.

COIN COUNTING—THE BPN IN ACTION

In our studies of BPN mechanics, we have devised a simple small experimental system for coin counting. The idea is to present the network with a mixture of quarters, dimes, nickels, and pennies, and let it predict the resulting total. Since this total is known exactly, a measure of the success of the network can be gained. At the same time a traditional polynominal fit for the data was computed and compared with the true value.

A training set of 20 combinations of coins and their corresponding totals was prepared and used to train a three layer ($4 \times 9 \times 1$) BPN. The input layer of four cells represents the quarters, dimes, nickels, and pennies data. The middle layer is set at $2n + 1 = 9$ cells. (The figure of $2n + 1$ for the number of elements of the middle layer of a BPN derives from a mathematical mapping theorem by

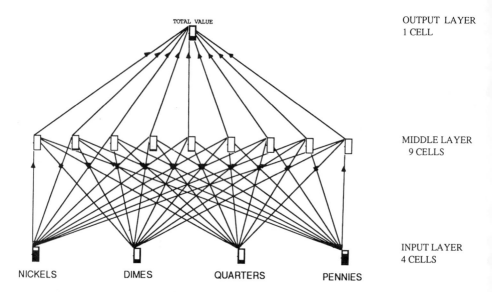

FIG. 7.5 Three-layer BPN for coin counting. This shows the actual interconnections between the cells, utilizing MACBRAIN software.

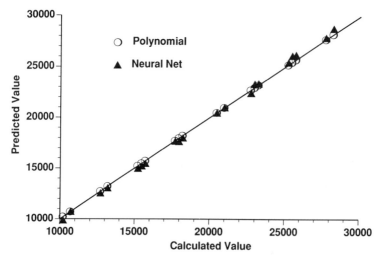

FIG. 7.6 Neural net vs. polynomial fit. These data have 0% error in the training set.

Kolmagorov, much advanced in NN theory but still open for review (Giroso and Poggio, 1989)). The single output cell represents the total value. A schematic of this network is shown in Fig. 7.5. After training, the network was presented with previously unseen coin combinations. It was able to give reasonably good values, although a second-order polynomial fit was more accurate (Fig. 7.6).

To test the effect of noise in measured data on the performance of NN and polynomial fits, two more sets of training data were created. In one, for each coin combination (input layer) in the training set, three values (output layer) were generated by randomly setting the value between −10 and 10% of its actual value. In the second, the error range was −30 to 30% of actual value. Neural nets were trained on these data sets and second-order polynominal fits were made. When these NNs were presented with new coin combinations, the valuations showed little or no deterioration from the 0% valuations while the accuracy of the polynominal fit degraded as "noise" increased (Fig. 7.7 and Table 7.3).

One interesting feature of the NN was the ability to reverse train—given one total, the appropriate compositions could be found with good accuracy. This "dissolution" would defeat traditional statistical approaches.

The results demonstrate some of the characteristics of a NN, such as its ability to learn, to see through noise, and its inability to count accurately. We note that for small "toy" datasets (small N, see above) there is not much difference in results by using traditional statistics, or the BPN. For these small amounts of

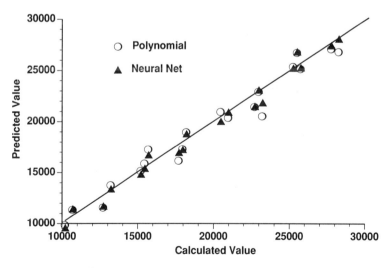

FIG. 7.7 Neural net vs. polynomial fit. These data have 30% error in the training set.

TABLE 7.3 Sample of Coin Data, 0% and 30% Error

	Pennies	Nickels	Dimes	Quarters	Value	Polynomial	Corr. Coef.	Neural Net	Corr. Coef.
0% error	750	750	250	250	13250	13250	1	13079	0.999
	500	500	1000	500	25500	25500		26158	
30% error	750	250	250	750	17204 23250 26153	20599	0.989	21935	0.995
	750	500	0	500	13591 15500 18879	15913		15475	

data, there was no practical speed difference in the two methods, so if you already have a good traditional solution there is no need to pursue another. We were happily encouraged by the predictive power of the BPN in both directions and also its immunity to noise.

SUMMARY

The above experiment in coin counting shows that NNs do work and are robust and easy to apply. Their limitations do not diminish their applicability to many problems that defeat traditional computers in the area of pattern recognition. The example of apples and bananas show how simply and usefully they may work. We look forward to more exciting developments in this field, and in particular to the availability of the first true NN chip which will build in silicon the paradigms we have discussed, and make real-time, large dataset handling practical.

REFERENCES

Anderson, J. A. and Rosenfeld, E. 1988. *Neurocomputing, Foundations of Research*. MIT Press, Cambridge, MA.

Carpenter, G. A. 1989. Neural network models for pattern recognition and associative memory. *Neural Networks* 2: 243–257.

Caudill, M. 1987. Neural Networks Primer, Part I. *AI Expert Magazine* 12: 46–52.

Caudill, M. 1988a. Neural Networks Primer, Part II. *AI Expert Magazine* 2: 55–61.

Caudill, M. 1988b. Neural Networks Primer, Part III. *AI Expert Magazine* 6: 53–59.

Durbin, R. and Rumelhart, D. E. 1989. Product units: A computationally powerful and biologically plausible extension to backpropagation networks. *Neural Computation* 1: 133–142.

Giroso, F. and Poggio, T. 1989. Representative properties of networks: Kolmagorov's theorem is irrelevant. *Neural Computation* 1: 465–469.

Hecht-Nielsen, R. 1988. Neurocomputing: Picking the human brain. *IEEE Spectrum* (March): 36–41.

Hopfield, J. J. 1982. Neural networks and physical systems with emergent collective computational abilities. *Proc. Nat. Acad. Sci.* 79: 2554–2558.

Kohonen, T. 1982. Self-organized formation of topologically correct feature maps. *Biological Cybernetics* 43: 59–69.

LeCun, Y., Boser, B., Denker, J. S., Henderson, D., Howard, R. E., Hubbard, W., and Jackel, L. D. 1989. Backpropagation applied to handwritten zip code recognition. *Neural Computation* 1: 541–551.

Pineda, F. J. 1989. Recurrent backpropagation and dynamical approach to adaptive neural computation. *Neural Computation* 1: 161–172.

Rumelhart, D. E., Hinton, G. E., and Williams, R. J. 1986. Learning internal representations by error propagation. In *Parallel Distributed Processing,* Vol. 1, (D. E. Rummelhart and J. L. McClelland, Ed.), pp. 318–326.

Stubbs, D. F. 1988. Neurocomputing. *MD Computing* (June): 1–12.

White, H. 1989. Learning in artificial neural networks: A statistical perspective. *Neural Computation* 1: 425–464.

8

Descriptive Analysis and Concept Alignment

Michael O'Mahony

University of California
Davis, California

CONCEPT ALIGNMENT IS NECESSARY FOR DESCRIPTIVE ANALYSIS

The descriptive analysis of flavor is a widely used technique in the sensory analysis of foods. There are several standard techniques available (Cairncross and Sjöström, 1950; Caul, 1957; Amerine et al., 1965; Stone et al., 1974; Meilgaard et al., 1987). These use a system whereby particular sensory characteristics of a food are identified and defined for trained panelists, using physical standard stimuli. Each characteristic is given an agreed name or label for purposes of communication. In this way, an ad hoc language is created for communication about the sensory characteristics of the particular food product. Various scaling techniques are used to measure the strengths of these characteristics.

If a group of untrained judges were to list the sensory characteristics of a food without any such training or standardization, there would be little agreement among the judges. First, untrained judges do not have a common language with which to communicate sensations; a given sensation would be given different descriptive labels by different judges. This is rectified in training by agreeing on a set of descriptive labels. The second reason poses greater problems; untrained judges do not have an agreed-upon system of categorization. For example, a sensation that may be categorized as "salty" by one judge may be categorized as something other than "salty" by another. This is similar to the phenomenon of a reddish-orange color being categorized as orange by one judge and red by another. In terms of those who study the information processing that takes place in the brain for such categorizations (Miller and Johnson-Laird, 1976), the sensory concept of saltiness for one judge is not exactly aligned with that of the other judge. A sensation may fall within one judge's concept of saltiness, while for another judge with a slightly different saltiness concept, the sensation may fall outside his concept. Obviously this could lead to a great deal of mis-communication. Judges may disagree about the presence or absence of a given sensory characteristic of food, not because one judge detected it and the other did not, but because the judges had a different set of concepts. This would be analogous to using a set of instruments that had not been calibrated.

For a panel of judges to be used as an analytical tool for descriptive analysis, the judges will need to agree on which sensations should and which should not be included within a given concept. There must be an alignment of that concept among the judges. For proper communication among panelists, they will need to have all of their required sensory concepts aligned with those of the other judges. Thus, the training of judges is essentially a process whereby judges have their concepts aligned. The lengthy procedures of tasting, discussion, presentation of standards and practice in identifying various sensory characteristics boil down to techniques to bring about concept alignment and the establishment of an agreed descriptive label for each concept.

Any process of alignment of judges' conceptual structures will involve the modification, splitting, or joining of existing concepts, and the creation of new concepts. Because of this, it is as well to study the mechanisms of sensory concept formation. The better this is understood, the more efficient the process of training for descriptive analysis can be made. Progress is beginning to be made in this area (Civille and Lawless, 1986).

MECHANISMS OF SENSORY CONCEPT FORMATION

In this examination of the formation of sensory concepts, let us, for simplicity, first consider a color concept. Let us choose "red." Red is not a single color. A whole set of quite distinguishable colors are given the label "red"; we call them the various shades of red. For convenience, we organize colors into groupings of similar hue under a common name or label. Thus, there are many colors that we call "red." There are many that we call "green" or "blue" or "brown" or whatever. Other cultures may group colors in different ways. The choice of stereotypes for given colors and the way in which colors are grouped and organized is not purely arbitrary. It remains an area of active research (Miller and Johnson-Laird, 1976). Suffice to say that a given person learns within his culture to attribute the label "red" to a particular set of colors. The person has an idea or concept of what redness is supposed to be and if a color falls within his concept, he attributes to it the label "red." It has redness.

How is such a concept formed? According to current theory of sensory concept formation (Hull, 1920; Pikas, 1966; Nelson, 1974; Miller and Johnson-Laird, 1976; Millward, 1980), the formation of a sensory concept involves two processes: abstraction and generalization. For the first process, a concept like redness is first abstracted from a set of red and nonred objects. A child learning to name colors learns that a certain grouping of hues is given the label "red"; others are given other labels. Gradually he forms an abstract concept of redness and learns to assign the label "red" to those colors which fall within the concept. This process should not be underestimated; it is an extremely complex piece of information processing (Miller and Johnson-Laird, 1976). At the time of writing, there is no computer program written for the formation of an abstract sensory concept like redness; it is too complex. Furthermore, children learn to name colors, forming the appropriate color concepts, rather later than would be expected from their other cognitive development. Some cultures never manage to form abstract color concepts; they might say that the sky is like blood or fire but they could not say that the sky, fire, and blood have redness. Yet abstraction of a concept for an adult in western culture is so simple as to be at the level of a party game. Presumably this is because by the time he reaches adulthood, he has already developed the software in his brain for this task.

The second part, generalization, refers to the fact that the concept is broadened beyond the stimuli from which it was extracted. If this generalization step did not take place, a person would never be able to label a new shade of red that he had not seen before as "red." This idea of abstraction and generalization can now be applied to sensory concepts other than colors, such as "sweet," "salty," "sour," "crunchy," "crumbly," "fruity," etc.

Having understood that concepts are formed by a process of abstraction and generalization, it is now possible to examine the techniques used in descriptive analysis protocols for eliciting and controlling concept formation. Do they form appropriate concepts for the particular descriptive analysis at hand? Are the concepts aligned among the judges? For this, the concepts will have to be measured.

HOW TO MEASURE A SENSORY CONCEPT

A sensory concept is an abstract entity and as such cannot be measured. At the time of writing, too little is known about brain function to be able to indicate exactly what physiological or cell biological processes take place when a concept is being formed. Yet, the behavioral output that is the result of decisions made as a consequence of a given concept is susceptible to measurement. A judge can tell which stimuli give sensations that fall within or outside his or her concept. Accordingly, "concept" measurement involves judges reporting whether the sensations elicited by various stimuli fall inside or outside a given concept. Alternatively, it can involve the judge having to sort a set of stimuli into their various conceptual groupings or categories. Both procedures will give evidence of the judge's conceptual structure.

The Problem of Response Bias

The measurement of a sensory concept for a single judge involves the judge examining a set of stimuli and categorizing them as to whether they elicit sensations that fall within or outside a given concept. However, there is a measurement problem here; it is the problem of response bias.

Response bias can best be understood by considering sensory difference testing. It is the vital cognitive problem that must be solved for all sensory difference tests. Therefore, it is worthwhile deviating from the main discourse on concept measurement to understand the appropriate aspects of response bias and criterion variation and how they relate to difference testing. We will then return to concept measurement.

The notion of response bias in a judge caused by criterion variation is embedded in signal detection theory (Green and Swets, 1966). Generally, when a judge is comparing two food stimuli in a sensory difference test, the foods will be very similar; they will be difficult to distinguish. If they were easily distinguishable, no formal difference testing would have been required. In fact, in the case where stimuli are easy to distinguish, criterion variation has an insignificant effect and the following argument does not apply.

The judge finds it hard to determine whether two almost identical stimuli are the same or different. The difference will be very slight, and it will be difficult for the judge to determine whether he is detecting a real difference or imagining it. Again, it is important to remember that this only applies to small differences. When the judge is uncertain and trying to answer the question as to whether he really detects a difference, a second question is implied in the judgment: How great a difference must there be before the judge believes the difference is not due to his imagination? This is not a sensory question, it is a cognitive question.

The judge's response will depend on his willingness to commit himself and say that the difference is real. In essence, the judge selects a given degree of difference as his criterion for saying that the difference is no longer imaginary and can be reported as real. Again, this criterion of difference is not a sensory function; it is cognitive. It is set arbitrarily and can vary. If the judge is biased towards detecting a difference, he will be more willing to say that two stimuli are different. He will have a more lax criterion of difference and commit himself to declaring that smaller differences constitute a real difference. On the other hand, should he be feeling cautious, he will be less willing to commit himself to saying that the two stimuli are different; he will have a stricter criterion and will need to experience a greater difference between the sensations elicited by the two foods before he will be willing to commit himself to declaring a difference.

The criterion can vary during the course of an experiment; the judge's opinion as to what is a real and what is an imaginary difference is not at all stable for such fine discriminations. Naturally, this variation must not be allowed to happen. It would mean that a judge could often unwittingly change his mind and report two stimuli as being different when shortly beforehand he had reported them as the same. The judge's ability to distinguish would not have changed, he would merely have changed his criterion. There are two main approaches for overcoming criterion variation (O'Mahony, 1990): signal detection measures and forced choice procedures. The former will be discussed first.

The strategy in signal detection measurement is to allow the criterion to vary and arrange for the judge to make the same judgments at several criterion levels ranging from strict to lax. There are various methods for doing, this but the most economical is to arrange for the judge to make judgments using several criteria simultaneously. In effect, it is like asking the judge whether he would report a difference if he were using a very strict criterion, whether he were using a fairly strict criterion, a fairly lax criterion, a very lax criterion, etc. The wording of the question would not be exactly as above. It translates into asking whether the judge would say whether he is sure that there is a difference, whether he would say there is perhaps a difference, whether the stimuli might perhaps be the same, whether he is sure the stimuli are the same, etc. If the judge reports he is sure there is a difference, he is using a strict criterion; if he reports that there is a difference but he is not sure, he is using a more lax criterion. Such a response scheme with responses graded in terms of sureness has the effect of forcing the judge to use several criteria simultaneously. The number of sureness levels chosen will depend on the situation. Two levels of sureness ("sure" vs. "not sure") giving four response categories ("different sure"; "different not sure"; "same not sure"; "same sure") are adequate for many purposes (O'Mahony, 1979b). Three levels of sureness ("sure"; "not sure"; "I don't know but I will guess") giving six response categories can also be straightforward for judges to handle (O'Mahony, 1972). It is desirable to have at least two levels of sureness (four response categories) (Brown, 1974), and there should not be a "don't know" category or else the less confident judges will not be forced to push themselves into differentiating the stimuli.

Having obtained a set of responses for different criterion levels for replicate sets of the two food stimuli, it is then possible to calculate an index of how sensitive the judge is to the difference. There are several indices that can be used: d' (pronounced "d-prime"), P(A) (Green and Swets, 1966), and the R-index (Brown, 1974). For those not familiar with psychophysics, it is worth briefly mentioning what these measures represent. It is beyond the scope of this article to explain signal detection theory in any detail, but any modern introductory text in experimental psychology should provide a brief introduction, and Green and Swets (1966) give a more thorough treatment.

Imagine that two foods are being sampled for difference testing; d' is the difference between the mean flavor strengths of the two foods. Each food will vary slightly in perceived flavor strength when tasted because of physiological factors inherent in the judge, like sensory adaptation (sensory "fatiguing"). The variation in perceived flavor strengths of the two foods can each be theoretically described by frequency distributions. If these frequency distributions are assumed to be normal, with equal variance, then d' is defined as the difference between the mean flavor strengths of the two foods, taking as the unit of measurement the standard deviation obtained for the frequency distributions. If the means of the two frequency distributions are 1.75 standard deviations apart, then $d' = 1.75$.

To the food scientist, this may seem an obscure measure, but it is commonly used in psychophysics. It is also used in the theoretical modeling of difference testing which has culminated in the work of Ennis (1990), where the frequency distributions used are multivariate, to reflect the multivariate nature of the sensory characteristics of food.

P(A) is another signal detection measure; it is exactly related to d' and can be understood as follows. One commonly used method of computing d' is to plot what is called a Receiver Operating Characteristic (ROC) curve (Green and Swets, 1966). This is a plot of the proportion of times a judge correctly identifies the stronger of the two stimuli in the difference test as being stronger (a "hit") versus the proportion of times he (incorrectly) identifies the weaker of the two stimuli as being stronger (a "false alarm"). A judge with a given criterion level will produce a given proportion of hits and false alarms and thus a single point on the graph. If the judge now assumes a stricter criterion level, he will be more cautious about reporting a stimulus as actually being stronger, and so his proportion of "hits" and "false alarms" will be reduced; this will provide a second point on the graph. Several criterion levels will produce several points on the graph. These are best obtained by using "sureness" judgments as described above. A curve drawn through these points is called an ROC curve; it is arch-shaped. The greater the ability of the judge to distinguish the difference between the two foods (the greater d'), the taller the arch and the greater the area below the curve. The proportion of area under the curve can therefore also be used as a measure of how well the judge can distinguish the two foods and is denoted by P(A). It may seem an obtuse measure, but it is common in psychophysics. From P(A), which is easy to measure from the plotted ROC curve, it is a simple matter to compute d'. The use of d' requires some statistical assumptions, and the exact shape of the ROC curve tells the experimenter whether these assumptions have been fulfilled or not. If they have not, then P(A), which requires no statistical assumptions, is generally used as an alternative to d'.

Again, it is beyond the scope of this article to delve into signal detection theory too deeply and the reader can only be advised to read more in the literature. For present purposes, however, there are two main points to note. First, signal detection theory provides alternative measures of how well a judge can discriminate between two foods. Second, these measures are quite independent of any variation in the judge's criterion level; this is because d' and P(A) are obtained using an ROC curve. A judge with a given set of criterion levels will provide a set of points with which to plot an ROC curve. The curvature of the curve determines d' and P(A). A judge with a given sensitivity to the difference between two foods will always produce the same curve; it will arch to the same extent. His criteria may vary from experimental session to experimental session, and thus the points on the curve obtained for the various criterion levels will also vary. However, they will all fall on the same ROC curve with the same curvature giving the same values of d' and P(A). Thus, one set of points may be obtained during one session while a different set may be obtained in the next session. Yet, when either set of points are joined up, they will produce the same ROC curve, with the same curvature and thus the same signal detection indices of degree of difference. It is in this way that the measures are independent of the actual criteria chosen. It does not matter what criteria are chosen during an experimental session, as long as several are used. There need only be enough to provide sufficient points to produce the curve and measure its curvature. There are variations on this procedure, but all use essentially the same device for overcoming the effects of criterion variation.

The R-index (Brown, 1974) is a further signal detection index, yet is more simple to understand. It is merely an estimated probability value. It is the probability of distinguishing between the two foods under consideration. For example, if the two stimuli were A and B, it is the estimated probability of correctly choosing A, when A and B are presented in paired comparison. If the probability were 100%, A would be chosen all the time; the foods would be perfectly distinguishable. If the probability were at the chance level of 50%, then A and B would be perfectly indistinguishable. Intermediate values indicate partial success at distinguishing between the two stimuli. The higher the R-index value, the greater the degree of distinguishability. Remarkably enough, this simple measure can be shown to be identical to P(A) in many cases and can thus be included among the array of signal detection measures, with the advantage that its underlying mathematical model is well understood. The R-index is relatively easy to understand, which makes it eminently suited to food research. Its computation is simple and has been given in detail elsewhere (O'Mahony, 1979b, 1988).

The second approach to controlling variation in the criterion is to use the familiar forced choice procedures. Consider first a paired comparison task. Should two stimuli be presented to a judge and the judge asked whether they were the same or different, the question would be biased; it would be prone to criterion variation. How different do the stimuli need to be before they are reported as different?

However, if a directional forced choice question were asked, the problem would be solved. For example, if one stimulus contained more sugar and the judge were asked to indicate the sweeter of the two, the criterion problem would be solved. The criterion question is whether one of the two stimuli is sufficiently sweeter than the other to be reported as sweeter. By using the directional forced choice question, the experimenter is in effect telling the judge that one of the stimuli will be sufficiently more sweet to be called sweeter; all he has to do is indicate which one. The criterion question is, in effect, answered by the experimental situation. The procedure is equivalent to setting the judge's criterion so that he will report differences of this magnitude as different, always assuming he can detect them.

The other common forced choice procedures use the same strategy. The triangle test arranges for the judge to pick the odd sample; the criterion question for this test is whether any sample is sufficiently different from the other two to be reported as being different. Informing the judge that one sample will be different from the other two effectively answers this question. All the judge needs to do is indicate the odd sample. Again, this is equivalent to setting the criterion to that level of strictness that is just sufficient to differentiate one of the stimuli. For the duo-trio test, when the judge is choosing which of the unknown stimuli are the same as the standard, the implicit criterion question is whether one of the two unknown stimuli are sufficiently similar to the standard to be reported as such. By informing the judge what one of the unknowns will be, the criterion question is effectively answered and the judge sets his criterion accordingly.

Response Bias, Concept Measurement, and Signal Detection

We can now return to the problem of concept measurement, bearing in mind problems of response bias (criterion variation). If a judge considers a stimulus and is asked whether the sensation it elicits falls within the sensory concept under consideration, the response the judge gives will depend on his criterion for inclusion within the concept. Where is the line to be drawn between sensations that fall within the concept and those that fall outside? How different from the stereotypic stimulus for that concept (either a remembered or a previously given standard) does the test stimulus have to be before it is considered as no longer falling within the concept? To be in the concept or not to be in the concept: that is the criterion question!

Because setting a criterion is a cognitive function, the criterion can change in an uncontrolled way, and sometimes a stimulus will be placed within the concept, only to be placed outside on another occasion. The solution to the problem is the same as that for difference testing: forced choice or signal detection procedures.

Ishii and O'Mahony (1987a, 1991) chose the latter strategy, incorporating sureness judgments and computing a signal detection index. Judges were required to state whether a given stimulus fell within the concept or outside and state whether they were sure or not. Thus, judges used the responses: "within the concept-sure," "within the concept-not sure," "outside the concept-not sure," "outside the concept-sure." From these responses, they could compute a suitable signal detection index giving the degree of inclusion of the stimulus within the concept.

The index chosen was Brown's R-index (Brown, 1974; O'Mahony, 1979b, 1988). This has been described earlier as the probability of distinguishing between the two stimuli under consideration. For two stimuli, A and B, it is the probability of correctly choosing A when A and B are presented in paired comparison. A probability of 100% indicates perfect discrimination, 50% chance discrimination. Intermediate values indicate partial success at distinguishing between the two stimuli; the higher the R-index value, the greater the degree of distinguishability.

Now, this logic can be applied to measurement of inclusion of a given stimulus within the concept. An R-index of 100% indicates that the stimulus is perfectly distinguishable conceptually from the prototypical stimulus for that concept; it always falls outside the concept. An R-index of 50% indicates that it is conceptually indistinguishable; it always falls within the concept. Intermediate values indicate that the stimulus is only sometimes placed within the concept. The higher the R-index, the greater the tendency for the stimulus to be excluded from the concept.

It should be remembered that we are now talking about distinguishing conceptually. All stimuli may be perfectly easy to distinguish in terms of their sensory properties (R-index = 100%), but they may not be conceptually distinguishable. To take a color example, perfectly distinguishable shades of red will all fall within the red concept; they will be conceptually indistinguishable

(R-index = 50%). Red and black will be conceptually distinguishable (R-index = 100%). Red and a reddish-orange may have intermediate values indicating that they are partially distinguishable conceptually. The reddish-orange may sometimes be included within the red concept and sometimes not.

Concepts with Fuzzy Edges

Partial inclusion within a sensory concept can alter the way we visualize the concept. If the stimulus were either always within or always outside the concept, we could visualize its "probability of inclusion function" like a balloon. The rubber of the balloon marks the boundary of the concept. A stimulus is either within the balloon or outside it; it cannot be partially inside. But now replace the balloon with a cloud; it is thick in the center but at the edges it gradually thins out into nothing. Again a stimulus can always be in the concept (thick cloud at the center), always outside (outside the cloud), or partially inside (at the fuzzy thinner edges of the cloud). The analogy with the s-orbital in chemistry is obvious. There is even fuzzy set theory available for dealing with such categorizations (Keuning et al., 1986; Lincklaen Westenberg et al., 1989). R-index values are thus well able to represent the fuzziness of the concept's edges by giving graded values for the degree of inclusion of a range of stimuli.

There is one problem. The experiment must be designed in such a way that the judge reappraises the inclusion of a given stimulus each time he is presented with that stimulus, or else he may merely remember his last response to the stimulus and unthinkingly repeat it. This can be achieved if enough test stimuli are presented to the judge successively to cause him to forget his past responses to a given stimulus; this can be achieved within an experimental session without too much difficulty (Ishii and O'Mahony, 1987a, 1991). Naturally, the greater the number of stimuli, the greater the time interval between successive presentations of a stimulus, the greater the distraction due to other stimuli and the greater the chance of the judge forgetting. If there were very few stimuli in the experiment, a long time (perhaps days) would have to elapse between successive presentations of stimuli to allow forgetting. Yet, this would not allow the measurement of concept drift over periods of a few days, discussed later.

Group Concepts

Measurement of an individual judge's concept is not the only procedure for concept measurement. The jointly held concept for a sample of judges can be measured by simply asking each judge whether a list of stimuli fall within the concept or not. Judges will not all agree, and so a given stimulus will have a given probability of being included within the group's concept. The stimulus may not always be totally included, nor may it be totally excluded from the concept; it may be partially included. Partial inclusion of stimuli, once again, gives the concept fuzzy edges. Thus, the replicate tastings required for determining partial inclusion within the concept can be achieved either by repeated tasting for a given judge or by using a single tasting per judge with many judges (group concept). Ishii and O'Mahony (1990) used the latter method when comparing umami taste concepts held by a group of Japanese and a group of Americans; this will be discussed later.

Sorting

A further way of getting at a judge's conceptual structure is to require judges to sort a set of stimuli into groups that are conceptually the same. For colors, all the reds would be sorted into one category, the greens into another, the browns into another, etc. Differences in conceptualization between judges would be seen by the different ways in which judges sorted their stimuli. This technique was used to assess a psychophysical descriptive analysis protocol (Ishii and O'Mahony, 1987b; O'Mahony and Ishii, 1987); this is also discussed later.

There are more indirect approaches than simple sorting. There are various multivariate techniques (Schiffman et al., 1981; Beiber and Smith, 1986) that have been used to sort taste, smell, and food stimuli into multidimensional space (Yoshida, 1963; Schiffman and Erickson, 1971; Schiffman, 1974; Schiffman and Dackis, 1975, 1976; Schiffman et al., 1975, 1977, 1978, 1979, 1980, 1981; Schiffman and Engelhard, 1976). Stimuli that are placed close together in space are conceptually similar; those placed far apart are conceptually different. Even though the methodology of sorting and the multivariate techniques are very different, the results of both techniques are comparable in that they both identify sets of stimuli that can be grouped as being conceptually similar. For sorting, conceptually similar stimuli are grouped into categories which are explicity given by the judge. For the multivariate techniques, the categories need to be inferred from clustering in the data. The choice of method depends on the goals of the study.

Having examined some ways of making behavioral measurements to infer a judge's conceptual structure, it is now possible to apply these methods to a set of conceptual problems in sensory evaluation and psychophysics. We will consider some examples taken from our laboratory to illustrate the methods. We will consider problems of standardization and concept alignment for descriptive analysis; we will consider measurement of a taste concept for international marketing and finally examine a method of taste descriptive analysis used mainly by psychophysicists.

CONCEPT ALIGNMENT USING STANDARDS

As already discussed, for a panel of judges to be used as an analytical tool for descriptive analysis, the judges will need to have their concepts aligned. Current descriptive analysis techniques use the presentation of physical standards to define the various sensory concepts representing the sensory characteristics of the food to be assessed. Generally, a standard is given to define a given characteristic; for example, sugar may be given to define sweetness, citric acid to define sourness, etc. The question arises: how many standards are needed to define a concept so that it is aligned among judges? Although a single standard is often used, it may be asked whether this is sufficient. A single standard would define the center of the concept, but it would not define its edges. A single standard would allow abstraction, but there would be no control over the generalization step. Would judges generalize concepts to the same degree?

Accordingly, Ishii and O'Mahony (1987a) investigated the efficacy of a single standard in the alignment of a sensory concept among a group of judges. The idea of the experiment was simple. Judges were given a 300 mM NaCl solution to define the concept. They were told that this was a standard that defined the taste concept, that it was prototypical, that it was at the center of the conceptual space.

The judges were then presented with 14 stimuli and were required to judge whether they fell within or outside the concept. They were given various concentrations of sodium chloride (NaCl) including the standard (1500 mM, 800 mM, 300 mM, 100 mM, 50 mM); these vary not only in intensity but also in actual taste. They were also given 100 mM NaCl with added orange color and vanilla odor and another sample with yellow color and lemon odor; they were given mixtures of NaCl with potassium chloride (KCl), with citric acid, with monosodium glutamate (MSG), and with fructose in weak and in strong concentrations; they were given pure sodium citrate and KCl solutions. Judges were required to indicate whether each stimulus fell inside or outside the concept defined by the 300 mM NaCl and add whether they were sure or not of each judgment. From such data R-indices indicating the degree of exclusion/inclusion within the concept were computed, as discussed earlier. Each stimulus was tasted eight times; all the tasting took place in random order. An extra set of 300 mM NaCl stimuli was included to act as the referent "noise" stimulus for the R-index calculation. Thus, judges tasted 120 (15 × 8) stimuli in random order, with water rinses taken between the stimuli to minimize adaptation effects (O'Mahony, 1979a; Halpern, 1986).

The resulting R-indices calculated for each stimulus for each judge are given in Table 8.1. Values of 50% indicate that the stimulus was judged by that given

TABLE 8.1 R-indices Indicating the Probabilities of Distinguishing a Stimulus, Conceptually, from 300 mM NaCl

Stimuli	Judges											
	1	2	3	4	5	6	7	8	9	10	11	12
1500 mM NaCl	—	—	—	75.0	—	—	50.0	90.6	43.8	93.8	87.5	31.2
800 mM NaCl	—	—	—	93.8	87.5	97.7	50.0	92.2	50.8	93.8	87.5	38.3
300 mM NaCl	75.8	56.3	75.0	62.5	56.3	77.3	62.5	62.5	50.0	54.7	51.6	52.3
100 mM NaCl	98.4	—	—	—	87.5	—	62.5	92.2	96.1	93.0	87.5	74.2
50 mM NaCl	—	—	—	—	—	—	68.5	82.8	99.2	93.8	87.5	85.9
NaCl, orange color, vanilla odor	—	—	—	—	93.8	—	—	—	—	93.8	85.2	87.5
NaCl, yellow color, lemon odor	—	—	87.5	—	—	95.3	—	89.1	97.7	93.0	87.5	87.5
Sodium citrate	—	—	—	—	—	—	—	—	—	93.8	87.5	87.5
KCl	—	—	—	—	—	—	—	—	—	93.8	87.5	87.5
KCl/NaCl	—	81.3	68.8	93.8	93.8	71.1	56.3	—	88.4	82.0	71.1	52.3
MSG/NaCl	—	87.5	—	—	—	—	—	79.7	96.9	93.8	87.5	80.5
Citric/NaCl	—	—	—	—	—	95.3	87.5	95.3	—	93.8	87.5	87.5
Strong fructose/NaCl	—	—	—	—	—	91.4	—	—	—	93.8	87.5	87.5
Weak fructose/NaCl	—	—	93.8	93.8	93.8	89.1	56.3	96.9	75.8	90.0	87.5	80.5

The higher the R-index above 50%, the lower the probability of the stimulus falling within the concept. R-indices of 100%, representing complete exclusion from the concept, are represented by dashes.

judge to be always conceptually identical to the 300 mM NaCl standard. Values of 100% indicated by dashes in the table, represent the stimulus being conceptually completely distinguishable from the standard. Intermediate values indicate partial inclusion within the concept, making it fuzzy edged; the higher the R-index the less the inclusion within the concept.

A quick glance at the table will reveal that for the 12 judges, no judge had the same distribution of R-indices; no judge included or excluded the stimuli from his concept in the same way; no judge had formed the same concept from the 300 mM NaCl; this was a significant majority trend. Because earlier experimentation had indicated that the stimuli were all completely distinguishable for these judges, the differences between the judges were due to variations in conceptualization, not differences in sensory distinguishability. It can also be seen that for judges 1–9, who excluded some stimuli completely from the concept (R = 100%), no judges agreed on which set of stimuli should be excluded; this was a significant majority trend. Thus, presentation of a single standard did not produce a concept that was aligned among the judges; it did not control the generalization step in concept formation.

Ishii and O'Mahony hypothesized that presentation of more than one standard would do this. Certainly Civille and Lawless (1986) had quite independently come to the conclusion that use of more than one standard would be desirable. Standards could be given to indicate which sensations lie within the concept while further standards could represent sensations falling outside the concept. The former would indicate how much the sensation could differ from the standard and still be included within the concept; the latter would indicate how far the sensation should differ from the standard to be excluded from the concept. The latter would put limits on the generalization step.

To investigate this, Ishii and O'Mahony (1991) essentially repeated their experiment with a few variations. Instead of using a model system, the stimuli were now variations on the lemon-flavored soft drink Tang. A standard-strength Tang drink was presented to judges to define the concept. Other Tang stimuli were then presented to the judges to be assessed as being inside or outside the Tang concept, as defined by the standard. Again sureness judgments were required to allow computation of R-indices.

The stimuli presented were a stronger and a weaker Tang drink (Stronger Tang, Weaker Tang), which vary not only in intensity but actual flavor. Also, there was Tang with just detectable amounts of KCl or citric acid added (KCl Tang, Citric Tang), Tang with a just detectable lemon odor added (Lemon Tang), and Tang deepened in color slightly by the addition of a trace of red coloring (Deep Tang). These stimuli were more difficult to distinguish than those in the previous experiment. The reason for this was to make the differences between stimuli more like those encountered in a "real-life" setting where stimuli that differ only slightly are assessed by highly trained judges, who are able to distinguish the stimuli with ease. Consequently many judges failed the screening that ensured that only judges who could easily distinguish the stimuli were allowed to take part in the experiment. For every two judges who passed the screening, seven were rejected.

Because the Tang was a stronger stimulus than the NaCl solutions, more rinses were required between tastings, fewer stimuli were used and each stimulus was presented fewer times. The six stimuli along with the standard Tang, as the referent "noise" stimulus, were presented four times, each giving a total of 28 (4 × 7) presentations per judge.

The results of the experiment are given in Table 8.2 (section A). The results confirmed the prior experiment. In their fuzzy-edged concepts, no judge had the same distribution of R-indices, indicating that the Tang concepts for the nine judges were not aligned. This was a significant majority trend. Also, for the eight judges who excluded stimuli completely from the concept, only two (#4 and #6) had the same pattern of exclusion.

Judges were then recalled for a second session approximately a week later to see whether a Tang concept could be aligned between judges. This concept would include the Standard, Weaker, Stronger, and Lemon Tangs but exclude the KCl, Citric, and Deep Tangs. Seven standards were given instead of one. Along with the Standard Tang, three stimuli were presented as examples of stimuli falling within the concept to attempt to broaden the concept so it would include Weaker Tang, Stronger Tang, and Lemon Tang. To induce inclusion of Weaker Tang, an even weaker Tang stimulus was presented as an example of a stimulus falling within the concept. To induce inclusion of Stronger Tang, an even stronger Tang stimulus was presented as an example of a stimulus falling within the stimulus. To induce inclusion of Lemon Tang, a Tang drink with an even stronger lemon odor was presented. To exclude Citric Tang from the concept, a Tang drink with less added citric acid was presented to the judge as an example of a stimulus falling outside the concept because it was too sour. A Tang juice with less red coloring was also used to exclude Deep Tang. A Tang juice with added caffeine was used to investigate whether it would exclude the KCl Tang. The caffeine and KCl both produced off-flavors but not the same off-flavor. If the Tang with added caffeine caused exclusion of the KCl Tang, then a degree of generalization of flavor quality from the standards would be demonstrated. This would mean that one standard might be able to act for more than one related flavor. It turned out that this was successful.

The results of this experiment are given in Table 8.2 (section B). This time all the test stimuli were either always excluded or always included within the concept. The judges agreed and within the limits of measurement, the concept was aligned among the judges. Thus, the use of standards to define the edges of the concept by giving examples of sensations to be excluded as well as included within the concept was superior to using only one standard.

TABLE 8.2 R-indices Indicating Percentage Probabilities of Distinguishing a Tang Juice Stimulus Conceptually from Base Tang Juice after Presentations of One or of Seven Standards

	A One standard						B Seven standards					
Judge	Weaker Tang	Stronger Tang	Lemon Tang	KCl Tang	Citric Tang	Deep Tang	Weaker Tang	Stronger Tang	Lemon Tang	KCl Tang	Citric Tang	Deep Tang
1	50.0	50.0	62.5	—	62.5	—	50.0	50.0	50.0	—	—	—
2	50.0	62.5	—	—	50.0	—	50.0	50.0	50.0	—	—	—
3	62.5	50.0	62.5	—	87.5	87.5	50.0	50.0	50.0	—	—	—
4	62.5	87.5	—	—	—	—	50.0	50.0	50.0	—	—	—
5	78.1	87.5	87.5	87.5	56.25	87.5	50.0	50.0	50.0	—	—	—
6	87.5	87.5	—	—	—	—	50.0	50.0	50.0	—	—	—
7	—	62.5	—	—	—	—	50.0	50.0	50.0	—	—	—
8	—	84.4	—	—	90.6	—	50.0	50.0	50.0	—	—	—
9	—	—	—	—	—	—	50.0	50.0	50.0	—	—	—

The higher the R-index above 50%, the lower the probability of the stimulus falling within the concept. R-index values of 100%, representing complete exclusion from the concept, are represented by dashes.

There was an attempt, using a different set of judges, to see whether the same concept could be formed by a context effect. A single standard Tang was presented to define the concept. Then, under the same conditions as before, the Stronger Tang, Weaker Tang, and Lemon Tang were presented along with four other juices of very different flavor. The idea was that the contrast between the three test Tang drinks and the other four juices would force the three Tang drinks to be always included within the concept, by sheer strength of the contrast effect with the other four juices. This only worked for two of the 14 judges tested; only these two judged the three test Tangs to be conceptually indistinguishable from the standard. Although all judges excluded the four other juices completely from the concept, judges did not agree on the degree of inclusion of the three test Tang solutions within the concept. The contrast effect was not as effective at concept alignment as the use of a set of standards.

Thus the result was clear. Use of more than one standard controlled the generalization step and the concept was aligned between judges. Thus, during the concept alignment part of the training of judges for descriptive analysis, it may be necessary to present standards defining sensations that fall outside as well as within a given concept. It is reasonable to suppose that the greater the number of standards employed, the more precise the alignment obtained. The goals of the experiment will determine the degree of precision needed and the number of standards (if more than one) that are required.

CONCEPT STABILITY AND DRIFT

Is a sensory concept fixed or does it change over time? If a standard stimulus forms a given concept for a judge on one occasion, will it form the same concept for that judge on a later occasion? In other words, is the conceptualization system in the brain stable or do concepts drift over time?

This question was answered by Ishii and O'Mahony (1991) in a control experiment in their Tang juice study. A further set of 10 judges were required to perform the first part of the experiment where the one standard Tang stimulus was presented to define the concept; the experiment was performed as before and R-indices duly noted for each stimulus and each judge. Then, approximately 1 week later, the experiment was repeated on these judges to see whether the same set of R-indices were obtained for each judge. In fact, the experiment was a control experiment to check that the effect of multiple standards in the first experiment really had been caused by the multiple standards and not by mere practice. However, the experiment also enabled the study of concept drift.

The results are given in Table 8.3 Section A gives the data for the first

TABLE 8.3 R-indices Indicating the Percentage Probabilities of Distinguishing a Tang Juice Stimulus Conceptually from Base Tang Juice after Presentations of a Single Standard in Experiment II

Judge	A One standard						B Week Later Retest					
	Weaker Tang	Stronger Tang	Lemon Tang	KCl Tang	Citric Tang	Deep Tang	Weaker Tang	Stronger Tang	Lemon Tang	KCl Tang	Citric Tang	Deep Tang
1	50.0	50.0	—	—	87.5	—	50.0	50.0	—	—	75.0	50.0
2	50.0	50.0	—	—	62.5	50.0	50.0	75.0	—	—	—	75.0
3	62.5	50.0	—	50.0	—	—	62.5	62.5	62.5	62.5	—	81.25
4	75.0	62.5	75.0	—	—	—	81.25	12.5	—	—	90.6	—
5	—	37.5	—	—	—	—	—	50.0	—	—	75.0	—
6	—	50.0	—	—	—	—	75.0	50.0	—	—	—	—
7	—	50.0	—	—	—	—	75.0	50.0	—	—	—	—
8	—	50.0	—	—	50.0	87.5	—	62.5	—	—	—	87.5
9	—	75.0	87.5	—	—	—	87.5	71.9	81.25	87.5	78.1	84.4
10	—	87.5	—	—	—	—	—	—	—	—	—	—

The higher the R-index above 50%, the lower the probability of the stimulus falling within the concept.
R-index values of 100%, representing complete exclusion from the concept, are represented by dashes.
R-indices below 50% indicate that the stimulus is perceived as more like the standard than subsequent presentations of Base Tang.

session, section B for the second session. In both parts of the experiment, judges behaved in the same way as in the first part of the original experiment; the concepts were fuzzy edged and judges did not form the same concept from a single standard. Exceptions were judges #6 and 7, who formed identical concepts in both parts; it is hypothesized that this was a chance event. The remaining eight judges formed different concepts.

However, the most important result was that for all judges, the concept in the second part of the experiment was not the same as in the first part. This was a significant majority trend. Further evidence of concept drift over time was indicated by R-index values below 50% (judges #4 and 5). This indicated that these stimuli were considered to be more like the previously represented standard than the standard itself when presented later. This indicated a drift in the center of the concept away from the previously presented standard.

Examination of Table 8.1 giving the results of Ishii and O'Mahony's (1987a) original experiment with the salt taste concept, indicates a similar phenomenon. Unlike their Tang juice experiment, Ishii and O'Mahony presented the two sets of the 300 mM NaCl (standard) stimulus for assessment. One set served as the referent noise while the other served as a test stimulus equal in strength to the standard. Had the judges' concepts remained stable, the test 300 mM NaCl stimuli would have remained conceptually indistinguishable from the noise 300 mM NaCl stimuli, resulting in R-indices of 50%. Table 8.1 indicates that this only occurred for one of the dozen judges (judge #9), a significantly small minority. Also, other stimuli were sometimes chosen as more like the standard than the standard itself (see judge #7).

Ishii and O'Mahony (1987a) went on to repeat their salt concept experiment approximately 1–2 weeks later and found, as with the Tang experiment, that no judge had the same concept on both occasions.

Thus, there was a dynamic aspect to concepts; they changed over time. What the physiological mechanism is that causes such drift remains unknown. Yet, Skarda and Freeman (1987) indicated an interesting parallel in recordings of chaotic firing in the olfactory bulbs of rabbits deprived of water. The rabbit was thirsty, and if a specific smell stimulus was given importance by being paired with the presentation of water to drink (it became the signal for the availability of water), then that smell would elicit a given pattern of activity in the olfactory bulb. Each smell had its own pattern. Yet, over time the pattern for a given smell would drift. It would be pushing analogies too far to draw an exact parallel between this and concept drift, but at least both sets of data suggest a lability in the brain for the engrams representing sensory concepts.

If concepts are so labile, it might be asked whether the order of presentation of stimuli during an experiment might affect how a concept drifted. After their Tang juice study, O'Mahony and Ishii (1991) tried manipulating the order of presentation of stimuli to see whether they could induce the sensory concept to drift.

The experiment was similar to their previous experiments. A standard Tang juice was presented to define the concept. Then, along with this standard Tang, Tang juices of varying strengths were presented.

They found some evidence that if a run of Tang juices of high concentration were presented, the concept tended to drift so that it began to encompass stronger stimuli and exclude the weaker ones. If a run of weaker Tang juices were presented, the concept tended to drift in the opposite direction, to encompass the lower concentration stimuli and exclude the stronger stimuli. The results were not generally clear-cut, however. Of the judges tested, seven gave data that indicated the above trends albeit not always clearly; three gave evidence of an opposite effect. The data are suggestive but not conclusive.

DOES MSG BECOME SALTY AT HIGH CONCENTRATIONS?

Psychophysical taste description will be described in more detail later, but a brief psychophysical problem will be examined at this point. The actual taste of gustatory stimuli alters with concentration, as well as the intensity. This also happens with color. As a colored light gets brighter, it also changes its hue; this is called the Bezold-Brücke phenomenon. The gustatory analog of the phenomenon is well documented (Peryam, 1960, 1963; Bartoshuk et al., 1964, 1974; Dzendolet and Meiselman, 1967; Bartoshuk, 1968; O'Mahony and Stevens, 1975; O'Mahony and Bramwell, 1976; O'Mahony et al., 1976a,b; Cardello and Murphy, 1977; Murphy et al., 1981). The taste of MSG also changes with concentration; at higher concentration it has been reported as being more "salty" (Peryam, 1963; Bartoshuk et al., 1974). However, in these studies, the term "salty" was not defined by the use of standards and so it was difficult to interpret exactly what was meant by MSG becoming "salty." Thus, the statement that MSG becomes "salty" at higher concentrations is essentially meaningless. Yet, the quality of the taste does change.

If "salty" was defined, using at least one standard, then it would be possible to determine whether high concentration MSG stimuli would be classified as falling within such a salty concept. In the previously described experiment of Ishii and O'Mahony (1987a), which examined the efficacy of using a single standard to define a "salty" concept, an extra study was included. The procedure was the same as in the previous experiment. A 300 mM NaCl solution was used, once again, as the standard to define the concept; 300 mM could be argued to be a fairly typical salty stimulus. Then, a set of stimuli were presented to the judges to be classified as falling within or outside the defined concept, with sureness ratings to allow the computation of R-indices. The stimuli were: 80 mM, 300 mM, 1000 mM NaCl; 80 mM, 300 mM, 1000 mM MSG; 192 mM KCl; and 530 mM fructose. Each stimulus was presented five times with plenty of interstimulus rinsing to minimize adaptation effects (O'Mahony, 1979a; Halpern, 1986). The 40 stimuli (5 × 8) were presented to 10 judges in random order.

The results are shown in Table 8.4. It can be seen that the two stronger MSG concentrations were excluded from the concept (R = 100%) by a significant majority (9/10) of the judges. 80 mM MSG was excluded by 8/10 judges. As expected, fructose and KCl tended to be excluded. Incidentally, in the context of these stimuli, a significantly higher proportion of subjects had an R-index of 50% for the 300 mM NaCl stimulus than in the previous experiment; thus, there was less drift in the center of the concept for this study.

When 300 mM NaCl was used to define the salty concept, then higher concentrations of MSG were not included within the concept; according to this definition, they were not "salty." Of course, they may be included in a salty concept that was defined differently. It may be asked why MSG was described as being "salty" at high concentrations. It may be hypothesized that the reason was because the descriptive terms available to judges were restricted to the four words "sweet," "sour," "salty," and "bitter." There will be more about this restriction later.

CONCEPT MEASUREMENT IN MARKETING

The problem was simple. Japanese manufacturers of seasonings had coined the word "umami" to describe the taste imparted to foods when they were cooked in

TABLE 8.4 R-Indices Indicating the Probabilities of Distinguishing a Stimulus, Conceptually, from 300 mM NaCl

Stimuli	Judges									
	1	2	3	4	5	6	7	8	9	10
1000 mM NaCl	90	80	—	—	50	80	80	—	—	50
300 mM NaCl	50	60	52	76	50	50	50	96	50	50
80 mM NaCl	82	80	—	—	50	82	60	—	—	80
192 mM KCl	—	80	—	—	—	—	—	—	—	—
530 mM fructose	—	80	—	—	—	—	—	—	—	
1000 mM MSG	—	80	—	—	—	—	—	—	—	—
300 mM MSG	—	80	—	—	—	—	—	—	—	—
80 mM MSG	98	80	—	—	—	—	—	—	—	—

The higher the R-index above 50%, the lower the probability of the stimulus falling within the concept.

R-index values of 100%, representing complete exclusion from the concept, are represented by dashes.

certain broths. Seasonings are now available to impart the "umami" taste, and in advertising and marketing in Japan, the word "umami" is being used to describe their taste. Were the Japanese customers understanding correctly what was meant by the term umami? Were the manufacturers really communicating clearly with the consumers? There was a further question: Could American consumers be made to understand the term clearly enough for advertising purposes without undergoing a session of descriptive analysis training to align their concepts?

The Umami Taste

It is worth briefly considering the umami taste (O'Mahony and Ishii, 1985). A special feature of Japanese cooking is the use of broths, rather than spices, for seasoning. Three of these broths have primary importance in the emergence of the umami taste concept; these are: Kombu, made from the seaweed *Laminiara japonica* or sea tangle; Katsuobushi, made from dried flakes of the bonito fish; or Shiitake, made from the shiitake or black mushroom *(Lentinus edodus)*.

Kombu broth has long been used for seasoning. For example, rice for sushi is often cooked in this broth, rather than water, to add flavor. The active flavor ingredient of Kombu is MSG, which is now prepared and sold as a pure seasoning. The story goes that a professor from Tokyo Imperial University, Kikunae Ikeda, took a break from his research and settled down to a bowl of tofu, which had been seasoned with Kombu broth. He was struck by the way the broth had made the tofu taste so delicious. He investigated this further, and in 1908 he finally isolated a crystalline flavor compound from the broth, which was identified as MSG. Shortly after, it went into commercial production. The active principle in Katsuobushi is the histidine salt of inosinic acid. This inosinate, however, did not go into commercial production, but the sodium salt of the appropriate isomer, IMP (disodium-5'-inosinate), is now produced commercially. The active principle giving the umami taste to Shiitake broth is another ribonucleotide: GMP (disodium-5'-guanylate).

Measuring Umami Taste Concepts

The term umami is used in the taste literature (Kuninaka, 1965, 1981; Yamaguchi et al., 1971; Cagan et al., 1979; Yamaguchi, 1979; Yamaguchi and Kimizuka, 1979; Torii and Cagan, 1980; Yamaguchi and Takahashi, 1984; Kawamura and Kare, 1986). It can be defined precisely as the taste imparted by the three broths. To align judges' concepts of "umami" the three broths would be useful standards.

Because the term "umami" is used in sales and marketing in Japan, the question becomes one of whether the everyday concept that untrained Japanese consumers have of the umami taste resembles the concept systematically defined by the three broths. It is this systematically defined concept which is being referred to in sales and marketing. It is thus necessary to know whether the everyday concept of consumers is aligned closely to the systematically defined concept, to determine whether those involved in sales and marketing are communicating accurately when they use the term "umami" to consumers.

The experiment was simple (Ishii and O'Mahony, 1990). A group of one hundred Japanese were told to taste and swallow a set of food stimuli and categorize them as being "umami" or not. Judges rinsed with water between tastings to minimize adaptation effects (O'Mahony, 1979a; Halpern, 1986). From the data an umami concept for the group of Japanese was established in the way discussed earlier; disagreement between judges resulted in each stimulus having a specific probability of being included within the concept for that group of judges.

At first, judges performed the experiment after only being told that the umami taste was the taste of Kombu, Shiitake, and Katsuobushi broths. In this case, judges were classifying the foods in terms of their everyday concept of umami. After completing this part of the experiment, the judges repeated the experiment, except this time the term "umami" was defined by presentation of the three broth standards; judges wore noseclips to eliminate confusing olfactory cues from the broths. The judges then repeated the experiment, categorizing the foods as to whether they fell within the new umami concept, defined by the broths. So that they did not merely repeat their responses from the first part of the experiment, the concept defined by the broths was not called umami; it was called "brosu," the nearest approximation to the word "broth" in Japanese. The aim of the experiment was to determine how closely the everyday verbally defined concept resembled the concept systematically defined by the broths.

Fifteen food stimuli were presented for classification by each judge as to whether they tasted "umami" or not. The stimuli were easily distinguishable and covered a wide range of flavors that might be encountered in the marketplace; they were not the closely related and difficult to distinguish flavors used in the prior Tang juice experiment. This was because the aim of the study was to investigate consumer rather than expert behavior and consumers encounter a wide range of easily distinguishable foods. The stimuli were: NaCl, KCl, MSG, IMP, GMP, sucrose, Hunt's canned tomato paste, a noncommercial beef extract, Marmite (a concentrated yeast extract for spreading on bread, popular in Britain), soy sauce, Pripp's Sports Drink (a Japanese soft drink), a portion of a Kraft American Cheese slice, a Hershey's Chocolate Kiss, Karinto (sweet peanut cookie), and Himemaru (fried rice cracker).

The results are shown in Fig. 8.1. The percentages range from 100% to zero, indicating that foods fall completely inside or outside the concept and also partially inside the concept; the concept was thus fuzzy edged. The two sets of percentages were similar; they had significantly high correlation ($r = 0.98$). The data show that mere naming of the broths evoked a taste concept close to that

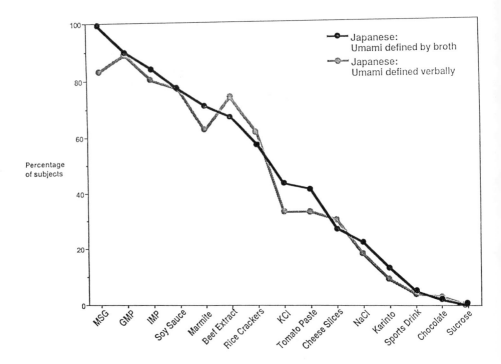

FIG. 8.1 Percentage of Japanese judges who categorized various foods as falling within the umami concept for different definitions of "umami."

obtained by systematic definition using the broths as standards. This simplifies the task of communicating the concept for measurement, sales, and advertising.

The experiment was then repeated on one hundred Californians known to be unfamiliar with the broths in their ordinary cuisine. Naturally, in the first part of the experiment, the umami taste was not defined verbally as the taste of the three broths; the judges would not have known their tastes. For Californians, the umami taste was defined as the taste of Chinese restaurant food or the taste of clear soup in Japanese restaurants. All judges had had experience of Chinese and Japanese style food in America; Japanese and Chinese restaurants are common in California. For the second part of the experiment the taste defined by the three broths was named "dashi."

The results for the Americans are shown in Fig. 8.2. The correlation, although lower for the Americans, was still significant ($r = 0.95$). Although the effect was not as great for the Californians, the concept evoked by mere verbal description was close to the concept defined by the broths. This indicates the potential for the use of the umami concept in American sales and advertising, at

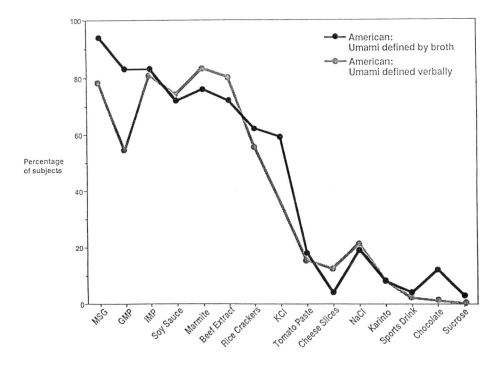

FIG. 2 Percentage of American judges who categorized various foods as falling within the umami concept for different definitions of "umami."

least in California. Perhaps there would be a less close correspondence in parts of the United States less familiar with oriental cuisine. Other verbal definitions of the umami taste, like "meaty" or "broth-like," might be appropriate here.

It is interesting to note that for Californians, when the concept was defined by giving the judges the three broths, 94% of judges included MSG within the concept. Yet, in additional testing with 32 Californians, when the reverse process of using solid MSG to define the concept was tried, the broths of Kombu, Katsuobushi, and Shiitake were only included by 63%, 41%, and 31% of the judges. It may be hypothesized that this lack of transitivity was caused by variations in concept breadth. The actual tastes of the three broths were all easily distinguishable yet were conceptually the same. The umami concept abstracted from these three standards would be expected to be comparatively broad, while the concept coming from the single stimulus, MSG, would be expected to be comparatively narrow. The taste of MSG has a greater chance of falling within a comparatively broad concept than have the different tastes of three broths of falling within a comparatively narrow concept.

TASTE PROFILING

What Is Taste Profiling?

At this point, a descriptive analysis technique design for analysis of taste sensations will be discussed. The technique is called "taste profiling" and has been used commonly by taste psychophysicists. Measurement techniques used in psychophysics have a habit of being adopted in sensory evaluation and versions of this technique have been used to assess food constituents like sweeteners, MSG, KCl, and sodium benzoate (Peryam, 1960, 1963; Acton et al., 1970; Acton and Stone, 1976; Du Bois et al., 1977, 1981; Kier, 1980).

Despite the fact that the technique is used mainly by psychophysicists, who have training in psychology, it appears to ignore any issues of concept alignment. Untrained judges are generally used, and very often no standards are given beforehand to define and align concepts. The method requires judges to partition any given taste sensation into estimated amounts of component "sweetness," "sourness," "saltiness," and "bitterness." The numerical estimates have been given in terms of a nine-point intensity scale (Peryam, 1960, 1963), percentages (Acton et al., 1970; Acton and Stone, 1976; Du Bois et al., 1977, 1981; Kier, 1980), or magnitude estimation (Smith and McBurney, 1969). The latter version of this technique is used widely by taste psychophysicists (Meiselman and Halpern, 1970a,b; McBurney and Shick, 1971; Bartoshuk et al., 1972, 1974a,b, 1978, 1988; McBurney, 1972; McBurney et al., 1972; McBurney and Bartoshuk, 1973; Bartoshuk, 1978, 1979; McBurney and Gent, 1978; De Simone et al., 1980; Murphy et al., 1981; Kuznicki and Ashbaugh, 1982). An additional term "other" has also been added for tastes that do not fit into the traditional four (Bartoshuk, 1974, 1975; Bujas et al., 1974; Bartoshuk and Cleveland, 1977; Kuznicki, 1978; Kuznicki and Ashbaugh, 1979; Settle et al., 1986) and for sensations other than taste (Gent and McBurney, 1978; Gent, 1979).

To the sensory analyst, used to developing a pragmatic descriptive language, using judges trained with standards, taste profiling would seem to involve two inherent assumptions. The first assumption is that judges need no training. The second is that judges can describe any taste in terms of a limited number of the descriptive labels: "sweet," "sour," "salty," and "bitter"; the assumption is that these labels are appropriate and sufficient. Because there are some seemingly tenuous assumptions associated with this method, and because of its potential for use in sensory evaluation, it would seem worth exploring its validity. Let us look at those assumptions further.

Exploration of the Assumptions Behind Taste Profiling

Taste profiling, as generally used, employs no procedures for concept alignment; judges are used without prior training. There have been some notable exceptions to this. Smith and McBurney (1969) when developing the method used standards; unfortunately, those who followed were not as careful. Riskey et al. (1982) used standards when repeating an experiment of Meiselman and Halpern (1970b), who did not use standards, and obtained very different results.

The rationale for not using prior training with standards is that it may be asssumed that judges already understand what is meant by the terms: "sweet," "sour," "salty," and "bitter." It is assumed that in everyday language, judges have these four concepts sufficiently aligned. This assumption is a little surprising. There are certainly many examples of taste "confusions" like the "sour-bitter" confusion (Myers, 1904; Pangborn, 1961; Amerine et al., 1965; Meiselman and Dzendolet, 1967; Robinson, 1970; Gregson and Baker, 1973; McAuliffe and Meiselman, 1974), which can be eliminated merely by prior presentation of standards to define the terms (O'Mahony et al., 1979). Untrained judges do not even have their concepts aligned for a simple concept like "salty"; this can be illustrated by a simple classroom demonstration. If students are given a set of NaCl solutions at different concentrations and required to describe their tastes; they will not correspond on the ones that are called "salty." This lack of alignment is not surprising when the results of Ishii and O'Mahony's (1987a, 1990a) experiments are considered, whereby concepts were not even aligned after the presentation of a standard.

The second assumption is that any taste can be described in terms of the descriptive labels: "sweet," "sour," "salty," "bitter," and "other" (if the latter is to be included). In terms of concept alignment, it means that the taste concepts representing these five labels must be aligned among the judges. If combinations of these five descriptive terms were to be used (31 possible combinations), it would imply that there were 31 taste categories that could be used consistently over judges based on their concept alignment.

At this point, we wish to avoid the controversy of whether taste is "analytic"—a mixture is perceived as a mixture of identifiable separate components, like two sounds played simultaneously—or whether taste as "synthetic"—a mixture is perceived as a new single sensation, like two colors mixed together. The synthetic/analytic nature of taste has been investigated (Erickson and Covey, 1980; Erickson, 1981, 1982; O'Mahony et al., 1983) and will not be pursued further. Suffice to say that it is vitally important for understanding what the judge is perceiving during taste profiling. If taste is synthetic, there will need to be 31 taste concepts to represent the 31 separate unitary blends of taste. This is a high number, and the taste profiling task will be very difficult to learn, stretching the judges' information processing channel capacities (Garner, 1952; Beebe-Center et al., 1955; Attneave, 1959). If taste is analytic, the taste categories using combinations of descriptive labels will be easily identifiable subcategories of the major five concepts; yet there will still be 31 of them. In fact, research indicates a partial synthetic and partial analytic nature for taste (Erickson and Covey, 1980; Erickson, 1981, 1982; O'Mahony et al., 1983).

Using combinations of these five descriptive terms assumes that judges have their concepts aligned and that the terms "sweet," "sour," "salty," and "bitter" are appropriate. Why were these four chosen?

The four were chosen because of a controversial assumption that there are four basic or primary tastes. This widely discussed issue has been thoroughly reviewed by Erickson (Erickson and Schiffman, 1975; Erickson, 1977, 1981, 1982, 1983, 1984a,b, 1985a,b; Erickson and Covey, 1980; Schiffman and Erickson, 1980) as well as many other authors (Frings, 1954; Andersen, 1970; McBurney, 1974; Price and De Simone, 1977; Dethier, 1978; Kuznicki and Ashbaugh, 1979; McBurney and Gent, 1979; Faurion, 1983; Scott and Chang, 1984; O'Mahony and Ishii, 1987). The issue is confused by a lack of differentiation between physiological and psychological levels of argument as well as by a lack of definition of "primacy." The number of primary tastes could refer to the number of taste receptors, the number of transduction mechanisms, the number of neural codes, channels, or fiber types, or the number of gustatory cortical processes, areas, or cell types. In the absence of sufficient evidence, it would seem unwise to assume a prior specified number of any of these physiological entities. Certainly it would seem that there are more than four taste transduction mechanisms (Kinnamon, 1988). Considerations of some type of "psychological primacy" also suffer from lack of definition.

It may be hypothesized that the choice of these particular four tastes is more dependent on language and culture than physiology. There is a need to communicate about food, and an appropriate language is developed for this purpose.

The main effect of the primacy issue has been the creation of traditions. It has often resulted in the number of stimuli used in an experiment being restricted to four. It has restricted the number of labels used for taste concepts to four (or five, with an "other" label), despite reports of difficulties with such a restriction (Murphy, 1971; O'Mahony and Thompson, 1977; Schiffman and Erickson, 1980) and despite the tendency for subjects to use more labels when the restriction is lifted (O'Mahony et al., 1976a, 1976b; O'Mahony and Thompson, 1977).

There have been several protocols adopted by psychophysicists when they have required judges to describe tastes in terms of "sweet," "sour," "salty," and "bitter." Harper et al. (1966) required judges to pick only one of these labels to describe a given taste, while others (Soltan and Bracken, 1958; Robinson, 1970) required a single label to describe the predominant taste. A "tasteless" category has been added, requiring subjects to categorize stimuli as "sweet," "sour," "salty," "bitter," or "tasteless" (Bartoshuk et al., 1964; Dzendolet and Meiselman, 1967; Meiselman and Dzendolet, 1967) or predominantly so (Fox, 1954; Hoover, 1956; Soltan and Bracken, 1958). Cardello and Murphy (1977) extended this approach by not only requiring one of these five quality descriptions but also a magnitude estimate (Stevens, 1957) of that quality. Bartoshuk et al. (1969) used the same technique, extending the categories with the addition of a category not described by these headings. Cardello (1978, 1979) did a similar thing with two extra categories: "indistinct or vague" for weak sensations that could not be recognized and "complicated tastes" for strong sensations that could not be classified as one of the four "primaries."

From these methods, taste profiling (Peryam, 1960, 1963; Smith and McBurney, 1969) was a logical extension. The technique also has some roots in a technique of color naming. For this, descriptions are limited only to combinations of the terms "red," "yellow," "green," and "blue," with an indication of relative strength (Jameson and Hurvich, 1959; Boynton et al., 1964; Boynton and Gordon, 1965; Jacobs and Gaylord, 1967; Kintz et al., 1969). Again, with this technique, no standards were given beforehand to align concepts. The reported success with this technique is not surprising. The few judges used in these studies were only required to label relatively simple spectral colors. It is not difficult for any person who has mixed paints as a child to describe orange (red, yellow), turquoise (blue, green), and mauve (red, blue) fairly consistently and quantitatively. Yet, how successful would this technique have been with the many shades of brown? For more general naming of colors, color chips or color standards are required (Munsell, 1976).

The color concepts of untrained judges only constitute an approximate tool. Furthermore, the constant communication regarding colors in western culture would ensure greater commonality for color labels than for taste labels; this is certainly true for preschool children (O'Mahony et al., 1978).

What follows is an account of three studies that investigated the efficacy of taste profiling. It is especially difficult to validate an introspective method. How can one get into a judge's head to know what he is experiencing? Yet, a consideration of concept formation and concept alignment provides some indirect ways of tackling this problem.

Consistency Tests

A test that gives consistent results does not necessarily give valid or "true" results. There is a difference between validity and consistency (repeatability, reliability). For example, a consistent result may be consistently invalid. It is sobering to know that judges will respond consistently when asked to describe days of the week or U.S. states in terms of "sweet," "sour," "salty," or "bitter" (O'Mahony, 1983). It is not difficult to see why Friday and Saturday are "sweet" while Monday is "bitter." It is not surprising that Californians call California "sweet," although it is puzzling why Texas should be "salty." With any such introspective method, there is a problem with validity. It is naive to assume that a judge has perceptions just because he gives appropriate responses. For difference tests we do not ask a judge whether he can detect a difference; we ask him to prove it by demonstrating his ability to pick the odd sample or pick the sample that matches the standard. Any introspective method must be approached with great caution and with an awareness of all the other factors that might be causing the judges to respond.

Although consistency (repeatability and reliability) is not the same as validity, consistency is worth examining. A valid measure will be consistent. An invalid measure may or may not be consistent. A consistent measure may or may not be valid. An inconsistent measure will be invalid.

If judges had their concepts perfectly aligned with a set of concepts that was appropriate for the stimuli to be described and an agreed set of labels for those concepts, the following propositions regarding consistency would apply:

1. Separate stimuli would be given separate descriptive labels by a given judge.
2. A judge would give the same descriptive label to a stimulus presented twice, given no extraneous effects like cross-adaptation.
3. Judges would agree on the label for a given stimulus.

Further, should judges possess an aligned conceptual structure appropriate for mixtures as well as single solute stimuli and if tastes do not blend:

4. Judges would give stimuli containing mixtures of solutes a descriptive label which was the composite of the single component solutes.
5. The descriptive labels would reflect the number of solutes present in the stimulus.

These points were tested in a simple experiment (O'Mahony et al., 1990). Twenty-eight judges tasted a series of "basic" taste stimuli and their mixtures: NaCl, fructose, citric acid, quinine hydrochloride (QHCl), and the binary mixtures NaCl/fructose, NaCl/citric acid, citric acid/fructose, as well as two "nonbasic" stimuli: MSG and KCl. These nine stimuli were presented once each with the exception that MSG, KCl, and the NaCl/fructose mixture were presented twice each, giving a total of 12 presentations. Rinses were taken between stimuli to minimize adaptation effects (O'Mahony, 1979a; Halpern, 1986). Untrained judges were required to taste these stimuli and describe their taste in terms of any combination of the descriptive labels: "sweet," "sour," "salty," "bitter," "other"; this amounts to 31 possible descriptions. Control experiments had shown that judges could easily distinguish between the stimuli, so any errors in labeling were caused by the labeling process itself rather than difficulties in distinguishing stimuli.

To test the first proposition, the data were examined to determine whether any of the nine separate stimuli were given the same descriptive label (facsimile labeling). Some stimuli were presented twice (MSG, KCl, NaCl/fructose); if they were given the same label, this was not scored as a "facsimile labeling," although such stimuli did have twice the chance to be scored as a facsimile. Because there were nine stimuli there was a total of eight possible facsimile labelings per judge. Facsimile labeling occurred for 22/28 judges, a significant majority of judges.

For the second proposition, the data were examined to see whether the three stimuli presented twice (MSG, KCl, NaCl/fructose) were given the same descriptive label on the second presentation as on the first (a consistent response). At least one inconsistent response was given by 24/28 judges, a significant majority. Eleven judges gave at least two inconsistent responses; three gave all three.

For the third proposition, the data were inspected to see whether there was agreement among the judges on the use of descriptive labels for the stimuli. This was done by inspecting, for each stimulus, the number of judges who used a unique label, a label not used by any other judge. The picture is complicated by the three stimuli that were presented twice; to allow simple comparison, the first and second presentations of such a stimulus were treated as separate stimuli. All 12 stimuli had at least one unique descriptive label, a significant majority; 10/12 had at least two; 5/12 at least three. It should not be assumed that a unique label for a stimulus implies that all other descriptive labels were the same; they were not. There was a range of agreement and disagreement between judges for various stimuli. Some of the best agreement was as follows: citric acid/fructose (called "sour-sweet" by 23/28 judges), QHCl ("bitter," 22 judges), fructose ("sweet," 20 judges), NaCl ("salty," 19 judges), KCl ("salt-bitter," 12 judges; "bitter," 9 judges), citric acid ("sour," 11 judges; "sour-bitter," 9 judges). It could not be argued from this data that judges had their concepts aligned.

For the fourth proposition, the data were examined to see whether the descriptive labels given to binary mixtures were the sums of the labels of their components. There were four presentations of binary mixtures (NaCl/fructose given twice). Only 4/28 judges managed to do this for all four mixture stimuli, a significant minority; nine judges did not manage it for any of the four mixture stimuli.

For the fifth proposition, the data were examined to see how the descriptive labels reflected the number of solutes in the stimuli. No judge fulfilled this requirement for all 12 stimuli. Four did so for 11 stimuli and 13 judges managed it for at least 9 stimuli; four managed it for less than half the stimuli. At least the mean number of words used per subject for single solute stimuli was less than for binary mixtures for a significant majority of the judges (23/28).

Thus, use of a taste profiling approach for descriptive analysis did not fulfill the consistency conditions required, should judges have had their concepts aligned. Yet, when the judges were given a few minutes of training with standards to align their concepts and descriptive labels, all five conditions were fulfilled without error.

Concept Sorting

Ishii and O'Mahony (1978b) approached taste profiling from a different direction, one of appropriateness. They argued that a taste descriptive language is merely a means of communicating the judge's taste conceptual structure. Taste descriptive terms are essentially labels for taste concepts. The system for labeling the concepts must not cause any confusion in the communication of these concepts. Thus, if a judge's taste concepts could be measured, it would be easy to see whether the labels used to communicate them did so clearly and without confusion. For example, every taste concept should have a different label. A simple way to measure a judge's set of taste concepts is by the third method discussed earlier: sorting.

Just as color chips can be sorted into their conceptual categories (various shades of red together, shades of green together, etc.), so can tastes. Instead of color chips for stimuli, boxes full of filter papers that had been soaked in a given solution and then dried were used. Each box contained filter papers impregnated with a given taste; these could be tasted as often as wished. Sufficient rinsing was performed between tastings to prevent adaptation effects (O'Mahony, 1979a; Halpern, 1986).

One hundred seventy Americans and 209 Japanese were tested; these groups were chosen because of prior evidence of differences in descriptive strategies (O'Mahony and Ishii, 1986). To get the general idea of sorting, judges first sorted color chips. Then they sorted a set of 12 boxes of filter paper taste stimuli into their taste conceptual categories. The stimuli were: NaCl (weak and strong), LiCl, MSG (weak and strong), IMP, GMP, citric acid, caffeine, sucrose, aspartame, KCl, and sodium benzoate.

The numbers of categories obtained and the stimuli that were categorized together varied from judge to judge. Generally, but by no means always, MSG, IMP, and GMP were placed together; NaCl and LiCl were placed together, as were sucrose and aspartame. Citric acid and caffeine formed separate categories. KCl and sodium benzoate tended to form separate categories while also tending to share categories with other stimuli. However, for the purposes of this experiment, the actual categorization was not important.

Once the judges had finally decided on the conceptual categories present, they were asked to give each one a descriptive label for its taste. No words were supplied or suggested so as not to bias the choice of words (O'Mahony and Thompson, 1977). Not unexpectedly, each conceptual category was given a separate label. Such labels, however, were useless for communicating the taste categories because judges rarely shared the same label for a given category.

Then, to test the taste profiling approach, judges were asked to label their conceptual categories using any one of the 31 possible combinations of the labels: "sweet," "sour," "salty," "bitter," and "other". To correspond with general taste profiling practice, none of the judges had had any prior training to align their concepts. What happened was that, using this system of labeling, the number of labels obtained tended to be less than the number of categories. That is, the number of conceptual categories that would be inferred from the number of labels, tended to be less than the actual number of conceptual categories present; there was a shortfall. For the Japanese, the mean number of categories that could be inferred from the labels was 4.9, whereas the actual mean number of categories was 5.7. For the Americans the mean values were 5.3 and 5.7. In both cases, the shortfall in numbers was significant. A more telling way of describing the data is as follows. The reductions were caused by approximtely half the Japanese (a mean shortfall of 1.6) and a third of the Americans (mean shortfall 1.7). These are unacceptably large proportions of judges whose descriptive labels do not communicate their conceptual structure without a distorting shortfall.

Thus, although the 31 combinations of terms available as concept labels far outnumbered the number of conceptual categories sorted (generally 4–7; max. 10), different categories were given the same label. The restricted number of labels did not provide a wide enough selection of appropriate labels for each taste conceptual category to have a separate label. Thus, separate categories were given the same label. Confusions could thus occur in the communication of those taste concepts, because of more than one concept having the same label. The descriptive method distorted communication of the conceptual structure. The method is faulted.

Taste Profiling with Concept Alignment

The final experiment gave a generous chance to taste profiling. What if judges were trained with standards to define the terms "sweet," "sour," "salty," and "bitter" and any combination of these terms? Then, would the taste concepts generated and aligned by this training still be adequate for the description of stimuli like MSG, sodium benzoate, or KCl? Would the sensations elicited by these latter stimuli fall within the concepts generated during training? Naturally, the concepts generated and aligned during training would be arbitrary. Yet, if very broad concepts were generated, broader than those generally used by untrained judges, then the chances of sensations elicited by stimuli like MSG or KCl falling within the concepts would be maximized. Accordingly, O'Mahony et al. (1990) performed the following experiment, which required a series of control experiments. For simplicity, the controls will not be detailed here; we will concentrate on the main study.

Five judges underwent training and concept alignment. They were given standards to define "sweet," "sour", "salty," and "bitter." They were also given standards to define mixtures of these four tastes in every possible combination. They were also tested extensively to ensure that they all had their concepts aligned. First, they were given solutions of NaCl, fructose, citric acid, and quinine hydrochloride to define the four "primary" tastes. Each stimulus was presented at a stronger and at a weaker intensity; each level of intensity had been previously intensity matched. After judges had learned to respond appropriately to these eight stimuli (i.e., call NaCl "salty," etc.), the next set of 11 standards was given. These were every possible combination of mixtures of two, three, and four stimuli at the lower intensity level. After judges had learned and were able to respond appropriately to all 19 (11 + 8) stimuli, a further set of 28 stimuli was added. These were mixtures made up in the same combinations of lower concentrations as before, except that in each combination, the mixture was made up with one of the stimuli having the higher concentration. When judges had learned to respond appropriately to all 47 (8 + 11 + 28) stimuli, they were presented with a further set of 12 binary mixtures. These were every possible combination of two stimuli with one of the stimuli being at the lower concentration and the other being at a trace concentration. Experiments indicated that these mixtures were distinguishable for the judges. Thus, judges had their concepts aligned using 59 standards. Judges were not considered to be trained until they were able to respond appropriately, without error, to 100 stimuli presented randomly and consecutively. Judges tasted and expectorated all stimuli with suitable interstimulus rinsing to prevent adaptations effects (O'Mahony, 1979a; Halpern, 1986). The training required for the judges was extensive, ranging from 13.5 to 67.75 hr; it took place over a period of 6 weeks.

Once the judges were trained and were shown to respond appropriately without error, they could be considered to be instruments that had been identically calibrated; their concepts were aligned. They were then told that they were to be given five further mixtures. This was a lie; they were given MSG, KCl, sodium benzoate, sodium bicarbonate, and sodium citrate.

The judges had had their concepts aligned, with agreed descriptive labels. If these concepts were appropriate for the five new stimuli, their sensations would fall within the concepts. Because the concepts were aligned, a given stimulus would fall within the same concepts for each judge. Thus, for each stimulus, judges would agree on the descriptive labels to be given to that stimulus. Because the concepts "sweet," "sour," "salty," and "bitter" were given the broadest possible definition, the chances of the sensations from these new stimuli falling within the aligned concepts were maximized. The chances for judges agreeing on the descriptions were thus maximized.

On the other hand, should the aligned concepts not be appropriate for the new stimuli, the sensations would not fall within the concepts. Judges would need to extend or generalize their concepts to accommodate them. There is certainly no guarantee that all judges would modify their concepts in the same manner. One judge may extend one concept to include a given new stimulus, a second judge may extend another quite different concept. Accordingly, the judges' descriptions would disagree for a given stimulus.

Consistency of response over judges for a given new stimulus would have indicated that the concepts were appropriate for that stimulus. However, for all five stimuli, the judges disagreed. They also complained that it was difficult to describe the tastes of the stimuli; it seemed their concepts were inappropriate for descriptive analysis of these new stimuli. A further point: when the final set of binary mixtures (with trace stimuli) were first presented, judges could not always respond appropriately to them. Thus, the responses that they had learned for the previous mixture stimuli did not generalize to mixtures with very different proportions.

It would seem that a set of concepts derived from four "primary" stimuli and their mixtures were not even appropriate for mixtures of these stimuli, made up in different proportions, let alone the five "nonprimary" stimuli. Again, the use of combinations of only these four descriptive labels would seem too restrictive for any general application.

The long training times in the experiment force a consideration of whether the judges were perceiving the tastes of the mixtures in a synthetic or in an analytic way. If taste were analytic, judges would have been working with four concepts, with each mixture stimulus falling into one or more of each. If taste were synthetic, the judge would have had to work with up to 59 separate concepts, each one for a unitary taste either from a single solute stimulus or a blended mixture. The long training times, confirming earlier work (O'Mahony and Buteau, 1982), certainly suggest some element of synthesis, as was found in earlier studies (Erickson and Covey, 1980; Erickson, 1981, 1982; O'Mahony et al., 1983).

Verdict on Taste Profiling

It would seem that methods that come under the general heading of taste profiling do not have wide applicability. Lack of training of the judges and restriction of the available descriptive terms to "sweet," "sour," "salty," "bitter" (and "other") causes problems. Even with suitable preliminary concept alignment, the system would seem to have limited applicability. Results, using this method, should be interpreted with caution with an eye to implied response demand characteristics.

It was useful to consider taste profiling because, even though it does not have wide use in the sensory evaluation of food, it provides an illustration of how a feasible seeming measurement technique can be faulted. It also illustrates how lack of clear thought can lead to faults in methodology, which can in turn lead to spurious results.

CONCLUSION

Before descriptive analysis, it is important for judges to have their concepts aligned with common descriptive labels. The degree of alignment will depend on the precision required. Concepts can be measured. For this, stimuli can be categorized as within or outside a concept (this might involve signal detection sureness ratings because of response bias), or stimuli can be sorted into their conceptual categories. These are not the only possible approaches. Once conceptual structure is measured, it can be used for a variety of reasons. Examples were given from this laboratory ranging from modification of training for descriptive analysis and assessment of descriptive techniques to problems of definition for sales and marketing. This is not a definitive list, nor is it the only approach to concept formation. The reader would do well to pursue the work of Civille and Lawless (1986) and of Solomon (see Chapter 9).

It seems that to try to describe a taste or any sensory characteristic of a food to a person who has never experienced the sensation is a little like trying to describe redness to a blind man. To describe a taste to people, one needs to give it to them and then label it. The more examples that are given to define the taste, the more precise will be the concept formed.

Finally, it is worth noting that scaling and descriptive analysis are essentially introspective techniques and as such are less objective than a response measurement technique like difference testing. For the latter, a judge is not merely asked whether he can detect a difference; he is required to demonstrate behaviorally that he can. This acts as a built-in validity check. Introspective techniques have no such behavioral demonstration component. The only way to use them is to validate them before use; this is equivalent to calibrating one's instrument. If, in descriptive analysis, a judge reports the presence of a given sensation, this report can only be trusted if that judge has demonstrated that he can reliably report the sensation when it is present and not report it when it is absent. To ensure that the judge is not having a momentary lapse when he is reporting the sensation, other equally reliable judges are used. The validity of the judge's report can be assessed from the reports of the other judges. Yet, this is only possible if the judges' concepts are aligned.

REFERENCES

Acton, E. M. and Stone, H. 1976. Potential new artificial sweetener from study of structure-taste relationships. *Science* 193: 584.

Acton, E. M., Leaffer, M. A., Oliver, S. M., and Stone, H. 1970. Structure taste relationships in oximes related to perillartine. *J. Agric. Food Chem.* 18: 1061.

Amerine, M. A., Pangborn, R. M., and Roessler, E. B. 1965. In *Principles of Sensory Evaluation of Food,* p. 267. Academic Press, New York.

Andersen, H. T. 1970. Problems of taste specificity. In *Taste and Smell in Vertebrates* (G. E. W. Wolstenholme and J. Knight, Eds), p. 71. J. & A. Churchill, London.

Attneave, F. 1959. *Application of Information Theory to Psychology.* Holt, Rinehart and Winston, New York.

Bartoshuk, L. M. 1968. Water taste in man. *Percept. Psychophys.* 3: 69.

Bartoshuk, L. M. 1974. NaCl thresholds in man: Thresholds for water taste or NaCl taste. *J. Comp. Physiol. Psychol.* 87: 310.

Bartoshuk, L. M. 1975. Taste mixtures: Is mixture suppression related to compression? *Physiol. Behav.* 14: 643.

Bartoshuk, L. M. 1978. The psychophysics of taste. *Amer. J. Clin. Nutr.* 31: 1068.

Bartoshuk, L. M. 1979. Bitter taste of saccharin related to the genetic ability to taste the bitter substance 6-n-propylthiouracil. *Science* 205: 934.

Bartoshuk, L. M. and Cleveland, C. T. 1977. Mixtures of substances with similar tastes. A test of a psychophysical model of taste mixture interactions. *Sensory Processes* 1: 177.

Bartoshuk, L. M., McBurney, D. H., and Pfaffmann, C. 1964. Taste of sodium chloride solutions after adaptation to sodium chloride: Implications for the 'water taste'. *Science* 143: 967.

Bartoshuk, L. M., Dateo, G. P., Vandenbelt, D. J., Buttrick, R. L., and Long, L. 1969. Effects of *Gymnema sylvestre* and *Synsepalum dulcificum* on taste in man. In *Olfaction and Taste III* (C. Pfaffmann, Ed.), p.436, Rockefeller University Press, New York.

Bartoshuk, L. M., Lee, C-H., and Scarpellino, R. 1972. Sweet taste of water induced by artichoke *(Cynara scolymus). Science* 178: 988.

Bartoshuk, L. M., Cain, W. S., Cleveland, C. T., Grossman, L. S., Marks, L. E., Stevens, J. C., and Stolwijk, J. A. J. 1974a. Saltiness of monosodium glutamate and sodium intake. *J. Amer. Med. Assoc.* 230: 670.

Bartoshuk, L. M., Gentile, R. L., Moskowitz, H. R., and Meiselman, H. L. 1974b. Sweet taste induced by miracle fruit *(Synsepalum dulcificum)*. *Physiol. Behav.* 12: 449.

Bartoshuk, L. M., Murphy, C., and Cleveland, C. T. 1978. Sweet taste of dilute NaCl: Psychophysical evidence for a sweet stimulus. *Physiol. Behav.* 21: 609.

Bartoshuk, L. M., Rifkin, B., Marks, L. E., and Hooper, J. E. 1988. Bitterness of KCl and benzoate: Related to genetic status for sensitivity to PTC/PROP. *Chem. Senses* 13: 517.

Beebe-Center, J. G., Rogers, M. S., and O'Connell, D. M. 1955. Transmission of information about sucrose and saline solutions through the sense of taste. *J. Psychol.* 39: 157.

Bieber, S. L. and Smith, D. V. 1986. Multivariate analysis of sensory data: A comparison of methods. *Chem. Senses* 11: 19.

Boynton, R. M. and Gordon, J. 1965. Bezold-Brücke hueshift measured by color-naming technique. *J. Optical Soc. Amer.* 55: 78.

Boynton, R. M., Schafer, W., and Neun, M. E. 1964. Hue-wavelength relation measured by color-naming method for three retinal locations. *Science* 146: 666.

Brown, J. 1974. Recognition assessed by rating and ranking. *Brit. J. Psychol.* 65: 13.

Bujas, Z., Szabo, S., Kovačić, M., and Rohaček, A. 1974. Adaptation effects on evoked electrical taste. *Percept. Psychophys.* 15: 210.

Cagan, R. H., Torii, K., and Kare, M. R. 1979. Biochemical studies of glutamate taste receptors: The synergistic taste effect of l-glutamate and 5'-ribonucleotides. In *Glutamic Acid: Advances in Biochemistry and Physiology* (L. S. Filer, S. Garattini, M. R. Kare, W. A. Reynolds, and R. J. Wurtman, Ed.), p. 1. Raven Press, New York.

Cairncross, S. E. and Sjöström, L. B. 1950. Flavor profiles—A new approach to flavor problems. *Food Technol.* 4: 308.

Cardello, A. V. 1978. Chemical stimulation of single human fungiform taste papillae: Sensitivity profiles and locus of stimulation. *Sensory Processes* 2: 173.

Cardello, A. V. 1979. Psychophysical exponents for single papillae: A comparison with whole-mouth exponents. *Percept. Psychophys.* 25: 510.

Cardello, A. V. and Murphy C. 1977. Magnitude estimates of gustatory quality changes as a function of solution concentration of simple salts. *Chem. Senses Flav.* 2: 327.

Caul, J. F. 1957. The profile method in flavor analysis. *Adv. Food Res.* 7: 1.

Civille, G. C. and Lawless, H. T. 1986. The importance of language in describing perceptions. *J. Sensory Stud.* 7: 203

De Simone, J. A., Heck, G. L., and Bartoshuk, L. M. 1980. Surface active taste modifiers: A comparison of the physical and psychophysical properties of gymnemic acid and sodium lauryl sulfate. *Chem. Senses* 5: 317.

Dethier, V. G. 1978. Other tastes, other worlds, *Science* 201: 224.

Du Bois, G. E., Crosby, G. A., Stephenson, R. A., and Wingard, R. E. 1977. Dihydrochalcone sweeteners. Synthesis and sensory evaluation of sulfonate derivatives. *J. Agric. Food Chem.* 25: 763.

Du Bois, G. E., Crosby, G. A., Lee, J. F., Stephenson, R. A., and Wang, P. C. 1981. Dihydrochalcone sweeteners. Homoserine-dihydrochalcone conjugate with low aftertaste, sucrose-like organoleptic properties. *J. Agric. Food Chem.* 29: 1269.

Dzendolet, E. and Meiselman, H. L. 1967. Gustatory quality changes as a function of solution concentration. *Percept. Psychophys.* 2: 29.

Ennis, D. M. 1990. Relative power of difference testing methods in sensory evaluation. *Food Technol.* 44: 114.

Erickson, R. P. 1977. The role of 'primaries' in taste research. In *Olfaction and Taste* VI. (J. Le Magnen and P. MacLeod, Ed.), p. 369. I. R. L., London.

Erickson, R. P. 1981. A new direction in taste psychophysics: Aristotle and Henning. Abstract: 3rd Annual Meeting: Association for Chemoreception Sciences, Sarasota, FL.

Erickson, R. P. 1982. Studies on the perception of taste: Do primaries exist? *Physiol. Behav.* 28: 57.

Erickson, R. P. 1983. Taste: A time for re-evaluation. *E. C. R. O. Newsletter.* No. 27: 281.

Erickson, R. P. 1984a. Öhrwall, Henning and von Skramlik; The foundations of the four primary position in taste. *Neuroscience and Biobehavioral Reviews* 8: 105.

Erickson, R. P. 1984b. On the neural bases of behavior. *Amer. Scientist* 22: 233.

Erickson, R. P. 1985a. Grouping in chemical senses. *Chem. Senses* 10: 333.

Erickson, R. P. 1985b. Definitions: A matter of taste. In *Taste, Olfaction and The Central Nervous System* (D. W. Pfaff, Ed.), p. 129. Rockefeller University Press, New York.

Erickson, R. P. and Covey, E. 1980. On the singularity of taste sensations: What is a taste primary? *Physiol. Behav.* 25: 527.

Erickson, R. P. and Schiffman, S. S. 1975. The chemical senses: A systematic approach. In *Handbook of Psychobiology* (M. S. Gazzaniga and C. Blake-more, Ed.), p. 303. Academic Press, London.

Faurion, A. 1983. Taste—A time for re-evaluation—Reply. *E. C. R. O. Newsletter* 28: 291.

Fox, A. L. 1954. A new approach to explaining food preferences. Abstract papers, ACS 126th Meeting, Div. Agric. Food Chem. p. 14A.

Frings, H. W. 1954. Gustatory stimulation by ions and the taste spectrum. Abstract papers, ACS 126th Meeting, Div. Agric. Food Chem., p. 14A.

Garner, W. R. 1962. *Uncertainty and Structure as Psychological Concepts.* John Wiley, New York.

Gent, J. F. 1979. An experimental model for adaptation in taste. *Sensory Processes* 3: 303.

Gent, J. F. and McBurney, D. H. 1978. Time course of gustatory adaptation. *Percept. Psychophys.* 23: 171.

Green, B. and Gelhard, B. 1989. Salt as an oral irritant. *Chem. Senses* 14: 259.

Green, D. M. and Swets, J. A. 1966. *Signal Detection Theory and Psychophysics,* John Wiley & Sons, New York.

Gregson, R. A. M. and Baker, A. F. H. 1973. Sourness and bitterness: Confusions over sequences of taste judgements. *Brit. J. Psychol.* 64: 71.

Halpern, B. P. 1986. What to control in studies of taste. In *Clinical Measurement of Taste and Smell* (H. L. Meiselman and R. S. Rivlin, Ed.), p. 126. MacMillan, New York.

Harper, H. W., Jay, J. R. and Erickson, R. P. 1966. Chemically evoked sensations from human taste papillae. *Physiol. Behav.* 1: 319.

Hoover, E. F. 1956. Reliability of phenylthiocarbamide-sodium benzoate method of determining taste classifications. *J. Agric. Food Chem.* 4: 345.

Hull, C. L. 1920. Quantitative aspects of the evolution of concepts. *Psychol. Monographs* 28: 1.

Ishii, R. and O'Mahony, M. 1987a. Defining a taste by a single standard: Aspects of salty and umami tastes. *J. Food Sci.* 52: 1405.

Ishii, R. and O'Mahony, M. 1987b. Taste sorting and naming: Can taste concepts be misrepresented by traditional psychophysical labelling systems? *Chem. Senses* 12: 37.

Ishii, R. and O'Mahony, M. 1991. The use of multiple standards to define sensory characteristics for descriptive analysis: Aspects of concept formation. *J. Food Sci.,* in press.

Ishii, R. and O'Mahony, M. 1990. Group taste concept measurement: Verbal and physical definition of the umami taste concept for Japanese and Americans. *J. Sens. Stud.* 4: 215–227.

Jacobs, G. H. and Gaylord, H. A. 1967. Effects of chromatic adaptation color naming. *Vision Research* 7: 645.

Jameson, D. and Hurvich, L. M. 1959. Perceived color and its dependence on focal surrounding, and preceding stimulus variables. *J. Optical Soc. Amer.* 4: 890.

Kawamura, Y. and Kare, M. 1986. *Umami. A Basic Taste.* Marcel Dekker, New York.

Keuning, P., Backer, E., Duin, R. P. W., Lincklaen Westenberg, H. W., and De Jong, S. 1986. Fuzzy set theory applied to product classification by a sensory panel. *Chem. Senses* 11: 622.

Kier, L. B. 1980. Molecular structure influencing either a sweet or bitter taste among aldoximes. *J. Pharm. Sci.* 69: 416.

Kinnamon, S. C. 1988. Taste transduction: A diversity of mechanisms. *Trends in Neurosciences* 11: 491.

Kintz, R. T., Parker, J. A., and Boynton, R. M. 1969. Information transmission in spectral color naming. *Percept. Psychophys.* 5: 241.

Kuninaka, A. 1965. Flavor potentiators. Oregon State Univ., 4th Symp. Foods, p. 515.

Kuninaka, A. 1981. Taste and flavor enhancers. In *Flavor Research, Recent Advances* (R. Teranishi, R. A. Flath, and H. Sugisawa, Ed.), p. 305. Marcel Dekker, New York.

Kuznicki, J. T. 1978. Taste profiles from single human taste papillae. *Percept. Motor Skills* 47: 279.

Kuznicki, J. T. and Ashbaugh, N. 1979. Taste quality differences within the sweet and salty taste categories. *Sensory Processes* 3: 157.

Kuznicki, J. T. and Ashbaugh, N. 1982. Space and time separation of taste mixture components. *Chem. Senses* 7: 39.

Lincklaen Westenberg, H. W., De Jong, S., Van Meel, D. A., Quadt, J. F. A., Backer, E., and Duin, R. P. W. 1989. Fuzzy set theory applied to product classification by a sensory panel. *J. Sensory Stud.* 4: 55.

McAuliffe, W. K. and Meiselman, H. L. 1974. The roles of practice and correction in the categorization of sour and bitter taste qualities. *Percept. Psychophys.* 16: 242.

McBurney, D. H. 1972. Gustatory cross adaptation between sweet-tasting compounds. *Percept. Psychophys.* 11: 225.

McBurney, D. H. 1974. Are there primary tastes for man? *Chem. Senses Flav.* 1: 17.

McBurney, D. H. and Bartoshuk, L. M. 1973. Interactions between stimuli with different taste qualities. *Physiol. Behav.* 10: 1101.

McBurney, D. H. and Gent, J. F. 1979. On the nature of taste qualities. *Psychol. Bull.* 86: 151.

McBurney, D. H. and Gent, J. F. 1978. Taste of methyl-α-D-mannopyranoside: Effects of cross adaptation and gymnema sylvestre. *Chem. Senses Flav.* 3: 45.

McBurney, D. H. and Shick, T. R. 1971. Taste and water taste of twenty-six compounds for man. *Percept. Psychophys.* 10: 249.

McBurney, D. H., Smith, D. V., and Shick, T. R. 1972. Gustatory cross adaptation: Sourness and bitterness. *Percept. Psychophys.* 11: 228.

Meilgaard, M., Civille, G. V., and Carr, B. T. 1987. *Sensory Evaluation Techniques,* Vol. 2. CRC Press, Boca Raton, FL.

Meiselman, H. and Dzendolet, E. 1967. Variability in gustatory quality identification. *Percept. Psychophys.* 2: 496.

Meiselman, H. L. and Halpern, B. P. 1970a. Human judgements of *Gymnema sylvestre* and sucrose mixtures. *Physiol. Behav.* 5: 945.

Meiselman, H. L. and Halpern, B. P. 1970b. Effects of *Gymnema sylvestre* on complex tastes elicited by amino acids and sucrose. *Physiol. Behav.* 5: 1379.

Miller, G. A. and Johnson-Laird, P. N. 1976. *Language and Perception.* Cambridge University Press, London.

Millward, R. B. 1980. Models of concept formation. In *Aptitude, Learning and Instruction, Cognitive Process Analyses of Learning and Problem Solving,* Vol. 2 (R. E. Snow, P. A. Federico, and W. E. Montague, Ed.), p. 245. Lawrence Erlbaum Associates, Hillsdale, NJ.

Munsell Book of Color. 1976. MacBeth Division of Kollmorgen Corp., Baltimore, MD.

Murphy, W. M. 1971. The effect of complete dentures upon taste perception. *Brit. Dental J.* 130: 201.

Murphy, C., Cardello, A. V., and Brand, J. G. 1981. Tastes of fifteen halide salts following water and NaCl: Anion and cation effects. *Physiol. Behav.* 26: 1083.

Myers, C. S. 1904. The taste-names of primitive peoples. *Brit. J. Psychol.* 1: 117.

Nelson, K. 1974. Concept, word and sentence. Interrelations in acquisition and development. *Psychol. Rev.* 81: 267.

O'Mahony, M. 1972. Salt taste sensitivity: A signal detection approach. *Perception* 1: 459.

O'Mahony, M. 1979a. Salt taste adaptation: The psychophysical effects of subadapting solutions and residual stimuli from prior tastings on the taste of sodium chloride. *Perception* 8: 441.

O'Mahony, M. 1979b. Short-cut signal detection measurements for sensory analysis. *J. Food Sci.* 44: 302.

O'Mahony, M. 1983. Gustatory responses to nongustatory stimuli. *Perception* 12: 627.

O'Mahony, M. 1988. Sensory difference and preference testing: The use of signal detection measures. In *Applied Sensory Analysis of Foods,* Vol. 1 (H. Moskowitz, Ed.), p. 145. CRC Press, Boca Raton, FL.

O'Mahony, M. 1990. Cognitive aspects of difference testing and descriptive analysis: Criterion variation and concept formation. In *Psychological Basis of Sensory Evaluation* (R. L. McBride and D. H. MacFie, Ed.), p. 117. Elsevier Applied Science, Barking, U.K.

O'Mahony, M. and Bramwell, A. 1976. How many taste changes are there available to signal tartaric acid taste thresholds? *I.R.C.S. Med. Sci.* 4: 68.

O'Mahony, M. and Buteau, L. 1982. Taste mixtures: Can the components be readily identified? *IRCS Med. Science* 10: 109.

O'Mahony, M. and Ishii, R. 1985. Do you have an umami tooth? *Nutrition Today* 20: 4.

O'Mahony, M. and Ishii, R. 1986. A comparison of English and Japanese taste languages: Taste descriptive methodology, codability and the umami taste. *Brit. J. Psychol.* 77: 161.

O'Mahony, M. and Ishii, R. 1987. The umami taste concept: Implications for the dogma of four basic tastes. In *Umami: A Basic Taste* (Y. Kawamura and M. R. Kare, Ed.), p. 75. Marcel Dekker, New York.

O'Mahony, M. and Stevens, J. 1975. Criterion points and qualitative taste descriptive categories along the taste continuum for citric acid solutions—Preliminary study. *I.R.C.S. Med. Sci.* 3: 56.

O'Mahony M. and Thompson, B. 1977. Taste quality descriptions: Can the subject's response be affected by mentioning taste words in the instructions? *Chem. Senses Flav.* 2: 283.

O'Mahony, M., Hobson, A., Garvey, J., Davies, M., and Birt, C. (1976a). How many tastes are there for low concentration 'sweet' and 'sour' stimuli?—Threshold implications. *Perception* 5: 147.

O'Mahony, M., Kingsley, L., Harji, A., and Davies, M. 1976b. What sensation signals the salt taste threshold? *Chem. Senses Flav.* 2: 177.

O'Mahony, M., Autio, J., Heintz, C., and Goldenberg, M. 1978. Taste naming by preschool children compared to colour naming: Preliminary examination. *IRCS Medic. Sci.* 6: 208.

O'Mahony, M., Goldenberg, M., Stedmon, J., and Alford, J. 1979. Confusion in the use of the taste adjectives 'sour' and 'bitter'. *Chem. Senses Flav.* 4: 301.

O'Mahony, M., Atassi-Sheldon, S., Rothman, L., and Murphy-Ellison, T. 1983. Relative singularity/mixedness judgements for selected taste stimuli. *Physiol. Behav.* 31: 749.

O'Mahony, M., Rothman, L., Ellison, T., Shaw, D., and Buteau, L. 1990. Taste descriptive analysis: Concept formation, alignment and appropriateness. *J. Sens. Stud.* 5: 71–103.

Pangborn, R. M. 1961. in discussion for Pilgrim, F. J. Interactions of supra-threshold taste stimuli. In *The Physiological and Behavioral Aspects of Taste* (M. R. Kare and B. P. Halpern, Ed.), p. 72. University of Chicago Press, Chicago.

Peryam, D. R. 1960. The variable taste perception of sodium benzoate. *Food Technol.* 14: 383.

Peryam, D. R. 1963. Variability of taste perception. *J. Food Sci.* 28: 734.

Pikas, A. 1966. *Abstraction and Concept Formation*. Harvard University Press, Cambridge, MA.

Price, S. and De Simone, J. A. 1977. Models of taste receptor cell stimulation. *Chem. Senses Flav.* 2: 427.

Riskey, D. R., Desor, J. A., and Vellucci, D. 1982. Effects of gymnemic acid concentration and time since exposure on intensity of simple tastes: A test of the biphasic model for the action of gymnemic acid. *Chem. Senses* 7: 143.

Robinson, J. O. 1970. The misuse of taste names by untrained observers. *Brit. J. Psychol.* 61: 375.

Schiffman, S. S. 1974. Physiochemical correlates of olfactory quality. *Science* 185: 112.

Schiffman, S. S. and Dackis, C. 1975. Taste of nutrients: Amino acids, vitamins, and fatty acids. *Percept. Psychophys.* 17: 140.

Schiffman, S. S. and Dackis, C. 1976. Multidimensional scaling of musks. *Physiol. Behav.* 17: 823.

Schiffman, S. S. and Engelhard, H. H. 1976. Taste of dipeptides. *Physiol. Behav.* 17: 523.

Schiffman, S. S. and Erickson, R. P. 1971. A psychophysical model for gustatory quality. *Physiol. Behav.* 7: 617.

Schiffman, S. S. and Erickson, R. P. 1980. The issue of primary tastes versus a taste continuum. *Neurosci. Biobehav. Rev.* 4: 109.

Schiffman, S. S., Moroch, K., and Dunbar, J. 1975. Taste of acetylated amino acids. *Chem. Senses Flav.* 1: 387.

Schiffman, S. S., Robinson, D. E., and Erickson, R. P. 1977. Multidimensional scaling of odorants: Examination of psychological and physiochemical dimensions. *Chem. Senses Flav.* 2: 375.

Schiffman, S. S., Musante, G., and Conger, J. 1978. Application of multi-dimensional scaling to ratings of foods for obese and normal weight individuals. *Physiol. Behav.* 21: 417.

Schiffman, S. S., Reilly, D. A., and Clark, T. B. 1979. Qualitative differences among sweeteners. *Physiol. Behav.* 23: 1.

Schiffman, S. S., McElroy, A. E., and Erickson, R. P. 1980. The range of taste quality of sodium salts. *Physiol. Behav.* 24: 217.

Schiffman, S. S., Reynolds, M. L., and Young, F. W. 1981. *Introduction to Multidimensional Scaling.* Academic Press, New York.

Scott, T. R. and Chang, F-T. T. 1984. The state of gustatory neural coding. *Chem. Senses* 8: 297.

Settle, R. G., Meehan, K., Williams, G. A., Doty, R. L., and Sisley, A. C. 1986. Chemosensory properties of sour tastants. *Physiol. Behav.* 36: 619.

Skarda, A. and Freeman, W. J. 1987. How brains make chaos in order to make sense of the world. *Behavioral and Brain Sciences* 10: 161.

Smith, D. V. and McBurney, D. H. 1969. Gustatory cross-adaptation: Does a single mechanism code the salty taste? *J. Exptl. Psychol.* 80: 101.

Soltan, H. C. and Bracken, S. E. 1958. The relation of sex to taste reactions. *J. Hered.* 49: 280.

Stevens, S. S. 1957. On the psychophysical law. *Psychol. Rev.* 64: 153.

Stone, H., Sidel, J., Oliver, S., Woolsey, A., and Singleton, R. C. 1974. Sensory evaluation by quantitative descriptive analysis. *Food Technol.* 28: 24.

Torii, K. and Cagan, R. H. 1980. Biochemical studies of taste sensation. IX. Enhancement of L-[^{3}H]glutamate binding to bovine taste papillae by 5'-ribonucleotides. *Biochem. Biophys. Acta* 627: 313.

Yamaguchi, S. 1979. The umami taste. In *Food Taste Chemistry* (S. Boudreau, Ed.), p. 33. Amer. Chem. Soc. Symp. Ser. No. 115, Washington, DC.

Yamaguchi, S. and Kimizuka, A. 1979. Psychometric studies on the taste of monosodium glutamate. In *Glutamic Acid: Advances in Biochemistry and Physiology* (L. J. Filer, S. Garattini, M. R. Kare, W. A. Reynolds, and R. J. Wurtman, Ed.), p. 1. Raven Press, New York.

Yamaguchi, S. and Takahashi, C. 1984. Hedonic functions of monosodium glutamate and four basic taste substances used at various concentration levels in single and complex systems. *Agric. Biol. Chem.* 48: 1077.

Yamaguchi, S., Yoshikawa, T., Ikeda, S., and Ninomiya, T. 1971. Measurement of the relative taste intensity of some 1-2-amino acids and 5'-nucleotides. *J. Food Sci.* 36: 846.

Yoshida, M. 1963. Similarity among different kinds of taste near the threshold concentration. *Jap. J. Psychol.* 34: 25.

9

Language and Categorization in Wine Expertise

Gregg E. A. Solomon

Harvard University
Cambridge, Massachusetts

CATEGORIZATION IN COGNITION AND PERCEPTION

Describing Wines

We laugh at the taster who describes a wine as *presumptuous* but not at the taster who describes it as *sweet,* disagree as we might with both. What is the difference? Both tasters' descriptions could be wrong, but wrong in very different ways. What one taster called sweet another might call dry. Here the correctness of the term would hinge on just where on a continuum each taster decided that sweetness ended and dryness began. But to call a wine *presumptuous* may be considered incorrect for a more profound reason; it could be argued that *presumptuous* is an entirely inappropriate term for describing a wine. Any wine. *Presumptuous* seems somehow a less sound attribute of a wine. It seems less fundamental, less scientific.

What then is the right way to describe a taste? Food scientists have long been interested in deriving the ultimate vocabulary of taste and smell. Conceivably, with such a vocabulary, expert judges could unambiguously represent the taste or smell of a food. To say the least, the advantages would be considerable. Information could, for example, be conveyed to other judges not present, or the sense of a volatile food could be compared as it changed over time. In industry, market analysts would have a greater knowledge of what consumers liked or did not like in their products and could better convey this information to experts in research and development.

But when we assess what it means for a description to be *right*, to what authority do we appeal? One of the principal problems in deriving this great fundamental lexicon of taste and smell is that groups of people vary; within groups, individuals vary; and individuals themselves even vary over time. Expert panels derive vocabularies for describing products that convey little to market analysts, let alone to naive consumers. I argue that the problem of group and individual differences is in fact no problem. It is part of the answer.

Are experts closer to the true way of describing nature than are novices? Are they more right in some absolute sense? We must consider an assumption underlying the search for the fundamental vocabulary of taste and smell: that such a vocabulary exists to be found. This is an assumption about the nature of perception. Is there a single, truest, way of classifying things in nature? In naming and labeling objects in the world, we classify them. When we say that an object is some type of thing (a sweet wine), we imply that it has something in common with others of that type. If I call a particular entity a *dog*, then obviously I imply that it has something in common with other *dogs*. We cannot help but create categories of things: *men* as opposed to *women, adults* versus *children, good* chili versus *bad* chili, *sweet* wines versus *nonsweet* wines. Categorization is a part of thinking. Obvious as the fact of naming may seem, how it is that we come by one name for an object as opposed to another is far from obvious.

Let us assume that some means of classifying wines are better than others because they better capture the way things are grouped in nature. What is it about the world that is captured by calling a wine *presumptuous?* Perhaps not much. The search for the fundamental odor and taste vocabulary would lie, in Plato's words, in attempting to "carve nature at its joints." There is structure in the world as we know it through our senses. Through a convergence of the food, social, and physical sciences we may yet arrive at a cross-cultural vocabulary that best capture basic and universal groupings of foods in the nature. There may actually be a discontinuity in nature between *sweet* and *nonsweet* foods. As evident as it appears to us that humans can be divided into *men* and *women,* so may there be equally self-evident discontinuities among foods. And so some vocabularies of classification would be truer than others.

There is another possibility: It may be that the vocabulary of one group cannot be said to be closer to the truth of the way things are in nature and so better than that of another group. Perhaps no single, most fundamental, scheme for classifying odors and tastes exists. There may be many equally valid ways of naming. *Presumptuous* may strike us as an odd term to use in describing and classifying wines, but it may not be worse in any absolute sense than calling a wine *sweet*. By this reasoning, the attempt to find a single vocabulary of taste and smell may be doomed from the start. Where does that leave us?

Enter experimental psychology. Individual and group differences in describing sense perceptions, far from being a problem, are in fact a means by which experimental psychologists infer the principles of thought and perception. By studying how people and groups differ and how they are the same, by studying patterns of behavior, we can address the assumptions underlying the search for a vocabulary of taste and smell. Comparisons of expert and novice wine tasters promise to tell us much about the extent to which groups and individuals differ. Knowing how and why experts and novices differ, we may also know something of how experts can best communicate to novices, and even of how novices become experts. But our first and central aim is to determine just how different people can be.

Expert-Novice Differences in Describing Wines

Expert and novice wine tasters certainly appear to be different. But demonstrating that difference reliably has proven to be quite difficult. Tasters differ in subtle ways; the methodological challenge is to set tasks difficult enough to expose meaningful differences between groups but not so difficult as to prove impossible. The linguist Adrienne Lehrer (1983) attempted to assess the abilities of wine tasters to describe wines. She had tasters describe three wines, then had other tasters attempt to match those descriptions to the wines about which they were written. Note that this method entails no assumptions about which descriptions are more true in an absolute sense. A description is considered correct only if it is matched to the right wine. This is known as referential agreement— subjects must agree on the reference of a term. Surprisingly, Lehrer found that tasters could not match descriptions to wines at a level above chance. There was no referential agreement.

Lawless (1984) duplicated Lehrer's method and found that not only could expert subjects match descriptions to wines, but they were significantly better at the task than were novices. Solomon (1990) recently confirmed the superiority of experts to novices in communicating information to each other about wines using a somewhat different methodology. Like Lehrer and Lawless, Solomon had his subjects describe a group of wines. However, in the second phase of his experiment, in which tasters attempted to match the descriptions to the wines, the tasters were presented with one wine at a time along with two descriptions. One of the two descriptions had actually been written about the wine presented and the other descriptions had been written about an entirely different wine. It was the subject's task to decide which of the two descriptions had been written about the wine at hand. This method allowed for more powerful analyses of the differences between groups and for more controlled comparisons. Solomon found that experts came to constitute something of a linguistic community. That is, experts could communicate information to each other at a level that novices could not share. Novices could not match wines with the descriptions written by other novices nor those of the experts, and the experts could not match the descriptions written by novices at a level above chance. But the experts could match the descriptions written by other experts. They showed greater referential agreement. In later studies, Solomon found that experts also outperformed the novices at discriminating among a series of very similar wines in a triangle test and at rank-ordering wines on a number of dimensions. These studies suggest that the greater precision that experts demonstrated in describing wines is, at the very least, associated with a more precise perceptual discrimination perform-ance.

In generalizing from these studies, we must be particularly cautious in interpreting the expert advantage at raw perceptual discrimination tasks. It is certainly possible that experts have greater initial discrimination abilities and that this given raw advantge leads to their superior performances on the other more complicated tasks. Their advantages at matching descriptions to wines could have resulted from their abilities to perceive the subtle differences in the stimuli that understanding descriptions requires. Though there is likely some threshold sensory sensitivity that is a prerequisite for improvement, it is not entirely accurate to say that experts are born and not made. Granted, their superior abilities could have led to their interest in becoming familiar with wines in the first place and then to their seeking to improve their abilities. Experts may in this way be something of a self-selected group. But innately given, raw perceptual acuity may not of itself account for all expert-novice differences. There is likely to be some influence of specialized linguistic experience and cognitive organiza-tion on differences between experts and novices.

Expert-Novice Differences: Evidence from Other Domains

Observers frequently comment on the seemingly remarkable abilities of chess experts to consider vast numbers of moves at a time. Studies by de Groot (1965) and Chase and Simon (1973) found that chess experts had no advantage in raw perceptual or cognitive capacity over novices. Experts did not see more, they saw better. They organized their perceptions in more meaningful ways and so could better act upon them. Consistent with anecdotal evidence, chess experts proved to be far better than novice subjects at reconstructing chess boards that had been taken from actual games. However, when subjects were asked to remember boards on which the pieces had been placed randomly, the experts actually recalled fewer pieces than did the novices. Chase and Simon went on to show that the expert recall superiority for the actual game boards was attributable to their recognizing the individual chess pieces as parts of larger, meaningful chunks. Experts considered groups of pieces together in terms of patterns of attack and defense. There were rule-governed associations within each chunk. Where a novice might memorize three or four things (individual pieces), an expert merely has to memorize one thing (a pattern, consisting of three of four pieces). When the pieces on the random boards were illegally placed, rule-governed chunks could not be created, and the expert recall superiority disappeared.

Chi et al. (1981) posited that expert physicists would show a similar superiority in cognitive organization. They demonstrated that a group of expert physicists, unlike novices, solved physics problems by considering the principles that lay beneath the problems, principles that related groups of problems. Chi et al. hypothesized that the representation of a problem is made in the context of its being categorized as a specific type of problem. "Thus, expert-novice differences may be related to poorly formed, qualitatively different, or nonexistent categories in the novice representation. In general . . . much of expert power lies in the expert's ability to quickly establish correspondence between externally presented events and internal models for these events" (Chi et al., 1981, pp. 122–123). By implication, expert wine tasters or food panelists may have certain cognitive or categorizing advantages in addition to any greater perceptual sensitivities they may possess.

Wine Categories

We can distinguish two ways by which experts may be said to differ from novices: They may differentiate more finely the individual perceptual features and dimensions of a wine, or they may possess more differentiated categories of wines. These two kinds of differentiation may, in fact, work in concert. Experts routinely identify far more attributes in wines and use more and more precise wine terms (Solomon, 1990). Such seemingly basic tastes as sour and bitter are regularly confused by novices (Amerine and Roessler, 1976; O'Mahony et al., 1979); they either use the terms interchangeably or use a more inclusive term such as "hard" or "biting" to suggest sour or bitter. The greater expert differentiation of categories of wine types is obvious to anyone who has had prolonged interaction with true experts. Where novices may only identify wines as being red or white or good or bad, experts further identify wines as cabernet sauvignons or pinot noirs or chardonnays, and some even name exact years of particular chateaux. The question is just what such categorization has to do with expert wine tasting performance.

Categorization allows us to make inferences. Rather than attend to every detail of every new object, we can consider instances of a class to be functionally equivalent, and so we are freer to attend to other aspects of the environment. It sufficed for our hirsute forebears to place large, fanged and clawed mammals in the *dangerous animal* category and so to allocate the rest of their limited attentional resources to surveying other aspects of surroundings and to getting to safety. Those creatures who spent too much their cognitive energies attending to every new animal as if it were unique failed to take advantage of some of the more obvious critical the benefits of categorization. As W. V. O. Quine said, "Creatures inveterately wrong in their inductions have a pathetic but praiseworthy tendency to die before reproducing their kind" (1969, p. 126). It would seem rather adaptive for an animal to be able to infer that because one saber-toothed tiger was dangerous, another saber-toothed tiger could be so characterized. Exemplars of the same category may be expected to be equivalent along certain dimensions.

Lawless (1985) has suggested one way that superior expert categorization could simplify the cognitive aspects of tasting: If a wine is of a given type, then ostensibly it has a greater chance of having certain characteristics. As experts taste a given type of wine, a particular Bordeaux for example, they know to look for hints, say, of raspberry or bell pepper. Knowing which characteristics have a higher probability of varying together would make the cognitive tasks easier. We thus find ourselves considering the implications for wine tasting of theories of categorization.

Prototype Theory

Traditional theories of categorization do not allow for judgments about the probability that a member of a category has certain attributes. According to the classical view (dating back to Aristotle), a category is defined by a set of necessary and sufficient features. The category *bachelor* is defined by the attributes: *human, male, adult,* and *unmarried.* An exemplar has all of the attributes defining the category—something either is or is not a member of a category. All exemplars of a category, therefore, ought to be judged as equally good instances of it. They ought to be, but they are not. The classical view cannot explain that a robin, for example, is judged to be a better example of the category bird than is an ostrich. Nor can it account for the observation that subjects are faster at deciding that a robin is a member of the category bird than at deciding that an ostrich is a bird. Such phenomena are called prototypicality effects.

Posner and Keele (1968) demonstrated there to be a relationship between category structure and prototypicality effects. Subjects were trained to learn categories of abstract dot patterns that had been systematically distorted to varying degrees from prototypes. The subjects' abilities to classify new patterns were highly correlated with the degree to which the patterns varied from the prototypes, though the subjects had never actually seen the prototypes. Furthermore, when presented patterns in a later recognition test, subjects were most likely to claim that they had seen the prototype, though they had not. Posner and Keele (1968) interpreted their results as indicating that the subjects actually abstracted a prototype and made judgments based on it.

Rosch and Mervis (1975) extended implications of the prototypicality effects by building on Wittgenstein's (1953) notion of family resemblances. Wittgenstein asserted that the classical view is inadequate to account for all kinds of categories. He argued that, for example, the category *game* cannot be defined by a set of necessary and sufficient features. He proposed, instead, that the category is characterized by a number of features that are typical of the group, but all of the features are not necessarily true of any one instance.

Rosch and Mervis (1975) further demonstrated that judgments of prototypicality do indeed derive from the family resemblance structure of the category. They created categories of letter strings with predetermined family resemblance structures. Those instances that were judged by subjects as most prototypical of a category shared more of the features common to other instances of the category and fewer of the features common to instances of contrasting categories. Most of the Smith brothers are bald, have big noses, and wear glasses. No single brother has all of these characteristics. And not all of the Smith brothers have beards, not all wear glasses, not all have big noses, not all are bald, yet they are all still identifiably members of the same family even though no two share all of the same qualities. There need not be any *necessary* features of the category, there is no single features that all exemplars must possess to be considered members of the category. Rather, there are a number of features that are *common* to the category. Those Smith brothers with more of the features common to the Smith family will be judged as more prototypical, as better examples of the family, than will those brothers with fewer of the features. By this reasoning, the probability of our judging a wine to be central example of a given wine category is related to the number of features it possesses that are common to other members of the category. Conversely, knowing that a wine is a member of a given category, a given family of wines, a wine expert could conceivably make probabilistic judgments about whether the wine had a given attribute based on his or her knowledge of the attributes common to exemplars of the category.

LINGUISTIC SYSTEMS

Linguistic Relativity

How do novices acquire expert knowledge of the families of wines, and how, if at all, is such knowledge accomodated into their conceptual systems? Can expert wine talk lead novices to expert wine perceptual categories? We return to our consideration of the nature of perception. The acquisition of perceptual categories has been most thoroughly investigated in the domain of color recognition and color terminology. The impetus for this line of research was largely provided by the provocative writings of Edward Sapir (1921) and Benjamin Lee Whorf (1956).

> The categories and types that we isolate from the world of phenomena we do not find there because they stare every observer in the face; on the contrary the world is presented in a kaleidoscopic flux of impressions which has to be organized in our minds—and that means largely by the linguistic system in our minds. We cut nature up, organize it into concepts, and ascribe significances as we do, largely because we are partners to an agreement to organize it in this way—an agreement that holds throughout our speech community and is codified in the patterns of our language. (Whorf, 1956, p. 213)

Whorf claimed, for example, that syntactic differences between English and Hopi lead native Hopi speakers to acquire a conception of time different from that of native English speakers. In Standard European languages, the major grammatical divisions are between objects (nouns) and actions (verbs). In Hopi these are both represented by duration. The major grammatical principle in Hopi is between the manifest and the nonmanifest. They have a different world view; they understand the world differently.

Whorf's linguistic relativist position holds that differences between languages will be paralleled by nonlinguistic cognitive differences in the native speakers of the languages. He further claimed that these differences are determined by the differences in the languages. If this hypothesis is true, then the search for a single, cross-linguistic vocabulary of taste and smell is misguided. There is no single appreciation of nature that guides the naming and organizing of our perceptions.

Color Terms: Evidence Against Extreme Linguistic Relativity

Brown and Lenneberg (1954) put the Whorfian hyopthesis to the test. Their challenge was to find a domain in which languages differed in the way they carved it up, but in which there was some way of measuring cognitive or perceptual ability independent of language. They chose to study color memory, for languages appeared to differ wildly in their color lexicons and color memory could be measured independent of naming. Brown and Lenneberg were thus concerned with whether the color terms of a language group determined how native speakers categorized their color perceptions. They found that people performed better on recognition tasks for colors that were highly codable. Codability was their measure of the degree of agreement on the naming of a color within a linguistic community. Colors for which there was less naming agreement proved to be more difficult to recognize in a memory task. Brown and Lenneberg, assuming that different languages would give rise to different codability scores, inferred that members of different linguistic communities would have different constructions of reality. There was support for Whorf's hypothesis.

Subsequent research in the color domain, however, has refuted an implication of Brown and Lenneberg's work: that a native speaker's language will determine his or her cognitive structures. Berlin and Kay (1969) found a striking cross-cultural regularity in the categorization of colors. Though language groups varied in the number of color terms that they used, the particular colors for which terms existed in a language followed a predictable sequence that depended on the number of color terms that existed in the language. For example, if a language had only two color terms, they would be terms roughly translating to "black" and "white." If the language contained a third term, it would be "red," and so on. This scale does not necessarily indicate the sequential evolution of color terms through which a language group proceeds. However, Berlin and Kay did find the complexity of color vocabulary to correlate with a measure of cultural and technological complexity and so suggested such a sequential evolution to be plausible.

Note that these terms do not necessarily match our meanings of the same color terms: They are expanded applications. The term we translate as "black" from a language with only two colors, black and white, does not necessarily match what we mean by "black." Rather, their term "black" would more accurately correspond to what we roughly refer to as the darker or cooler colors. Bousfield (1979) criticized this research, asserting that we cannot know that the color terms in a language mean the same thing to members of another language group that they do to us. Yet, as long as we define meaning in terms of referential agreement, agreement within a community about the reference of a term, we have no fears of Bousfield's criticisms. We need not consider whether your phenomenological experience of blue means the same thing as my experience of blue. If you call an object blue and I call it blue, then that is sufficient. We agree to agree. And indeed, across cultures people agree as to the best examples of colors, called "focal colors." They agree, for example, as to which of an array of different shades of yellow is the best example of yellow. Across cultures, people agreed the particular shade of yellow called focal yellow was a better example of the category yellow than were nonfocal yellows (Ticonderoga pencil yellow is an example of focal yellow).

In a series of experiments further undermining the Whorfian implications of Brown and Lenneberg's study, Heider and Olivier (1972) and Heider (1972) demonstrated that members of different language groups showed similar recognition memory performance for colors despite the fact that the languages contained different arrays of color terms. Like English speakers, members of the Dani tribe of New Guinea, whose language contains only two color terms, "mili" and "mola," were better at recognizing focal colors than nonfocal colors. Even though they had no term that corresponded exactly to what we call "yellow," they were much better at remembering focal yellow than they were at remembering nonfocal yellows. Focal yellow and nonfocal yellow would both be called *mili* by the Dani. Because they coded both focal and nonfocal colors identically, codability clearly could not have determined these prototype effects.

In the debate over whether members of different languge groups differ arbitrarily in how they carve up the color world, greater emphasis has been placed on perceptual predispositions. Attempting to explain Berlin and Kay's discovery of the regularity of color terminologies across languages, Kay and McDaniel (1978) found the selection of color terms for a language to follow nicely from divisions of the wavelengths of the focal colors. Furthermore, studies of primate visual systems actually have provided neurophysiological support for the psychological reality of focal colors (De Valois and Jacobs, 1968). It appears that certain colors, focal colors, are more memorable not because we have better terms for describing them, but because our physiological equipment makes them more memorable. And because they are more memorable, we may be more apt to have better terms for them.

This is not to claim that language has no effect at all on the ways we categorize colors. There do appear to be effects of language on the categorization of nonfocal colors. There is certainly some room for relativity in deciding where one category ends and the other begins. Kay and Kempton (1984) have shown that subjects' estimations of which of three colors is more different from the others, a nonlinguistic cognitive task, can be influenced by the way their language systems categorize the colors. Subjects were presented with three color chips at once: a chip the color of a nonfocal blue, a chip the color of a nonfocal green, and a middle chip. The wavelength of the middle chip was near the border between blue and the green. When the middle chip fell on the green side of the border, that is, when it was named green by subjects, it was judged by English speakers as being more different from the blue than from the other green, even if physically it was not more different. Similarly, when the middle chip was coded as blue, it was judged to be more different from the green chip than from the other blue chip. Native speakers of the Mexican Indian language Tarahumara do not have separate words for blue and green; both colors are coded as *siyóname*. When estimating differences between the colors, the Tarahumara speakers were just as likely to say that the green was more different from the middle chip as they were to say that the blue was more different. The results appear to indicate that, albeit in a small way, the way colors are named affected the way that native English speakers judged the physical similarities of the colors. Colors of the same name were judged to be more similar whether or not they actually were. It is at the extreme borders of categories that linguistic relativity appears to have some validity. Admittedly, this finding, important though it is, is a far cry from the grand claims of the Whorfian hypothesis.

Categorical Perception

Once seen as a promising domain in which to study the relativity of perception, much of the work in color categorization has instead pointed to the existence of perceptual universals. The Kay and Kempton findings notwithstanding, the color literature indicates that we are predisposed to making certain categorizations, if we make them at all, according to universal rules for category creation. Let us now turn to studies in another modality. Much of the work in psycholinguistics has also concerned the influence of language on perception. Is the influence of language so great that people would vary in their abilities to make acoustic discriminations?

Early investigations of phonemes indicated that we may perceive speech signals as falling into discrete categories. Phonemes, our basic units of speech, are roughly the oral equivalent of the letters of the alphabet, or more correctly, of the symbols used in dictionaries to guide pronunciation. The phoneme [æ], for example, is the oral equivalent of the English "a" sound in the word "bat." The phonemes [b] and [p] are called bilabial stop consonants because they are produced by an explosion of air from between the lips and by an activation of the vocal cords. In fact, the time of vocalization relative to the time air is expelled from between the lips, known as voice onset time (VOT), is what distinguishes the syllable [ba] from [pa]. An English speaker would vocalize 40 msec after expelling air in order to make a [pa] sound (a VOT of +40) and would vocalize at about the same time as expelling air in making a [ba] (a VOT of 0).

Lisker and Abramson (1970) synthesized a number of syllables that varied only along VOT. Despite the fact that the synthesized speech signals fell along a continuum, subjects' category judgments about whether a speech signal was a [ba] or a [pa] changed abruptly at a VOT of +25 msec. Moreover, the only disagreements between subjects concerned speech signals whose VOTs occurred at about +25 msec, the boundary between the phonemes.

Liberman et al. (1957) found that subjects were better able to discriminate between speech sounds that fell on different sides of the phonemic category boundary than they were between speech sounds that fell on the same side. This phenomenon is known as categorical perception. Speech signals sound more different when they are on oppsosite sides of the category boundary than do sounds that both lie on the same side—even if the latter two signals are in fact more physically different. A native speaker of English cannot tell the difference between a speech sound with a VOT of 0 msec and a sound with a VOT of –40 msec, but can tell the difference between a speech sound with a VOT of 0 msec and a sound with a VOT of +30 msec. It appears that our language processing systems somehow affect the way we perceive the world.

Researchers were led to question just how innately determined were the specific category boundaries and the general phenomenon of categorical perception. How much variation can there be between members of different language groups? In light of such questions, developmental studies took on great importance. Eimas et al. (1971) demonstrated that even prelinguistic, one-month-old infants show categorical perception of certain phonemes. They further found that infants discriminate speech sounds into categories that are phonemic for at least some known languages. Moreover, rhesus monkeys (Morse and Snowden, 1975) and chinchillas (Kuhl and Miller, 1975) have been shown to make some categorical distinctions on speech segments that correspond to human phonemic boundaries. Our shared categorical perceptions may be innately given to the extent that our auditory systems are similar to those of chinchillas and rhesus monkeys.

But phonemic boundaries cannot be entirely innately determined. Cross-cultural studies have shown that subjects from different language groups have different boundaries at which they make certain categorical distinctions (Lisker and Abrahamson, 1970). And speech sounds are more discriminable for subjects for whom the sounds were separated by a categorical boundary than they are for subjects whose languages do not operate with such a boundary. For example, there is no phonemic boundary between the speech sounds for [r] and [l] in Japanese, so native speakers have great difficulty discriminating between them (Miyawaki et al., 1975). It has also been shown that very young infants cannot make all of the categorical discriminations that adults can. Butterfield and Cairns (1974) suggested that we are predisposed to learn to make certain cuts in speech world, but that some distinctions require experience in a specific language before they are acquired. So different languages *can* affect the way we categorize the world, even to the point that they can influence our judgments as to whether stimuli are the same or different.

More recent studies have indicated that categorical perception is not quite as simple as was first believed. Findings on within-category discrimination have indicated that distinctions within phonemic categories are not completely suppressed. A bilabial stop consonant with a VOT of 0 msec and a bilabial stop consonant with a VOT of –40 are both considered to be [b]s, but subjects can be trained to tell them apart. Pisoni (1973) posited that the reason the earlier categorical perception tasks (Liberman et al., 1957, 1967) found subjects unable to use within-category acoustic information to make discrimination judgements was that the memory demands of the task were too high. Pisoni et al. (1982) used a simplified task that reduced the attentional and cognitive processing demands on the subjects, the cognitive load as it were. They demonstrated that adult English-speakers could, in fact, be trained in a short period of time to make category-like discriminations of speech stimuli that are considered to be the same phoneme in English.

These studies indicate that though we may be predisposed to making certain perceptual categories, and though these categories are in part mediated by linguistic experience, the information that is not directly relevant to categorization is not completely lost. Rather, because certain distinctions are less important in normal day-to-day functioning (such as in understanding speech), subjects must pay greater attention if they are to use this less salient information in discrimination tasks. Experts may have less of their cognitive resources tied up in the mundane aspects of discrimination tasks. Thus, trained expert subjects, subjects for whom there may be a lesser cognitive load, can learn to make distinctions that are not innately determined, distinctions that would elude a novice.

One way that cognitive load may be reduced for experts is through the process of automaticity. Simply, some tasks that initially require an individual to pay great attention eventually become mindlessly rote. For example, novice car drivers usually find that shifting gears requires tremendous attention. Experienced drivers can shift virtually without thinking; shifting has become automatic. Experts can therefore attend to other aspects of driving. Aspects of taste and odor discrimination tasks may be more automatic for expert tasters than for novice tasters, and so the experts may be able to pay greater attention to more subtle discriminations.

PERCEPTUAL SYSTEMS AND EXPERTISE

The Role of Experience in Reducing Memory Load

Olfaction may not be as different a modality as was first suspected. Yet, it is not at all clear just what *causal* role our linguistic abilities may play in wine expertise. Gibson (1966) place greatest emphasis on the roles of experience and the environment in perceptual descrimination. There is a tremendous amount of information in the world, and our sense organs are designed so to retrieve it.

According to this view, the perceptual system slowly becomes attuned to the information that is in the sense array. Gibson believed that perceptual learning entails increasingly more selective attention. Simply through greater experience, an individual naturally comes to make more differentiated discriminations. Experts, by reason of their greater experience in a domain, come to perceive aspects of a stimulus array that are not obvious to the novice.

> The gentleman who is discriminating about his wine shows a high specificity of perception, whereas the crude fellow who is not shows a low specificity. A whole class of chemically different fluids is equivalent for the latter individual; he cannot tell the difference between a claret, burgundy, and chianti; his perceptions are relatively undifferentiated. What has the first man learned that the second man has not? . . . he has learned to taste and smell more of the qualities of wine, that is, he discriminates more of the variables of chemical stimulation. If he is a genuine connoisseur and not a fake, one combination of such variables can evoke a specific response of naming or identifying and another combination can evoke a different specific response. He can consistently apply nouns to the different fluids of a class and he can apply adjectives to the differences between fluids. (Gibson and Gibson, 1955, p. 35)

With experience, an individual learns which aspects of a perceptual array warrant attention and which ought to be disregarded as irrelevant. In contrast, the novice is overwhelmed by the sheer mass of information present in the environment. Gibson, despite his abhorrence of all things representational, allowed that language could influence perceptual learning by highlighting those features of the environment that are most salient. Verbal cues could serve to direct attention to salient aspects of the environment and so facilitate perceptual learning. If expert wine descriptions are better, it follows that perceptual learning for experts should be better facilitated.

The Role of Language in Reducing Odor Memory Load

It has long been known that verbal labels can faciliate recognition memory performance for lexical and for visual items. If subjects can name something, it is much easier for them to remember it. In addition to directing attention to salient aspects of a sensory array, verbal cues might also reduce memory load, thereby freeing attention for discrimination. Experts might therefore make finer perceptual discriminations in part because they have good names for odors and tastes.

Is there a relationship between odor memory and descriptions? Daniel and

Ellis (1972) had demonstrated that stimulus codability was highly correlated with recognition performance for visual form. The more easily named was the visual form, the better one could remember it. There appeared, however, to be much less of a systematic matching of verbal labels to odors. Surprising research by Engen and Ross (1973) suggested that olfactory perception was unique among the modalities in that there appeared to be no relation between language and memory for odors. They found that neither verbal labeling nor familiarity with an odor improved memory performance. They presented subjects with a series of odors and a list of common names for the odorants. Another group of subjects was presented with a list of associations to the odorants. For example, *antiseptic, Lifesavers,* and *root beer* were association labels for odors whose common name labels were *alcohol, wintergreen,* and *licorice.* Subjects were instructed to match the odors with the correct labels. A final group received no labels at all. In the forced-choice recognition task conducted 3 months later, the subjects who had received the common name labels performed no better than the subjects who had received no labels at all. In fact, they found that performance was worse for subjects who had coded odors with the associations labels.

Engen and Ross (1973) suggested that odors may be experienced as "unitary perceptual events." That is, we may perceive olfactory stimuli with fewer features than we do other stimuli, therefore we tend to encode them more holistically. Later work (Davis, 1977; Lawless, 1978) tested more directly whether it was the paucity of features in the olfactory perceptual array that led to such poor initial recognition performance or whether the phenomenon was peculiar to olfaction. Lawless (1978) presented subjects with a set of odors, a set of simple, free-form shapes, and a set of pictures clipped from travel magazines. The subjects were subsequently tested for recognition memory. The recognition rates for the complex pictures were 100% for the first month but then dropped steadily to 81% over the next 3 months. The performance rates for the odors and simple shapes dropped 15% during the first month and then flattened out. Because the free-form shapes, assumed to be perceptually simple, were remembered in a manner similar to that of odors, Lawless contended that it is the feature-impoverished nature of odors that leads to their encoding as unitary events, rather than some property peculiar to olfaction.

Lawless also found support for the finding that while odors are often judged as familiar, such recognition is independent of verbal associations. Subjects often reported that they had smelled an odor before but they could not produce its name. Lawless and Engen (1977) had reported that in this *tip-of-the-nose* state the subjects could provide virtually no information about the name of the odor itself. However, subjects could name similar odors, indicating that they had some information about the odors when in this state.

Rabin and Cain (1984) argued that verbal labels could indeed facilitate odor recognition performance. They argued that the associational labels Engen and Ross had instructed their subjects to learn were inappropriate. In coding a peer's associations to an odor, their subjects had attempted to use labels that were not meant to have communicative accuracy. Furthermore, they argued that, because the forming of verbal-odor associations proceeds so very slowly, the previous findings in the odor literature had confounded the relative difficulty of label retrieval with a lack of any verbal-odor association at all.

In their own studies, Rabin and Cain found memory for odors to be strongly associated with label use. They presented subjects with a set of odors one at a time. Subjects rated the familiarity of each odor and attempted to generate a label for it. The results of a series of forced-choice two-alternative recognition memory tests indicated that memory for odors was strongly associated with rating of familiarity, the accuracy of the label, and the consistency of label use by an individual at the time of presentation and during subsequent recognition tests. It should be noted that odors that were badly encoded were still recognized at a better-than-chance rate, though not as high a rate, indicating that subjects could make use of information that is independent of label use. Other studies have since supported the contention that good descriptions can improve recognition performance (Walk and Johns, 1984; Lyman and McDaniel, 1986). Good descriptions would then appear to be an inextricable part of being an expert taster, both because, in a Gibsonian sense, the descriptions serve to direct a taster's attention to salient aspects of the sensory array, and because, in a cognitivist sense, the descriptions may serve to reduce memory load and free attention for subtle discrimination.

HIERARCHICAL SYSTEMS

The Basic Object Level

So how shall a wine be described? How is wine best named? Brown pointed out some years ago (1958) that an object can be referred to in many ways. "The dime in my pocket is not only a *dime*. It is also *money*, a *metal object*, a *thing*, and, moving to subordinates, a *1952 dime*, in fact, *a particular 1952 dime* with a unique pattern of scratches, discolorations, and smooth places" (p. 14). Yet, one name seems truest. He suggested that this level of truest classification was determined by that object's "usual utility." That is, we tend to name objects according to the usual way in which we interact with them. We seek in conversation to convey the appropriate amount of information, neither too little to be of use, nor so much that we either waste effort or confuse our conversational partners with unnecessary detail. We refer to objects in such a way as to convey, to highlight, that sense of it that is pertinent to the interaction. I would not say to a friend "Here's a 1952 dime for the phone," nor would I say "Here's a metal object for the phone." I'd say "Here's a dime for the phone."

Rosch and colleagues (1976) built upon Brown's line of reasoning, proposing a hierarchy of categorizations. An object is a member of a number of categories running from the most inclusive, superordinate levels (e.g., furniture, vehicle) to subordinate, more specific levels (e.g., kitchen chair, sports car). They emphasized the importance of correlations of features in the perceptual world at what they called the "basic object level." They defined the basic object level as the most abstract level of classification at which a number of behavioral factors converged. Essentially, this is the level of classification at which categories are maximally differentiated. Naming objects at this level (implicitly classifying them) communicates in the optimal way: It is a balance of the amount of information conveyed about an object and the cognitive effort associated with different levels of naming. The argument turns, in part, on the assumption that naming objects at more subordinate levels is generally a less economical expenditure of effort. For example, the object I see before me I am likely to call a guitar. True, it is also a *Japanese, Takamine classical guitar made in 1976*, but to call it so is to expend more cognitive effort with little gain in information. When I refer to the object as a *guitar* I imply a level of classification; I implicitly contrast it with *trumpet* or *violin* or *piano*. Each member of the category *guitar* shares many features with other guitars relative to the number of features it does not share with other guitars and relative to the number it shares with trumpets and pianos. To call a guitar a *1976 Takamine classical guitar* is to contrast it with, say, *1975 Takamine classical guitars*. At this more subordinate level, there are relatively fewer features that are typical of 1976 Takamine classical guitars but not of 1975 Takamine classical guitars. Exemplars of two different categories at this level of classifications share many more features relative to the number that they do not share. Naming an object at this subordinate level entails a fineness of differentiation that requires greater effort. Similarly, my naming the object before me *musical instrument*, a more superordinate level, is also not optimal. Musical instruments share fewer features with other musical instruments relative to the number of features that they do not share. At this level, differences between members of categories are more salient than are similarities.

It is generally the case that naming the object before me a 1976 as opposed to a 1975 Takamine classical guitar is to invoke unnecessary discriminations between categories, but there are those for whom finer distinctions between categories are meaningful. To a guitarist or a guitar dealer or a lutier, the differences between guitars of different years might be of great importance. For most of us, *guitar* is a basic object level category. But for lutiers, *guitar* may not convey appropriate distinctions between categories. Hence, Brown's term "level of usual utility is" synonomous with Rosch's "basic object level." For lutiers, the level of usual utility, the basic object level, is a level that would be subordinate for most people. For expert wine tasters, a basic object level categorization might be *1975 Chateau Margaux* as opposed to *1976 Chateau Margaux*. For the novice, the basic object level might be *wine* as opposed to *soda*.

Ethnobiology

We return to consideration of relativity in our carving up of nature. Lawless (1988), sympathetic to the notion of hierarchies of classification, cautions that odor naming tends to be idiosyncratic and that different cultures may utilize very different, nontranslatable, classification schemes based on their different experiences. There may yet exist regularities across cultures. Working in the branch of anthropology called ethnobotany or ethnozoology, Berlin (1978) derived a taxonomic hierarchy through his cross-cultural studies of folk biological classifications. One might have supposed that the classification schemes between pre-scientific cultures might also have been nontranslatable, that systems of classifying nature might vary wildly and even unpredictably. Instead, Berlin found a "generic" level of classification roughly corresponding to Rosch's basic object level. Cultures around the world tend to name and group plants and animals in the same ways. Berlin proposed that there is a level of categorization at which gross perceptual and behavioral discontinuities in the biological world are most easily perceived: " . . . the basic principles of classification of biological diversity appear to arise directly out of the recognition by man of groupings of plants and animals formed on the basis of such visible similarities and differences as can be inferred from gross features of morphology and behavior" (p. 10). In his words, there is a level at which classes "cry out to be named" (p. 11).

It is worth noting that categories are first learned and named by children at this generic or basic level. Children, for example, first learn the name *dog*. Later, they learn names of more subordinate categories like *schnauzers, chihuahuas,* and *poodles* and more superordinate categories like *mammals* and *vertebrates*. Might not novice to expert development parallel in some ways child to adult development? The relativity among classification schemes might be accounted for by the fact that cultures or communities with specialized needs, such as expert communities, make further classifications based on less perceptually obvious sensory features. There may be features and groupings that we are predisposed to consider as salient in the classification of foods and beverages. But experts are led by need to go beyond the generic level.

I suggest that, given a broad array of wines, we would all, as wine novices, be initially disposed to making certain divisions. As the child moves from the basic object level to more subordinate and superordinate classifications, so too may the novice. As the novice enters the community of wine experts, the more specialized demands of the field require a finer "level of usual utility." And as the child gains knowledge of greater families of categories, of superordinate groupings of objects, so too may novices, with increased knowledge of enology, come to appreciate underlying associations among wine categories.

Basic Object–Level Wine Features

How might wines be categorized at a basic object level? *Red* versus *white? Californian* versus *French* versus *Australian* versus *Chilean?* Or simply *wine* versus *soda* versus *water?* Moreover, what are the features typical of wine categories? Pleasantness would seem to be a universally acknowledged feature. Good/bad is the most common interpretation of dimensions yielded by multi-dimensional and factor analyses (Rabin, 1986). Moreover, what could be more basic to an animal's experience of food or drink? It may, evolutionarily, be basic to survival. Experts also appear to regard this hedonic dimension as most important. There are many score cards used by expert judges for rating wines. The bottom line for a wine, however the individual dimensions are weighted, is always a measure of how good it is.

Sweetness may be another, not altogether unrelated, dimension. It was the only dimension in Solomon's (1990) studies of wine tasters for which the novice rankings of wines agreed at a rate above chance. And it was only when Lehrer (1983) chose a set of white wines varying widely in sweetness that her subjects were able to match descriptions to wines. Many of the novices in Solomon's (1990) studies showed a tendency to refer, however vaguely, to sourness or bitterness. This tendency makes some evolutionary sense if we consider that sourness and bitterness are often associated with unripe or poisonous foods. This appears to be the extent of novice describing. As Lawless (1984) had previously noted, novices very rarely make any references to aromatics. The list so far is short, but it clearly will not suffice for us to accept the dogma of the four basic tastes blindly (O'Mahony and Ishii, 1986). Let scientists look to the features that novices actually identify. They should take care to distinguish between the features identified by novices and those identified by experts. Therein may lie answers about basic object–level features.

CONCEPTUAL SYSTEMS

Between Relativism and Biological Reductionism

The work of Ann Noble and her colleagues (1987) shows great promise in uncovering the chemical bases of certain grape types and certain descriptors. But classification for novices, at a basic object level, may be altogether different from that of experts. And indeed, as Noble discusses, expert wine categories are not solely determined by grape type. Pinot noirs can vary extraordinarily. Though expert category formation may not be arbitrary, neither is it necessarily predictable solely from perceptual information. We make categories and identify features because doing so enables us to make inferences about exemplars in cognitively economical ways. We may uncover the chemical characteristic that gives rise to the green pepper nose in a cabernet sauvignon, but it is entirely another matter to say that cabernet sauvignon will necessarily be an expert wine category. There is room for tremendous variation in expert classifications. If, for example, when presented with the task of categorizing a set of wines, experts create the category "Pauillac," they surely will make their categorization on the basis of characteristic features. Indeed, there may be scientific justification for creating the category, some underlying chemical or physiological truth uniting all its members. Most Pauillacs, for example, may have common growing conditions and soil types; the vintners may share enological techniques. This may lead to a real association of sensory features. Expert wine categories are grounded in facts, if they are to be of use, so they are not arbitrary. But even if novices could make the requisite preceptions, they would not necessarily be able to predict, just from the sensory characteristics in the array, that a set of wines would come to constitute the category Pauillac. Pauillacs may cohere as the category Pauillac due to higher-order social, as well as perceptual, factors. Given the realities of the wine world, given historical and social forces, experts may first come into contact with wines *as* members of certain categories. They drink Pauillacs *as* Pauillacs. Perhaps it is after having formed the category that experts come to know which features define the category. It is then that they come to know that a particular correspondence of sensory characteristics has relevance for a given categorization.

The same could be true in other domains. I could play for an uninitiated listener the following three classic jazz recordings: the 1946 bebop piece "Yardbird Suite" featuring Miles Davis on trumpet, the 1959 modal piece "So What," also featuring Miles Davis, and the 1961 modal piece "Stolen Moments" with Freddie Hubbard on trumpet. Were I to present "Yardbird Suite" *as* a bebop tune and "So What" and "Stolen Moments" *as* modal jazz, my listening friend would likely be drawn to the similarities in the harmonic structures, phrasing, and orchestration of the modal pieces and to their dissimilarities to the frenetic phrasings of the bebop recordings. Conversely, I could also present "Yardbird Suite" and "So What" *as* Miles Davis recordings and "Stolen Moments" *as* a Freddie Hubbard recording. In that case, my impressionable friend would probably listen more attentively to the contrast between the distinctively fragile tone of Miles Davis and the rounder, more openly trumpety sound of Freddie Hubbard. Were I to play for my friend a fourth recording and instruct him to classify it properly, he would likely attend to the harmonic structure and rhythms if he had listened to the previous recordings *as* bebop or modal pieces, and he would attend to the tone of the trumpet if the albums had been presented as featuring trumpet players. Each scheme for classification would be made on the basis of sensory characteristics, but the selection of a categorization scheme would not have been *determined* by perception.

A Cognitive Models Approach

Lakoff (1987) argued for the importance of a larger context in which to place the act of categorization. He advocated that categorization may not only be hierarchical, but that different properties acquire relevance for categorization only in the context of the particular categorization schema adopted. To say which of two women is more of a *real* woman is to assume a particular schema. Phyllis Schlafly and Gloria Steinem would probably disagree on who was more of a real woman; each might be considered more prototypical than the other depending on the classifying schema, or "cognitive model," adopted. Similarly, a wine is more than a concatenation of features. To say which of two wines is more a *real* cabernet sauvignon is also to assume a particular cognitive model of which attributes are relevant to the category. A Frenchman and a Californian, schooled in different styles of winemaking, might disagree on which style resulted in the truer cabernet.

Within the context of a particular cognitive model, wine tasters may show prototype effects in their judgments of category membership. Our Frenchman and Californian might yet agree as to which of the several wines were more typical of Napa Valley cabernets and which were less. Lakoff's point is that when we categorize, we make judgments about which features are important. When we consider "stays at home and minds the kids" to be a heavily weighted feature of the category "woman," we imply a model of categorization. And when we describe wines, we imply a level of classification and a purpose to the classification. Are we trying to describe the features of a typical Pauillac or of an ideal Pauillac. It could make a difference in what is considered a feature. To call a wine *sweet* or *presumptuous* or *balanced* or *mischievous* or *sensuous* or just to call it *good* is to say what is to count as a feature, it is to imply a categorization scheme, it is to imply a greater context of cognitive and cultural forces driving the categorization.

Conclusion

Is *sweetness* a more valid descriptor than is *presumptuousness?* Certainly, some descriptors are more readily understood than are others. One reason could be that we tend naturally to consider certain sensory characteristics as more salient for categorization, whether because of innate predispositions or because of Gibsonian perceptual learning. Sweetness may reflect classification at a basic object level. It is a better descriptor of a wine in that its reference is more readily understood by novices. But the criterion for validity is whether a term will communicate information to members of a particular linguistic community. Assuming that it does convey some information among members of a particular expert community, that the members agree on the sensory qualities to which it refers, then *presumptuous* may be a valid wine term. It may in fact represent description at a different level of classification than *sweetness*.

In addition to the salience of the qualities being referred to, the ease with which descriptors are understood is likely to be a result of the aptness of the associations or metaphors used. The referent of *sweet* is obvious. We may be disposed to identifying the taste quality, and we all have had experience in using the term to refer to foods. A particular tasting group may agree as to the reference of the term *flinty*. The particular quality may taste nothing like flint, but the group has agreed to a standard: The term communicates information. It is quite another question to ask whether a term will communicate information to those outside of a particular group. Perhaps outsiders can abstract some sense of the referent from the metaphorical sense of flinty. Perhaps not. But we must distinguish between the question of whether a term communicates anything to anyone and the questions of whether some metaphors communicate to a wider range of people.

All tasters may initially describe wines at a basic level of categorization most directly guided by the predispositions of their perceptual systems. It may be possible to approximate a generic level vocabulary of taste and smell. This is not the same thing as to say that there exists, for experts and novices, a single fundamental vocabulary. A basic object level or generic level vocabulary of novices need not be the vocabulary of experts. Experts, given their functioning in more specialized worlds, worlds of different social, communication, and perceptual demands, may have greater needs to construct more differentiated categories and more superordinate categories. The categories may be based on perceptual features, but higher conceptual and cultural factors may also influence how and which further perceptual characteristics are to determine categorization. Presumptuous may reflect a particular concatenation of characteristics that are relevant to some expert domains.

Sweetness and *presumptuousness* are equally correct if the wine tasters, in using those terms, are attempting to communicate to members of a community that agrees as to what sensory characteristics *sweetness* and *presumptuousness* refer. And they are equally correct if the features communicated are appropriate to the purpose of the description, that is, if the wine categories implied by the terms are appropriate to their communities' needs. A vintner engaged in quality assurance and a naive consumer looking for a bottle of wine with dinner would seek to make different distinctions between categories of wines. Used correctly, *sweetness* and *presumptuousness* might both be apt descriptions of wines, communicating relevant and meaningful information. Used incorrectly, *sweetness* and *presumptuousness* might communicate more about the wine tasters themselves.

REFERENCES

Amerine, M. A. and Roessler, E. B. 1976. *Wines: Their Sensory Evaluation.* W. H. Freeman, San Francisco.

Berlin, B. 1978. Ethnobiological classifications. In *Cognition and Categorization* (E. Rosch and B. B. Lloyd, Ed.). Erlbaum, Hillsdale, NJ.

Berlin, B. and Kay, P. 1969. *Basic Color Terms: Their Universality and Evolution.* University of California Press, Berkeley.

Bousfield, J. 1979. The world seen as a colour chart. In *Classifications in Their Social Context* (R. F. Ellen and D. Gleason, Ed.). Academic Press, London.

Brown, R. W. 1958. How shall a thing be called? *Psychol. Rev.* 65: 14–21.

Brown, R. W. and Lenneberg, E. H. 1954. A study in language and cognition. *J. Ab. Soc. Psychol.* 49: 454–481.

Butterfield, E. C. and Cairns, G. F. 1974. Discussion summary: Infant reception research. In *Language Perspectives: Acquisition, Retardation, and Intervention* (R. L. Schifelbusch and L. L. Lloyd, Ed.). University Park Press, Baltimore.

Chase, W. G. and Simon, H. A. 1973. Perception in chess. *Cog. Psychol* 4: 55–81.

Chi, M. T. H., Feltovich, P. J. and Glaser, R. 1981. Categorization and representation in physics problems by experts and novices. *Cog. Sci.* 5: 121–152.

Daniel, R. G. and Ellis, H. C. 1972. Stimulus codability and long-term recognition memory for form. *J. Exp. Psychol.* 93: 83–89.

Davis, R. G. 1977. Acquisition and retention of verbal associations to olfactory and abstract stimuli of varying similarity. *J. Exp. Psychol.: Human Learning and Memory.* 3: 37–51.

de Groot, A. 1965. *Thought and Choice in Chess.* Mouton, The Hague.

De Valois, R. L. and Jacobs, G. H. 1968. Primate color vision. *Science* 162: 533–540.

Eimas, P. D., Siqueland, E. R., Jusczyk, P., and Vigorito, J. 1971. Speech perception in infants. *Science* 171: 303–306.

Engen, T., Kuisma, J. E. and Eimas, P. D. 1973. Short-term memory of odors. *J. Exp. Psychol.* 99: 222–225.

Engen, T. and Ross, B. M. 1973. Long-term memory of odors with and without verbal descriptions. *J. Exp. Psychol.* 100: 221–227.

Gibson, J. J. 1966. *The Senses Considered as Perceptual Systems*. Houghton Mifflin, Boston.

Gibson, J. J. and Gibson, E. J. 1955. Perceptual learning: differentiation or enrichment. *Psychol. Rev.* 62: 32–41.

Heider, E. R. 1972. Universals in color naming and memory. *J. Exp. Psychol.* 93: 10–20.

Heider, E. R. and Olivier, D. C. 1972. The structure of the color space in naming and memory for two languages. *Cog. Psychol.* 3: 337–354.

Kay, P. and Kempton, W. 1984. What is the Sapir-Whorf hypothesis? *Amer. Anthro.* 86: 65–79.

Kay, P. and McDaniel, C. D. 1978. The linguistic significance of the meanings of basic color terms. *Language* 54: 610–646.

Kuhl, P. K. and Miller, J. D. 1975. Speech perception by the chinchilla: Voiced-voiceless distinction in alveolar plosive consonants. *Science* 190: 69–72.

Lakoff, G. 1987. Cognitive models and prototype theory. In *Concepts and Conceptual Development* (U. Neisser, Ed.). Cambridge University Press, Cambridge, England.

Lawless, H. T. 1978. Recognition of common odors, pictures and simple shapes. *Percept. Psychophys.* 24: 493–495.

Lawless, H. T. 1984. Flavor description of white wine by expert and nonexpert wine consumers. *J. Food Sci.* 49: 120–123.

Lawless, H. T. 1985. Psychological perspectives on wine tasting and recognition of volatile flavours. In *Alcoholic Beverages* (G. G. Birch and M. G. Lindley, Ed.), Elsevier Applied Science Publishers, London.

Lawless, H. T. 1988. Odour description and odor classification revisited. In *Food and Acceptability*. (D. M. M. Thomson, Ed.) Elsevier Applied Science Publishers, London.

Lawless, H. T. and Engen, T. 1977. Associations to odors: Interference, mneumonics, and verbal labeling. *J. Exp. Psychol: Human Learning and Memory*. 3: 52–59.

Lehrer, A. 1983. *Wine and Conversation* Indiana University Press, Bloomington.

Liberman, A. M., Cooper, F., Shankweiler, D., and Studdert-Kennedy, M. 1967. Perception of the speech code. *Psychol. Rev.* 74: 431–459.

Liberman, A. M., Harris, K. S. Hoffman, H. S., and Griffith, B. C. 1957. The discrimination of speech sounds within and across phoneme boundaries. *J. Exp. Psychol.* 54: 359–368.

Lisker, L. and Abramson, A. 1970. The voicing dimension: Some experiments in comparative phonetics. *Proceedings of Sixth International Congress of Phonetics Sciences, Prague, 1967.* Academia, Prague.

Lyman, B. J. and McDaniel, M. A. 1986. Effects of encoding strategy on long-term memory for odours. *Quarterly J. Exp. Psychol.* 38A: 653–765.

Miyawaki, K. Strange, W., Verbrugge, R. R., Liberman, A. M., Jenkins, J. J., and Fujimura, O. 1975. An effect of linguistic experience: The discrimination of [r] and [l] by native speakers of Japanese and English. *Percept. Psychophys.* 18: 331–340.

Morse, P. A. and Snowden, C. T. 1975. An investigation of categorical speech discrimination by rhesus monkeys. *Percept. Psychophys.* 17: 9–16.

Noble, A. C., Arnold, R. A., Beuchsenstein, J., Leach, E. J., Schmidt, J. O., and Stern, D. M. 1987. Modification of a standardized system of wine aroma terminology. *Am. J. Enol. Vitie.* 38: 143–146.

O'Mahony, M., Goldenberg, M., Stedmon, J., and Alford, J. 1979. Confusion in the use of the taste adjectives 'sour' and 'bitter.' *Chem. Senses and Flavour* 4: 301–318.

O'Mahony, M. & Ishii, R. 1986. The umami taste concept: Implications for the dogma of four basic tastes. In *Umami: A Basic Taste,* (Y. Kawamura and M. R. Kare, Ed.). Marcel Dekker, New York.

Pisoni, D. B. 1973. Auditory and phonetic memory codes in the discrimination of consonants and vowels. *Percept. Psychophys.* 13: 253–260.

Pisoni, D. B., Aslin, R. N., and Hennessy, B. L. 1982. Some effects of laboratory training on identification and discrimination of voicing contrasts in stop consonants. *J. Exp. Psychol.: Hum. Percept. Perform.* 8: 297–314.

Posner, M. I. and Keele, S. W. 1968. On the genesis of abstract ideas. *J. Exp. Psychol.* 77: 353–363.

Quine, W. V. O. 1969. *Ontological Relativity and Other Essays.* Columbia University Press, New York.

Rabin, M. D. 1986. The role of experience in olfactory discrimination. Doctoral dissertation, Yale University, New Haven, CT.

Rabin, M. D. and Cain, W. S. 1984. Odor recognition: Familiarity, identifiability, and encoding consistency. *J. Exp. Psychol.: Learning, Memory and Cognition* 10: 316–325.

Rosch, E. and Mervis, C. B. 1975. Family resemblances: Studies in the internal structure of categories. *Cog. Psychol.* 7: 573–605.

Rosch, E. and Mervis, C. B., Gray, W. D., Johnson, D. M., and Boyes-Braem, P. 1976. Basic objects in natural categories. *Cog. Psychol 8:* 382–439.

Sapir, E. 1921. *Language.* Harcourt Brace, New York.

Solomon, G. E. A. 1990. The psychology of expert and novice wine talk. *Amer. J. Psychol.* 103: 514–534.

Walk, H. A. and Johns, E. E. 1984. Interference and facilitation in short-term memory for odors. *Percept. Psychophys.* 36:508–514.

Whorf, B. L. 1956. *Language, Thought, and Reality: Selected Writings of Benjamin Lee Whorf.* MIT Press, Cambridge, MA.

Wittgenstein, L. 1953. *Philolsophical Investigations.* Macmillan, New York.

10

Individual Differences in Taste and Smell

David A. Stevens

Clark University
Worcester, Massachusetts

INTRODUCTION

Many sources of individual differences in sensory studies have been identified. These include gender, menstrual status, genetic endowment, age, and personality as defined by various psychological tests. Some of the dependent variables in which individual differences have been found in responses to odorants, tastants, and oral irritants include absolute and differential sensitivity, perceived quality, hedonic ratings, identification, rate of salivation, and relative sensitivity of receptor loci.

INDIVIDUAL DIFFERENCES AS SOURCES OF ERROR

Individual differences are often considered as annoying sources of variance in research. Statisticians refer to individual differences as "nuisance" variables.

They receive this designation since, if unrecognized, they may obscure the effects of treatments under investigation and bias interpretations of effects.

Effects on Description

The inferences that follow from an investigation are in part dependent upon the descriptive statistics employed since some characterization of the data is necessary to think about them. Those statistical descriptions are valid only to the extent to which they appropriately characterize the data. Typically, a measure of central tendency is used to represent a sample or population, and the arithmetic mean is the statistic of choice. This statistic is appropriate only if the distribution of scores in the group to be characterized is symmetrical and unimodal. Indeed, all measures of central tendency mislead if distributions are bi- or multimodal. While reduction of data by a measure of central tendency is usually necessary for description, generalization, and interpretation, its use where unwarranted results in inaccurate description and thus invalid generalization and interpretations. If there are systematic individual differences within the groups being described, a measure of central tendency may be an invalid statistic. For example, about 45% of the general population is wholly or partially anosmic to androstenone. Since people are either relatively sensitive or insensitive to the compound, the population mean threshold does not describe well human sensitivity to androstenone as it does not represent the threshold concentration for most people; rather, the mean represents the threshold for relatively few people.

Figure 10.1 shows data collected by Sontag (1978). Thirty subjects rated how well they liked coffee sweetened with sucrose over an 8-week period. Each point represents five replications. Solid lines are results from the first week, dashed lines, from the last week. The graph in the lower right corner shows the mean responses for the panel. The other graphs show the responses of eight of the panelists. The quadratic functions for the group clearly fail to represent the diverse functions that describe the responses of the individual subjects. Again, the mean does not well represent the individuals within the group.

Effects on Statistical Power

Individual differences affect the efficiency of tests to evaluate experimental treatment effects, i.e., the differences produced by the manipulation of the independent variable, which might be concentration of sweetener or processing time. In both parametric and most nonparametric statistical tests, the standard by which effects are measured is within-group variance, i.e., the variation found within groups that have had identical experimental treatment. Since the scientist believes that behavior is determined, people treated identically should respond

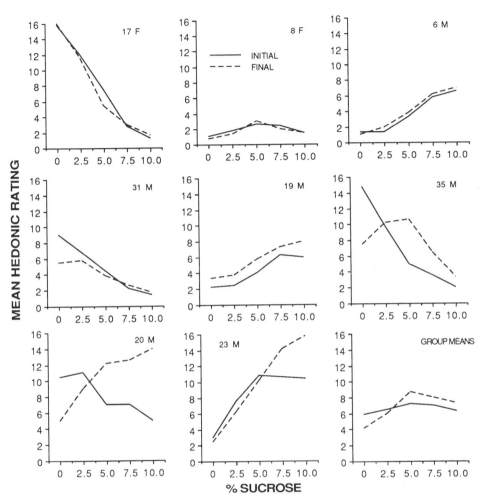

FIG. 10.1 Mean hedonic responses to coffee varying in added sucrose for the first and eighth week, of testing for the mean of 30 subjects (lower right graph) and for eight individuals. *(Redrawn from Pangborn's (1981) report of Sontag's (1978) data.)*

identically. The extent to which they do not is viewed as experimental error. This type of error is logically defined. Since all subjects given a particular treatment are presumably treated identically, responses to that treatment should be identical. If they are not, there must be error.

The extent to which treatments produce differences over and above the differences found within groups of people treated the same provides the *t* and *F* ratios used to test treatment effects and the measures of correlation used to assess degree of association. Thus, the larger the differences within a group receiving the same treatment relative to differences due to treatment effects, the less likely one is to detect, and to determine as statistically significant, an effect of treatment. This error, the failure to detect differences when they exist in the population being sampled, is called a Type II or *beta* error. The power of a statistical test is the probability of not making a Type II error. Within-group differences thus affect the power of a test.

The power of a test is determined by several factors. The power can be increased by lowering the probability criterion for rejecting the null hypothesis, usually set at 0.05. This is usually unsatisfactory as it increases the likelihood of declaring that differences exist when they do not (a Type I error). Power can be raised by increasing the sample size. This would be appropriate if the sources of the error variance do not interact with the treatments of interest, i.e., the effects are consistent over all levels of the experimental treatment. Otherwise there will be errors in generalization. A confounding of relevant variables will occur regardless of sample size if relevant sources of variance contribute to the data and are not identified. For example, if there were a sex difference in the perceived intensity of a flavorant, an increase in the number of men and women in the sample would not remove the effects of that difference between individuals, and the sex difference would remain a confounding variable. Thus the best method to increase power is the identification and removal of unwanted sources of within-group variance.

Removal of Error

The removal of sources of within-group variance can be accomplished two ways. One way is to hold the source of the variable constant. This method is often used. Research is conducted in environments in which all controllable variables are held constant, with especially trained and selected panels. Homogeneity is emphasized. Data so collected *should* be relatively free of contamination by nuisance variables. However, generalization from the results of such studies can be risky. The researcher cannot know if similar results would be found under different conditions. The results can be generalized only if the variables being manipulated and those held constant do not interact. Some time ago a worker in olfaction spoke of his dismay that others were not replicating his findings. Later, he remarked that he had spent years getting a panel of judges that would provide reliable data by replacing "bad" judges with "good" ones. He apparently controlled well for nuisance variables within his group of judges, but at the expense of then using a highly selected group that differed from the population in important aspects. Generalizations made from its data were inconsistent with data from more general groups. This lesson was also learned the hard way by psychologists who generated psychological principles using only 60-day-old male rats, of a single inbred strain, tested in a wide variety of ways in a very limited number of situations. We now know that some of the specific "universal" law they discovered are not at all universal. Clearly the experimenter is in a bind. Holding all variables constant except the independent variable ensures a powerful test, but if some of those controlled variables interact with the manipulated variable, valid generalizations to a population may be impossible. One must hold only the irrelevant, noninteracting variables constant. It is reasonable to select only subjects without respiratory infections for a study of odor identification; it is not reasonable to select only subjects of one gender if one wishes to generalize to both women and men.

The second method is to incorporate variables contributing to individual differences into the analysis, rather than holding them constant. Those variables thought to be responsible for individual differences are treated as independent variables along with the treatment variables. Not only will the effects of those individual differences be removed from the within-group variance, but their general effects and interactions with the experimental treatment can be evaluated. There is less risk of unwarranted generalization, and appropriate conditional statements can be made. If the individual differences are found to be trivial, and their isolation found not to affect appreciably the within-group variance, there is no additional cost to the investigator. Their effects can be pooled with the within-group variance to make a more powerful test. In analysis of variance, generalized randomized-block and split-plot experimental designs accomplish this. Isolation of the effects of individual differences can only be done, of course, if the sources of individual differences can be identified.

Identification of sources   Generally there are two strategies in identifying individual differences. The first is to inspect the data after they are collected. If the inspection reveals bi- or multimodal or unsymmetrical distributions, the data are examined for the identification of subpopulations responsible for the modes or lack of symmetry. However, it is not necessary that individual differences be revealed by the shape of distributions. Two subpopulations having reliably different means can sum to produce a symmetric, unimodal distribution. The shape of a distribution is a good but not infallible sign of subpopulations being contained within it.

Figure 10.2 shows the frequency distributions for two samples ($n = 200$) randomly drawn from normal populations, having means of 9.3 and 10.8, respectively, and the distribution obtained from the combination of them. Each of the two samples have standard deviations of about 1.5. The means of these distributions are about 14 standard errors apart, and the probability of their coming from the same population is less than one in ten thousand ($t = 9.2$, $df =$

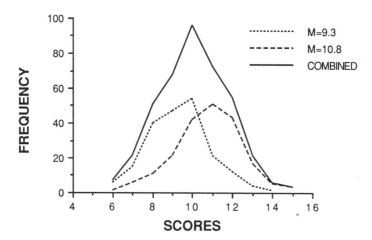

FIG. 10.2 Normal frequency distributions (*n* = 200, SD ~ 1.5) with means = 9.3 and 10.8, and the distribution of the combined groups.

398). Yet, when the two groups are combined, the distribution is not bimodal. The shape of the distribution gives no hint that it is made up of two different groups.

Figure 10.3 shows two groups with the same standard deviations, but with means a little over two standard deviations apart. Now, the distribution of combined groups shows two modes. There *is* bimodality, but even with group means two standard deviations (or about 30 standard errors of the mean) apart, the bimodal distribution hardly looks like the classic textbook examples of bimodal distributions with the antemode half the height of the modes. Inspection of distributions will not always show the existence of subpopulations.

Another way to identify sources of individual differences is to use prior knowledge about the phenomenon or theory (generalizations based on prior knowledge) to suggest (or in a critical test of theory, require) that particular systematic individual differences exist. For example, one knows that there are effects of dentition on many chemoreceptive phenomena, and includes dental status as an independent variable.

INDICATORS OF UNIDENTIFIED RELEVANT SOURCES OF VARIANCE

Not all workers view individual differences as nuisance variables. Some view them as important to study. Individual differences are seen as indications of relevant causal variables, i.e., factors that vary across people and importantly

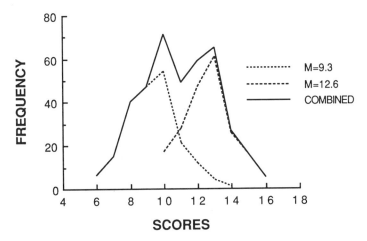

FIG. 10.3 Normal frequency distributions ($n = 200$, SD ~ 1.5) with means = 9.3 and 12.6, and the distribution of the combined groups.

affect the responses under investigation. This view holds that these variables must be accounted for in order to understand fully the phenomenon under investigation. In fact, individual differences are often the source of principal hypotheses from this viewpoint. The individual differences are seen as evidence for differences in mechanisms underlying sensory phenomena of interest.

Self-Attribution

Some recent work using a classification of people from self-attribution theory of personality is an example of a theory-driven search for an explanation of some of the individual variation found in psychophysical research. Here, an explanation of individual differences developed in the study of emotion was applied to sensory studies.

Following the theory of emotion of James (1884), Bem (1967, 1972) has argued that people determine their own moods and feelings the same way they determine them in others—by assessing their actions and context, and inferring their mood using that information. For example, people feel happy because they are smiling in a situation in which happiness is typical. They perceive their smiling, know that smiling is associated with happiness, and thus attribute their smiling to happiness.

The traditional view of emotion is that some emotional stimulus elicits feelings, which are followed by appropriate actions. For example, meeting a bear in the woods (stimulus) evokes fear (feeling), resulting in a look of panic, change in heart rate, and running from the bear (actions). James's view was that this traditional sequence is backwards. He theorized that emotional feelings follow the actions evoked by the emotional stimulus. For James, then, one sees the bear (stimulus), looks panicked, experiences increased heart rate, runs (actions), and then registers the feelings of fright through attribution. The feelings are the *result* of actions, not the cause. When this theory of emotion has been tested, there have been individual differences in the extent to which people respond the way the theory predicts they should. Laird (Laird and Crosby, 1974; Laird and Berglas, 1975) has produced a revision of this general self-attribution theory which accounts for some of the individual differences seen in the tests.

Laird argues that people differ in the extent to which they use different types of cues in making attributions about themselves. Self-produced cues are those that arise from an individual's actions and personal properties (the smiling in the first example above). Situational cues are those that come from the environment, and include normative expectations, that a rose will have a pleasant floral odor, for example. The latter kind of cues seem to be used by everyone, but there are reliable differences in the extent to which people use self-produced cues in determining their own mood, feelings, and abilities. When induced to perform the muscle movements associated with smiling and frowning, some people report strong emotional fluctuations corresponding to the manipulations of facial expression, and some do not. Those whose mood is affected by manipulation of facial expression are called self-produced cuers, and those who are relatively unaffected by the manipulations are called situational cuers.

This idea has been tested in a number of different situations. Self-produced and situational cuers differ in expected ways in tests of self-impressions (Kellerman and Laird, 1982), feelings of romantic love (Kellerman et al., 1989), and suggestibility with regard to feelings (Kellerman and Laird, 1982).

A particularly interesting effect was found with placebos (Duncan and Laird, 1980). Individual differences in reactions to placebos are often found in experimental studies. Some subjects show a "reverse placebo" effect; that is, they report their symptoms increase rather than decrease. In Duncan and Laird's study, subjects were tested to determine their fear of snakes and electric shock. Then they were given placebos identified as either "arousers" or "relaxers" and exposed to a snake and an electric shock generator (but not shocked), and tested again. The self-produced cueing subjects showed reverse placebo effects; they felt *more* fear of the snakes and shock when told they were receiving a relaxing drug, and showed *less* fear when told the drug was an "arouser." However, the situational cuers showed the usual positive placebo effects: less fear with the "relaxer" and more fear with the "arouser." These differences were predicted by Laird's self-attribution theory. Presumably, self-produced cuers used bodily cues (e.g., visceral responses to the snake) to a greater extent than situational cues (the information about the placebo). When they realized that their visceral responses had not changed after taking the "relaxer" placebo, they concluded that they must be *even more fearful* of snakes than they believed initially.

Perception of oral irritation   The stimuli for taste, smell, and flavor perception and the hedonic responses to those experiences include both self-produced and situational cues. The former include the mouth movements associated with chewing, salivation, swallowing, and sniffing. The latter include specific and normative expectancies about the stimuli, e.g., information contained on a product's label and through personal or cultural prior experience.

Since there are typically individual differences in the perception of taste, smell, and flavor, and there are both self-produced and situational cues involved in their perception, and there are individual differences in the extent to which these cues are used, it seemed reasonable to hypothesize that the two are related.

This idea was first tested using the Private Body Consciousness (PBC) test (Miller et al., 1981). People are classified as having high or low PBC on the basis of their answers to questions on how sensitive they are to changes in body temperature, internal tensions, heart rate, dryness of mouth and throat, and hunger contractions. Stevens (1990) compared data from subjects classified as having high and low PBC while doing a series of studies with Lawless on the sensory effects of capsaicin and piperine (the irritants in red and black pepper). In one test eight places in the mouth were separately stimulated with filter paper treated with capsaicin (3.2 μg) and the subjects periodically reported magnitude estimations of the intensity of the burning irritation it produced.

The intensity of the irritation produced by the irritant averaged over all oral loci is plotted separately for subjects with high and low PBC in Fig. 10.4. There

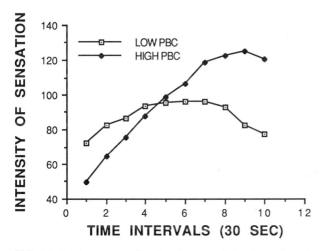

FIG. 10.4 Mean normalized estimates of the intensity of the burning irritation of capsaicin plotted as a function of time for subjects with high or low private body consciousness. *(From Stevens, 1990.)*

was a much flatter function for perceived intensity over time for the low PBC subjects than for the high PBC subjects. The flatter function for the low PBC subjects is what would be expected if they were less sensitive to the bodily responses that accompany oral irritation by pepper than those with high PBC. Differences between high and low PBC subjects were also found for various oral loci stimulated.

Figure 10.5 shows the intensity or irritation reported for each of the eight loci when stimulated with capsaicin. High PBC subjects reported more intense sensations from the tip and side of the tongue than the low PBC people. Similar results were found when piperine (320 µg) was the irritant.

When the two kinds of subjects were compared on the effects of various tastants on oral irritation (Stevens and Lawless, 1986), there were also differences. The results, presented in Fig. 10.6, show that generally tastants suppress irritation by piperine to a greater extent for low PBC subjects than for high PBC subjects.

Flavor perception   To study individual differences in the perception of flavor of a prepared food, a factorial study was done in which stimuli that should provide self-produced and situational cues were independently manipulated (Stevens et al., 1989; Stevens, 1990). The subjects were given samples of a chicken soup having sodium concentrations of 0.276% (normal), 0.420% (high), or 0.564% (very high). These samples were presented with three kinds of verbal information; they were told the soup had "less than normal flavoring," "more

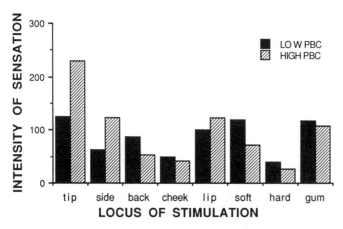

FIG. 10.5 Mean normalized estimates of the intensity of irritation produced by capsaicin applied to tip, side, and back of tongue, inside of cheek, inside of lower lip, soft and hard palates, and lower gum, by subjects with high or low private body consciousness. *(From Stevens, 1990.)*

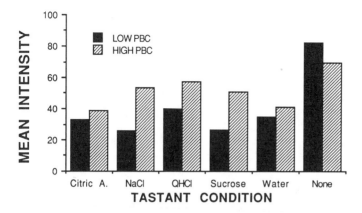

FIG. 10.6 Mean normalized estimates of the intensity of the irritation produced by pipeline with solutions of various tastants, or no tastant, in the mouth by subjects with high or low private body consciousness. *(From Stevens, 1990.)*

than normal flavoring," or were told nothing. The sodium concentration was assumed to provide self-produced cues and the verbal information was assumed to be a situational cue. The soups were tasted and rated on several attributes using 150 mm line scales.

The subjects were classified using the PBC test, and also for cuer type by utilizing a facial manipulation test (Duncan and Laird, 1980). For the latter test, the subjects were shown slides of four signed works of "art" (actually randomly placed geometric forms), framed with titles, and identified as being the work of an artist well known in Europe. Then, while viewing the slides, the subjects were told to contract the muscles near the corners of their mouth and extend their eyebrows, or to contract the muscles above their eyebrows by drawing the eyebrows down and together and to contract the muscles at the corners of the jaw by clenching the teeth. These manipulations produced smiles and frowns, respectively. Two of the picture titles had a positive emotional tone (e.g., "dancing") and two had negative emotional tone (e.g., "rage"). The emotional tone of the facial manipulation was always the opposite of that of the picture title. While posed and viewing the picture, the subjects filled out a mood adjective check list adapted from the Nowlis-Green Adjective Check List (Nowlis, 1965). The subjects were given a cover story to account for this test. They were told that we were interested in flavor perception, and that memory, muscular tension, and mood all affect perception, and it was necessary to obtain measures of these variables in order to control for them in our work.

The subjects were scored on the basis of the extent to which their mood was consistent with the facial manipulation. The subjects falling in the upper third (*n* = 29) of the distribution were classified as self-produced cuers, and those in the lower third (*n* = 31) as situational cuers. Thus, the self-produced cuers were those whose mood was most determined by facial manipulation, and the situational cuers were those whose mood was not.

Analysis of the judgments of a soup's overall flavor showed that the self-produced cuers had a steeper psychophysical slope (0.92, using log-log transformed data) than did the situational cuers (slope = 0.52,) for the intensity of the soup's flavor. The self-produced cuers, then, were more sensitive to the changes in sodium concentration than were the situational cuers. The results are shown in Fig. 10.7.

When the subjects were classified by PBC, similar results were found for judgments of intensities of saltiness. These are presented in Fig. 10.8. The psychophysical slope for the high PBC group was 1.17 (using log-log transformed data), while it was 0.84 for the low PBC group. Again the groups differed in sensitivity in the expected direction.

Figure 10.9 shows the hedonic responses of the two groups to soup samples with the lowest sodium concentration as a function of the three kinds of verbal information given. Different placebo effects were found for the two groups. A positive placebo effect for hedonic ratings was found for the low PBC subjects; they liked the soup verbally labeled "flavor reduced" less than that labeled "flavor added." The opposite was found for the high PBC subjects; they liked the soup labeled "flavor added" less than that labeled "favor reduced." Presumably,

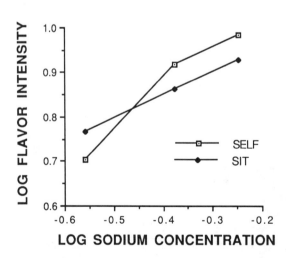

FIG. 10.7 Log mean estimates of overall intensity of flavor of chicken soup having three concentrations of sodium by subjects classified as self-produced or situational cuers. *(From Stevens, 1990).*

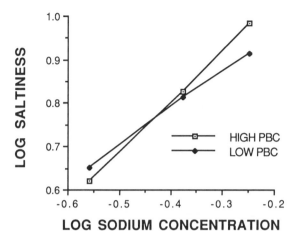

FIG. 10.8 Log mean estimates of saltiness of chicken soup having three concentrations of sodium by subjects with high or low private body consciousness. *(From Stevens, 1990.)*

since the sodium levels of the samples were independent of the verbal labels, the high PBC subjects heard that the soup had flavor added, could detect no difference in flavor, and concluded that such a group could not be as good as the others, showing a reverse placebo effect.

In another study (Hopmeyer and Stevens, 1989) judgments of the intensity of

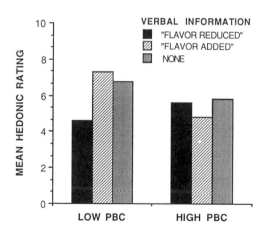

FIG. 10.9 Mean hedonic ratings of soup having "normal" sodium concentration by subjects with high or low private body consciousness when given three kinds of verbal information.

citric acid were compared for self-produced and situational cuers. In this study subjects were classified by the facial manipulation test described above. Twenty self-produced cuers and 20 situational cuers judged the intensity and color of 0.0028 M, 0.0056 M, and 0.0112 M citric acid solutions colored with 0.0001%, 0.0003%, or 0.0024% yellow food coloring in a factorial design. Again, the psychophysical slope for intensity of taste was steeper for self-produced cuers (slope = 0.74 using log-log transformed data) than for the situational cuers (slope = 0.50) (Fig. 10.10). There were no group differences in estimates of color intensity.

It is clear, then, that people's judgments of the intensity of oral irritation and at least some tastes vary with the extent to which those people report being sensitive to some bodily responses (PBC) and differentially use cues in assessing their mood (self-produced and situational cueing). These effects are not due to general differences in response functions. The groups did not differ in responses to variables not manipulated, for example, oiliness of soup, and they did not differ for all of the manipulated variables. In the soup study PBC classification showed differences only in saltiness and hedonics, and cue type showed differences only in flavor intensity.

Although this work identified variables contributing to individual differences in flavor perception, it did not identify the specific mechanisms underlying them. The differences in sensory responses between the self-produced and situational cuers may have their basis in the mouth movements used in tasting. It could be that proprioceptive stimuli (the sensations from oral and facial muscle move-

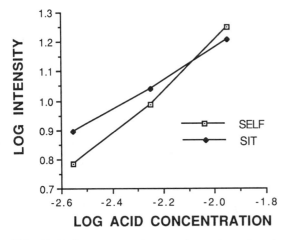

FIG. 10.10 Log mean estimates of taste intensity of three citric acid solutions by subjects classified as being self-produced or situational cuers.

ments) are the self-produced cues being used, or it could be that self-produced cuers have developed oral manipulations that result in a different flow of stimuli over the tongue. There is ample evidence that differences in stimulus flow affect the intensity of taste sensations.

This research is a example of how the recognition of individual differences contributes to the development of theory. Laird's theory (Laird and Crosby, 1974) of differential self-produced and situational cueing was developed because in earlier studies of self-attribution theory, all people did not respond to self-produced cues in a predictable manner. Also the individual differences of unidentified origin typically found in chemosensory psychophysical studies demanded explanation. Broad similarities in the kinds of cues used by people in the self-attribution of emotion and in flavor perception suggested that the same general explanation could account for differences in both domains. Rather than attributing individual differences in chemosensory perception to random error, they were investigated, and hypotheses developed from theory to account for some of them.

IDENTIFIED SOURCES

Often, however, the source of the variance, at least at one level of analysis, is known. Classification of subjects according to this known variable can then be used to study changes associated with that variable. The information acquired can be used to describe effects more completely, develop hypotheses, or test theoretical speculations.

If the source of variance is a general one, such as sex or age, it defines a number of variables. In addition to anatomical differences, gender defines, historically at least, differences in perceived expectancies, environments, opportunities, income level, interactions with children, and experience in food selection and preparation, to name only a few. As chronological age varies, so do sensory functions, cognitive functions, motor skills, etc. The design of a study using a general independent variable must include controls for variables correlated with the general one that might confound a more specific variable of interest. In a sensory study, nonsensory variables must eliminated as causal variables. Often this must be done indirectly, rather than directly. You cannot alter your subject's specific history, or the culture of 10 years ago, for example, but you may be able to show that particular confounding variables do not affect your results.

Age Effects

Food preferences change with age, and complaints about the flavor of food are not uncommon in the elderly. Sensory studies of changes in flavor perception done a dozen or so years ago were confounded by cognitive factors. It was not clear if the results were due to sensory differences, differences in people's concepts about foods, or some interaction of them. Accordingly, Stevens and Lawless (1981) examined the perceptual basis of flavor in three age groups: 18–25 years, 36–45 years, and 56–65 years (including a control for this variable).

Multidimensional scaling (MDS) was utilized. The subjects were given all possible pairs of 12 purees of unseasoned fruits and vegetables under red light (to control for differences in the foods' color) and asked to rate them on their similarity of flavor. Similarity ratings have the advantage of being free of experimenter bias; the subjects must generate the bases of the judgments rather than using some list of attributes provided by the experimenter. With MDS, the judged dissimilarities between the objects (purees) are plotted in multi-dimensional space, with the spatial distances corresponding to the dissimilarities. The number of spatial dimensions needed for an efficient solution presumably represents the number of perceptual dimensions utilized by the subject.

Interpretation of MDS analyses can be difficult since the experimenter cannot directly know from similarity judgments the characteristics the subjects were using. To assist in the interpretation, after the purees were rated, the subjects were asked to list the attributes they used in making their judgments. Then they tasted the purees again and rated them on the 10 most frequently named attributes. These ratings were used as dependent variables in multiple regressions over the coordinates of the MDS solutions. To control for a possible confounding of cognitive and sensory differences, we also had the subjects rate the flavors of the purees using the purees' names so that judgments from tasted foods could be compared with judgments of named foods.

Four-dimensional solutions efficiently described the judgments of each age group. Figure 10.11 shows the configurations projected onto two planes. The dashed lines are the vectors produced by the multiple regressions, and the interpreted dimensions are shown in bold type at the end of the vectors. For the younger group, sweetness and hedonics were highly correlated and considered a single dimension. However, for the middle group, hedonics and sweetness, now represented as a fruit-vegetable dimension, were now independent. Sweetness did not predict hedonic quality for the 36- to 45-year-olds. In the older group neither sweetness nor hedonics were principal dimensions. For them, intensities of attributes were important.

Importantly, there were no age differences in judgments based on food names. Concepts about food remain stable over age, and the group differences we found cannot be attributable to differences in ideas about the flavors of the purees.

These differences in flavor perception are interesting, for there are relatively small suprathreshold age effects for taste, except for bitterness, using model

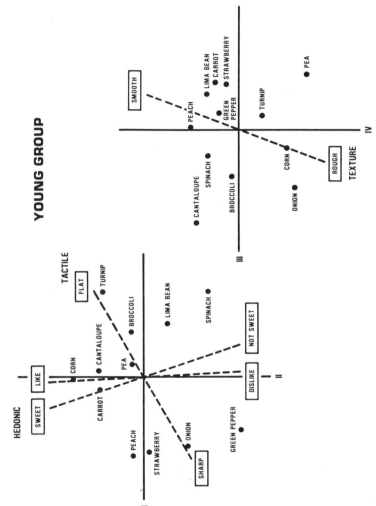

FIG. 10.11 Projections of MDS solutions for purees judged by three age groups. The axes are those produced by MDS. The dashed lines are attribute vectors produced by multiple regression, and the labels of interpreted dimensions are in bold type adjacent to the descriptors for the vectors. (*From Stevens and Lawless (1981) by permission of Academic Press Inc., London.*)

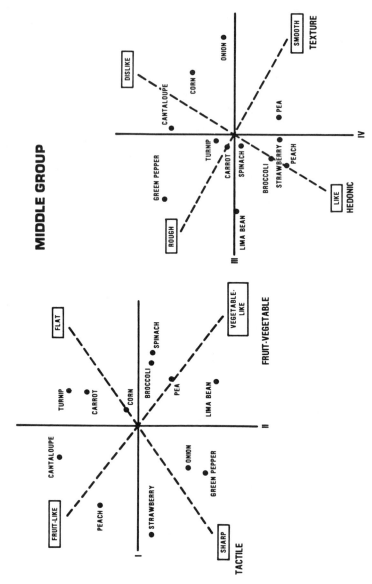

FIG. 10.11 *(cont.)*

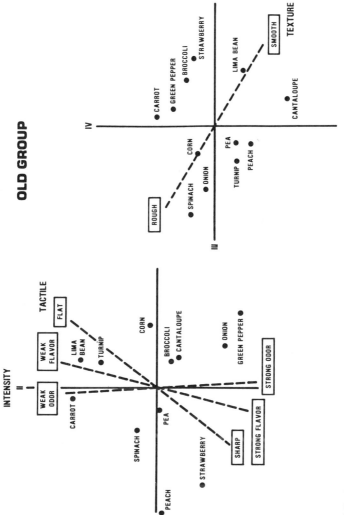

FIG. 10.11 (cont.)

solutions of tastants. For olfactory sensitivity, there are also relatively small effects up to age 65. But despite the lack of large psychophysical differences through late middle age, we found important differences in the dimensions used by people in discriminating foods. Apparently the way sensory information is used to produce the perceptions of flavor changes with age. It seems likely that this is more a result of experience, rather than chronology per se.

Specific Anosmias

Studies on specific anosmias are another example of utilization of a known source of variance. While individual differences in sensitivity to odors were recognized in the nineteenth century, they were not studied with a view toward olfactory receptor processes until Guillot (1948) suggested that specific anosmias were due to inherent defects in receptor cells and that the odors associated with them defined olfactory primaries.

Robert O'Connell reported on work using odorants for which there are large individual differences in sensitivity (see Chapter 4). In his work these individual differences provided a means to test existing ideas and formulate new ones about the mechanisms responsible for odor perception. There, relationships between individual's sensitivity to odorous compounds and relationships between the sensitivity to a compound and the odor it produces provided the data. Sensitivity to one odorant was related to sensitivity to another, and for those people insensitive to some compounds, the qualities of the odors of those compounds were different from those experienced by people sensitive to those compounds.

These findings, together with the fact that different kinds of molecules produce the same kinds of odors in sensitive people, argue for a multiple profile or multiple perceptual channel basis for olfactory perception. The former notion assumes that olfactory receptors respond to multiple characteristics of the odorant molecule, and the second that the olfactory percept is based on information from more than one channel.

Wysocki, Beauchamp, and colleagues have also been studying individual differences in sensitivity to androstenone. By studying the sensitivity of identical and fraternal twins, they have established that sensitivity to it is in part genetically determined (Wysocki and Beauchamp, 1984). Sensitivity also changes with age. Most children under 14 can smell androstenone. For those between the ages of 15 and 20 years, there is a decrease in the frequency of osmic males, but not females, who are generally osmic to the compound. For those over 21 years of age, there is a decrease in the frequency of osmic people of both sexes, with males continuing to be have a lower frequency of sensitive individuals (Dorries et al., 1989). However, for about half of those initially insensitive to androstenone, exposure to androstenone three times a day results in an increase in sensitivity (Wysocki et al., 1989). Thus, not only are there genetically determined differences in sensitivity to androstenone, but there are further individual differences in the effects of age and exposure on sensitivity to it.

Explanations of odor perception must account for these observations. Whether these individual differences as seen in specific anosmias, and in the age, sex, and experiential effects that interact with them, are the results of differences in the presence or nature of one or more receptors, differences in the receptors' environment (olfactory mucosa), or differences in the processing of the information provided by the receptors is not yet known.

CONCLUSION

Individual differences deserve the researcher's attention. Rather than being ignored or "obscured," as it were, by large sample sizes, individual differences should be identified and studied as independent variables. Do this to describe better the sensory phenomena under study and to understand better the phenomena, whether the concern is with underlying mechanisms or with general correlates, be they casual or not.

REFERENCES

Bem, D. J. 1967. Self-perception: An alternative interpretation of cognitive dissonance. *Psychol. Rev.* 74: 183.

Bem, D. J. 1972. Self-perception theory. In *Advances in Experimental Social Psychology*, Vol. 6 (L. Berkowitz, Ed.), p. 1. Academic Press, New York.

Dorries, K. M., Beauchamp, G. K., and Wysocki, C. J. 1989. Changes in sensitivity to the odor of androstenone during adolescence. *Develop. Psychobiol.* 22: 423.

Duncan, J. W. and Laird, J. D. 1980. Positive and negative placebo effects as a function of differences in cues used in self-perception. *J. Pers. Soc. Psychol.* 39: 1024.

Guillot, M. 1948. Anosmies partielles et odeurs fondamentales. *C. R. Acad. Sci.* 226: 1307.

Hopmeyer, A. and Stevens, D. A. 1989. Individual differences in psychophysical responses to chemical stimuli. *Chem. Senses* 14: 711 (Abs.).

James, W. 1884. What is an emotion? *Mind* 9: 188.

Kellerman, J. and Laird, J. D. 1982. The effect of appearance on self-perception. *J. Pers. Soc. Psychol.* 50: 296.

Kellerman, J., Lewis, J., and Laird, J. D. 1989. Looking and loving: The effects of mutual gaze on feelings of romantic love. *J. Res. Pers.* 23: 145.

Laird, J. D. and Berglas, S. 1975. Individual differences in the effects of engaging in counter-attitudinal behavior. *J. Pers.* 43: 286.

Laird, J. D. and Crosby, M. 1974. Individual differences in the self-attribution of emotion. In *Thought and Feeling: The Cognitive Alteration of Feeling States* (H. London and R. Nisbett, Ed.), p. 44. Aldine, Chicago.

Miller, L. C., Murphy, R., and Buss, A. H. 1981. Consciousness of body: Private and public. *J. Pers. Soc. Psychol.* 41: 397.

Nowlis, V. 1965. Research with the mood adjective check list. In *Affect, Cognition and Personality* (S. S. Tomkins and C. E. Izard, Ed.), p. 352. Springer, New York.

Pangborn, R. 1981. Individuality in responses to sensory stimuli. In *Criteria of Food Acceptance: How Man Chooses What He Eats,* (J. Solms and R. L. Hall, Ed.), p. 177. Forster Verlag, Zurich.

Sontag, A. M. 1978. *Comparison of sensory methods: Discrimination, intensity, and hedonic responses in four modalities.* M.S. thesis, University of California, Davis.

Stevens, D. A. 1990. Personality variables in the perception of oral irritation and flavor. In *Chemical Senses,* Vol. 2: *Irritation* (B. G. Green, J. R. Mason, and M. R. Kare, Ed.), p. 217. Marcel Dekker, New York.

Stevens, D. A. and Lawless, H. T. 1981. Individual differences in flavor perception. *Appetite* 2: 127.

Stevens, D. A. and Lawless, H. T. 1986. Putting out the fire: Effects of tastants on oral chemical irritation. *Percept. Psychophys.* 39: 346.

Stevens, D. A., Dooley, D. A., and Laird, J. D. 1989. Explaining individual differences in flavor perception and food acceptance. In *Food Acceptability* (D. M. H. Thomson, Ed.), p. 173. Elsevier Science, London.

Wysocki, C. J. and Beauchamp, G. K. 1984. Ability to smell androstenone is genetically determined. *Proc. Natl. Acad. Sci. USA* 81: 4899.

Wysocki, C. J., Dorries, K. M., and Beauchamp, G. K. 1989. Ability to perceive androstenone can be acquired by ostensibly anosmic people. *Proc. Natl. Acad. Sci. USA* 86: 7976.

11

Descriptive Techniques and Their Hybridization

Margery A. Einstein

SensTek, Inc.
Mercer Island, Washington

INTRODUCTION *(The Method and Its Ancestors)*

Descriptive analysis is the sensory method by which the attributes of a food material or product are identified, described, and quantified using human subjects who have been specifically trained for this purpose. It is a unique and highly specialized form of sensory analysis. This analysis can include all sensory parameters or it can be limited to certain aspects as in flavor or texture profiling. Proper use of the descriptive analysis method requires that the panel be carefully selected, trained, and maintained under the supervision of a sensory analysis professional who has extensive experience using this method.

A recent advertisement in *Food Technology* read, "You can't hide flat flavor!" It is possible to taste flat flavor, but how do you measure and correct it? For example, this problem is very clear when replacing natural with imitation ingredients. When imitation vanilla extract is substituted for aged bourbon vanilla extract, many of the background or complex character notes are missing. When evaluated using individual attribute analysis, little difference may be found. However, when a discrimination test like a triangle test is used, the samples are clearly different. This is when descriptive analysis is best applied. The trained panel will carefully identify and measure all the character notes in each sample, thereby documenting just how the two samples differ. If the panel has extensive experience evaluating vanilla extract, they can go one step further and document the exact type and strength.

Descriptive phases have long been used to evaluate the food we eat. There have been many attempts to formalize descriptive testing, most of which were product specific. After the Second World War, a need for a formal means of describing food arose. During the war and the postwar period, extensive information was collected on product preference and acceptability. Reliable methods already existed to determine if products or samples were different. Once identified as different, however, there existed no objective, reproducible, and reliable method to describe and document these differences

Around the same time, researchers at the Arthur D. Little laboratory had an interest in understanding the mechanism for seasoning in cooked food and applying this knowledge to product development. The Flavor Profile method was developed to meet this need as well as to provide research guidance to their clients for a wide variety of food products and materials (Caul, 1957, 1959).

Since flavor profiling was first developed, many attempts have been made to modify or improve upon this method. Some attempts were illegitimate since they represented changes that did not adhere to the essential elements of the descriptive analysis method. Genuine modifications resulted in development of the three additional systems: Texture Profile, QDA, and Spectrum, which are recognized today. The Texture Profile method was developed in part because textural attributes were not included in flavor profiling. It was developed in the early 1960s at the General Foods Technical Center by a team of researchers, under the leadership of Dr. Alina Szczesniak (Szczesniak, 1963; Szczesniak et al., 1963; Brandt et al., 1963). This system arose from the need to apply descriptive methods to the sensory analysis of food texture. They were able to apply their comprehensive knowledge of food rheology to the development and application of the Texture Profile method. Since that time, this method has been further developed. However, almost 40 years later, the foundation of rheological principles upon which the method was built is still applicable.

Quantitive descriptive analysis (QDA) resulted from the need to precisely quantify and statistically analyze certain applications of profile data (Stone et al., 1974, 1980; Stone and Sidel, 1985). Its introduction represented a major advance in the method and opened new opportunities for its use.

The Spectrum descriptive analysis method is the youngest of the descriptive analysis systems included in this review. It was developed by Gail Civille (1979) after years of experience employing descriptive analysis for a wide variety of food and nonfood consumer products. It blends several essential elements of descriptive analysis with the need to employ universal scales.

COMMON SYSTEMS *(The Parents and Grandparents)*

Four popular systems exist which utilize the descriptive analysis method. Information for the following brief synopsis of each of the systems was obtained from a summary of each method which has been prepared for publication by a special task group of ASTM Committee E-18 (ASTM, 1990). Since this review was based on information supplied by representatives of each system, it is as current as any information available at this time.

Flavor Profile Method

The Flavor Profile method is "based on the concept that flavor consists of identifiable taste, odor, and chemical feeling factors plus an underlying complex of sensory impressions not separately identifiable" (ASTM, 1990). The individual characteristics of product aroma, flavor, and aftertaste are identified, and then the intensity of each is assessed. The basic elements of the flavor profile method include amplitude or overall impression, identification of perceptible aroma and flavor character notes, intensity of each character note, order of appearance, and aftertaste.

The panelists are chosen because of their ability to discriminate odor and flavor differences (as specified by certain screening tests), to communicate their perceptions, and to possess specific personal criteria and skills. Training for flavor profiling includes basic and advanced instruction, interspersed with practice sessions.

To understand flavor profiling, one must understand how the panel operates. The panel leader is not only responsible for panel operation and sample preparation, but the panel leader conducts the panel discussion, records the data, and participates fully as a panel member. After the panel, the leader compiles and interprets the data. Another unique aspect of profiling is that the panel, which usually consists of four to eight members, works together to arrive at a consensus description. One may need several panel sessions to achieve consensus.

Quantitative Descriptive Analysis

Quantitative descriptive analysis (QDA) is based on the panelists' ability to verbalize product perceptions in a reliable manner. After screening, training, developing a language, and scoring the samples, the data are statistically analyzed (ASTM, 1990). The elements of QDA include panel agreement in development of a list of perceptible sensory attributes; the order of attribute occurrence; and relative intensity measurement for each attribute. This is followed by statistical analyses of the responses.

Panelist screening and qualification has three phases. They include familiarity and/or use of the product, ability to discriminate, and task performance. It should be noted that the QDA system specifies that panelists be users and likers of the product category. A typical panel is comprised of 10–12 individuals who have demonstrated their ability to work in a group. Panelists also must demonstrate that they can maintain their skills throughout the testing period. In this method, the panel leader does not participate in the panel. He or she coordinates the screening and training process, as well as being responsible for product preparation and handling.

Training follows the screening, qualification, and verification phases. It begins with the language development process, and its primary objective is the preparation of the product scorecard. The panelists work together to develop the language to describe the perceptible product attributes. The panel leader assists this process but does not actively participate unless the panel is experiencing difficulty. Panel agreement on each attribute is required. Order of occurrence is achieved by mutual agreement with resolution by the panel leader, if necessary. During the training sessions the panel also agrees upon the standard procedure to be used for sample evaluation. To assure that each product is used equally often by each panelist, a balanced block design is used. The QDA method uses six-inch horizontal line scales with word anchors at either end to measure each attribute. Specific attention in the data analysis is given to monitoring panelist performance and differentiation by attribute.

Test results are often graphically presented using the familiar spider plot. A spoke denotes each attribute. The distance of the attribute mean from the center of the plot along each spoke directly corresponds to the attribute intensity. The plot provides a visual presentation of product similarities and differences.

Spectrum Descriptive Analysis

The Spectrum descriptive analysis method is based on a detailed characterization of a product's sensory categories (ASTM, 1990). This characterization includes identification of the perceptible sensory attributes along with measurement of the intensity of each attribute. All intensities are reported relative to universal scales that enable comparisons not only within product groups but among all products tested. As with QDA, the data are statistically analyzed.

Panelists participating in the Spectrum method are selected based on six major criteria. These include acuity, rating ability, interest, availability, attitudes to task and products, and health. Panelist screening is extensive, resulting in selection of approximately 15 panelists, who are then ready to be trained for a variety of products.

Training, which consists of both orientation and practice sessions, is comprehensive. It includes exposing the panelist to the underlying dimensions of appearance, flavor, and texture, as well as providing the means to develop references, terminology, and scaling for product evaluation. The initial or orientation phase of training precedes extensive practice sessions where demonstrations are presented to expose the trainee to a wide range of products.

Initially the panelists work independently to develop a list of terms which describe all perceptible attributes. Subsequently, a series of references (which may include products) are provided by the panel leader. Training is followed by extensive discussion to determine which terms will be used on the evaluation ballot. Terms are positioned on the ballot in the order that they are perceived. Attribute intensities are measured using line scales which are anchored at either end by the terms none and extreme. One achieves familiarity with each scale by presenting a series of intensity references. This process, which is moderated by the panel leader, usually takes several sessions. Also during this time the evaluation procedure is established. At this point in the Spectrum method, two products are evaluated to validate both the procedure and panelist performance.

Once trained, 8–12 panelists participate in a panel session. At the outset of testing, several panel sessions are conducted to develop the terminology, references, ballots, and procedures appropriate for the test product. At least two and often three replications are included for each sample.

Texture Profile Method

This method is based on a systematic classification of textural properties, which is applied to the definition and description of each attribute. These are grouped into the following categories: mechanical attributes, geometrical attributes, and attributes related to moisture and fat. The sensory definitions for each attribute are derived from the physical and rheological definition (ASTM, 1990). Guidelines for training and further modifications and applications are reported by Civille and Szczesniak (1973) and Civille and Liska (1975).

In the original texture profile method, the attributes were measured using the five-point scale developed for the Flavor Profile method. Initially data were analyzed using consensus guidelines. Subsequent applications have resulted in use of line scales and statistical methods of analysis.

Summary

The systems which have just been briefly reviewed are to some degree proprietary. All but the Texture Profile method have been developed and are currently sustained by independent consulting groups. While they share several elements in common, each is unique and each is reliable, representing countless hours of experience and application. Furthermore, each has its strengths and weaknesses—although this latter point is always subject to considerable debate! Aside from the proprietary nature of these systems, several other factors exist that have complicated the universal application of the descriptive analysis method. First, it is hard to find detailed information about it. We cannot depend solely on the juried literature and other published information to learn about this method. Vast experience using descriptive analysis has been acquired by private industry. Regrettably, these data seldom find their way into the literature. Finally, descriptive analysis is very expensive. It takes time: time to train, time to maintain the panel, and then time to analyze the products and interpret the data. All this time, along with the cost of preparing product, references, and data analysis further increase the testing cost.

GENERIC DESCRIPTIVE ANALYSIS *(The Stepchildren)*

The four systems that have been described have been in use many years. While each has its own features, they share common universal elements. Examination of each reveals that some evolution has occurred since its inception. Nevertheless, there are certain properties that exist regardless of the specific system being used. When these universal elements are applied, the technique can be called generic descriptive analysis. Before examining the common traits or elements that exist in generic descriptive analysis, it is appropriate to present a theoretical example to demonstrate how descriptive analysis can be effectively used.

The strength of this method is to be able to measure and monitor small but meaningful changes over time. This is especially important when no other chemical or physical means exists to do this. Comparisons between samples should be made only after each sample has been evaluated independently. Analysis of a series of lemon mustard salad dressings demonstrates the usefulness of this method. Panel analysis results from formula one—a blend of traditional salad dressing ingredients-lemon, sugar, mustard, spices and oil—is shown in Figure 11.1. The perceptible attributes are listed along the x axis in the order that they occur, beginning at the y axis. Intensity is indicated by the height of each bar. While lemon and mustard are the predominant character notes, salty, sweet, sour, pepper and oil are also perceptible. Consumer testing of freshly prepared product indicates high acceptance of this product in a highly competitive market.

Despite the success of formula one, someone decided it was too expensive! The product was reformulated eliminating fresh lemon juice, reducing the sugar, and changing the spice/mustard blend. Figure 11.2 shows the results of descriptive analysis of formula two alongside formula one. In formula two, the lemon aromatics were reduced almost to the point of no longer being detectable, and the

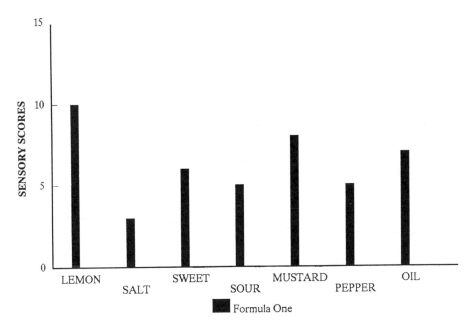

FIG. 11.1 Sensory intensity scores for the descriptive analysis of lemon mustard salad dressing (formula one).

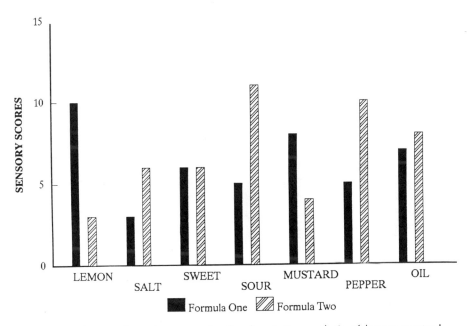

FIG. 11.2 Sensory intensity scores for the descriptive analysis of lemon mustard salad dressing (formula one and formula two).

salty taste was more perceptible even though the amount of salt in the formulation was not changed. Despite the decrease in sugar, there was no change in the amount of sweet perceived, while the sour taste became the strongest, most perceptible attribute in the formulation. Changes in the spice/mustard blend in formula two resulted in perceptible changes in both the amount of pepper and mustard. Clearly the differences between these two formulations could be detected by consumers, so if formula two was to be manufactured, it would be a replacement, not a substitution. Further consumer testing was necessary to verify that formula two enjoyed the same high acceptability rating as formula one.

Unfortunately, formula two failed to be well received by the consumer. Enthusiasm for formula one was high, and so the decision was made not to wait for the results of storage tests and immediately begin to distribute this product. Figure 11.3 shows the results from descriptive analysis testing after 3 months. All lemon aromatics were gone, and so was the pepper. A new character note which overwhelms the product description was detected. This attribute was described as soapy. Comparing formula one at zero time and three months, perceptible differences in the other attributes were also noted.

Descriptive analysis is the only way to describe and quantify the complex

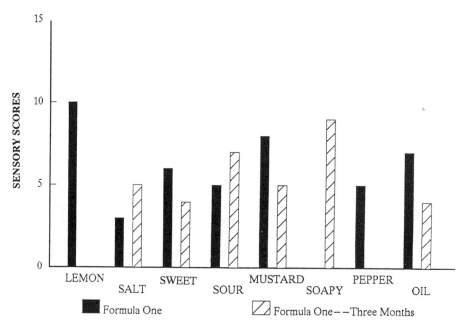

FIG. 11.3 Sensory intensity scores for the descriptive analysis of lemon mustard salad dressing at zero time and 3 months (formula one).

changes in perception found in these salad dressing formulas. Individual, independent attribute analyses, even when you are lucky enough to anticipate the perceptible attributes, do not permit as complete or as accurate a product description as can be obtained using descriptive analysis. Understanding the product profile is more complicated than tracking the individual intensity of a series of attributes. The appearance of new character notes (or attributes) and the changes in order of occurrence are two common phenomena that complicate product analysis. Attribute intensity can remain constant, and yet detectable differences can occur due to changes in the order of the attributes or the presentation of new ones.

Having seen how generic descriptive analysis is used, let us examine the unique elements of this method. The first feature of descriptive analysis is the ability to identify all the perceptions and then to describe and measure each. No other sensory analysis method is able to do this. One can achieve this through the use of a highly trained and maintained panel. The data is then compiled and analyzed. The method of analysis is to some degree dependent on the reasons or objectives for conducting the test.

Panel Selection and Training

Panel selection, including preliminary screening, is the first step in descriptive analysis. Training, however, is the key to successful employment of descriptive analysis. Some training may actually begin during the screening process. However, formal training usually begins with the orientation program. During this phase, the panelist gains knowledge of the senses and the basis for using human subjects as instruments. This is followed by examples of simple systems, which are examined to demonstrate the analysis procedure. A practice product, such as marmalade, or a juice drink is examined. The panelists learn to identify all the perceptible attributes and record them in the order that they occur. Next, samples are provided that vary in attribute strength. This provides an opportunity to measure intensity and to apply this measurement to some type of scale. The key aspect of this phase of the training is that the panelists are expected to identify all the sensory attributes and to come up with descriptive terms for each attribute. Subsequent panel discussion under the watchful eye of the panel leader results in the elimination of duplicate terms and the final list of attributes to be placed on the sample ballot. Once this has been accomplished, including a written definition for each term, references are provided and attached to various portions of each attribute scale. Now it is time to check panelist performance. One accomplishes this through a series of samples that include coded references and replicates. Often these results are revealed to the panelists for self-examination. This is later repeated to check actual performance. At this point, panelists who fail to replicate their own results or who cannot operate within the agreed upon definitions are excused. This usually concludes the formal training part of the program. Training is ongoing, however, and it is in this ongoing training or maintenance phase where some panels go astray. Prior to beginning, all aspects of panel training should be planned and written down in the form of a protocol. After training is complete, this protocol should be reviewed and revised, if needed, before further training is begun.

Screening The purpose of screening is to identify a group of prospective panelists who are expected to successfully complete the training phase. While success at acuity tests is important, the most important criteria for success as a panel member is interest and motivation. Only candidates who will ultimately be eligible to participate should be included in the screening process. The most effective method of screening is to design a series of exercises where the candidates are shown samples that may be references or product. Next they are asked to identify these same samples or references in a series.

These types of screening tests not only provide information about certain acuity skills, but also valuable information about the candidates' patience, persistence, motivation, cooperation, and ability to deal with abstract concepts. Information obtained during screening about behavioral skills is even more important than success at acuity tests.

Orientation An important aspect of panel training is the introduction or orientation phase. During this phase, the panel members learn to work as a group, gain valuable information about using their senses, and begin to learn to identify, describe, and quantify sample differences.

Training Training is a carefully designed series of exercises. One designs these exercises to teach, practice, and evaluate the panelists' performance. Specific information about reference standards and their role in training is discussed by Rainey (1986).

Also, during training, the panel members begin to understand the importance of developing and then following careful procedures for sample evaluation. When one panelist measures particle size after one chew and another after five chews, it is unlikely that good agreement about particle size will occur. Another important aspect of training is to practice identifying and recording the perceptible attributes in the order in which they actually occur. During training, the panel leader has many opportunities to observe and measure both individual and total panel performance. Group interaction is critical to make sure that all the attributes are understood and utilized in the same manner.

Maintenance Panel maintenance includes two tasks—ongoing training and monitoring performance. It is the key to successful use of the descriptive analysis method. Maintenance begins with training and is ongoing throughout the life of the panel. This includes review of orientation and training exercises, as well as implementation of new exercises designed to monitor individual and total panel scores.

Product Familiarity

Product orientation   Sample analysis begins once the panel is trained and its performance is validated. This process begins with specific product orientation, which usually includes an introduction to products within this product category including line extensions or competitive samples.

Evaluation procedures   A very important part of product training is the development of the sample preparation and the evaluation procedure. The panel leader directs the sample preparation while the panel develops the product evaluation procedure including identification and definition of the attributes. Sometimes only one difference exists between products: the order in which the attributes occur. When order is fixed on a ballot or scorecard, it is difficult to record changes in attribute order, and special provisions should be made for this. The evaluation procedure also must account for changes that occur with time and temperature. When evaluating a hot product, the sample changes so drastically as it cools that newly prepared samples from the same lot must be prepared and presented to the panel to complete the final portion of the analysis. The evaluation procedure must be built on what is actually observed by the panel and not, for example, dictated by the project leader.

Identification of terms   A fundamental principle of this method depends on the identification and the description of all the perceptible attributes in the sample by the panelists themselves. There are countless examples in the literature where the sensory method is claimed to be descriptive analysis. Instead, this is usually attribute analysis, where a series of terms or descriptors, which may or may not be perceptible in the sample, are submitted to the panelists for measurement. These terms are *presented* to the panel rather than being *identified or independently discovered* by them. Seldom are these attribute lists inclusive, and often they represent an ingredient or formula list.

Definitions   Much has been done but still much more must be done in the area of definitions and language development (Civille and Lawless, 1986). The same word or term may mean something different to each panelist. For example, "green" may be used to describe entirely different phenomena (e.g., "green" grass, "green" as in algae, "green" as in underripe).

There is no universal source for terms and their definitions. Most large descriptive analysis panels maintain libraries of terms, which include suggested definitions and references. In the past, much of this information has been proprietary and seldom published. Publication of the beer wheel (Meilgaard et al., 1982) and the wine wheel (Noble et al., 1987) introduced universal terms and references for these two product classes. More recently, lexicons have been published for warmed-over flavor (WOF) beef (Johnsen and Civille, 1986), catfish (Johnson et al., 1988), and peanuts (Johnsen et al., 1988). Texture terms, definitions, and scales were included in the first publication of the Texture Profile method (Brandt et al., 1963). Additional texture terms and scales were published by Muñoz (1986).

Ballot/scale The first system of descriptive analysis, the Flavor Profile method (Caul, 1957), did not provide for a ballot or scorecard. Instead, each panelist was expected to record all observations in the order in which they were perceived. It was later determined that ballots and scorecards are useful when one product is repeatedly analyzed or when statistics are appropriate for data analysis. Further advances now allow direct data input into the computer, so the paper ballot and its companion the digitizer will soon become relics. Use of ballots does have limitations, however, especially when the order of occurrence of one or more attributes changes drastically or new attributes are identified.

LEGITIMATE OFFSPRING *(The Grandchildren)*

A review of the literature indicates far more illegitimate than legitimate offspring of descriptive analysis methods. Short-cuts of the method often end up not being descriptive analysis at all. The two most common shortcomings are: (1) lack of adequate panel training and maintenance and (2) failure to identify all perceptible attributes.

However, there are excellent examples which represent legitimate hybridization of descriptive analysis techniques. These descendants, which represent valid or authentic applications of the method, have evolved from early roots in one or more of the basic systems and are probably best classified as utilizing a generic form of descriptive analysis. Unfortunately, to fully understand how descriptive analysis is used, it is necessary to retrieve many journal articles and piece together the actual methodology used.

Descendants

Lyon and co-workers at the USDA Russell Research Center, Athens, Georgia, have successfully used descriptive analysis to study poultry flavor and texture. Familiarity and experience with the major techniques (Lyon, 1980, 1987, 1988; Lyon and Lyon, 1989) have enabled this group to develop a modification of descriptive analysis that qualifies as a true descendant. Of special interest is research on poultry flavor changes. Panel data from this work resulted in the identification, description, and measurement of some of the character notes responsible for WOF in poultry. Day-old samples were described as bland. Recently, Lyon reported that maintenance of certain character notes, such as chickeny, meaty, brothy, and liver/organy, may be as important as the development of certain off-flavor notes when attempting to define and monitor warmed over flavor (Lyon, 1988; Ang and Lyon, 1990).

A modification of the flavor profile method was used by Lynch et al. (1986) to study flavor changes in ground beef packaged under different conditions. In this research, panel scores were recorded on a predetermined score card and the panel responses were averaged. As with the chicken, the ground beef attributes identified at zero time decreased before the stale or sour notes attributed to old ground beef were detectable.

The research findings of both groups demonstrate the value of using the descriptive analysis method to study the complex flavor changes that occur during storage. Frequently, when conducting shelf life or stability testing, one places emphasis on monitoring atypical or off-flavor to measure flavor deterioration. By using descriptive analysis, it is possible to track all the perceptible flavor changes. These changes often can be divided into three phases. The first phase is characterized by a decrease in intensity of the zero time or initial attributes. This is followed by a flat or bland phase where most attributes are low in intensity. Sometimes panelists want to actually add a descriptor such as bland at this point. The final phase is when the off-notes, such as rancidity or oxidation, are actually detectable. During all three phases the product is noticeably different in flavor, but it is not until the final phase that the change can actually be defined as off-flavor.

Catfish flavor has been effectively studied by Johnsen and Kelly (1990) using the descriptive analysis method. This team at the USDA Southern Regional Research Center (SRRC) reports that descriptive analysis panels have been useful when evaluating minced fish samples. Information obtained using both instrumental and sensory analysis data has been used to understand and control off-flavor—a problem that has long plagued catfish acceptability.

The lexicon of meat warmed-over flavors developed by Johnson and Civille (1986) laid the ground work for the application of the descriptive analysis method to research on meat flavor currently being conducted by the SRRC. This work, which is reported by Crippen et al. (1988), Spanier et al. (1990), and St. Angelo et al. (1987), shows how the descriptive analysis panel has been used to track the development or absence of WOF in meat that has undergone a variety of handling conditions and treatment. The strength of this work is that solid chemical, physical and sensory data are being used to unravel a complicated flavor issue.

Other researchers at USDA have provided important information on peanut flavor using descriptive analysis (Sanders et al., 1989). This research has helped guide harvesting, storage, and processing of peanuts and set both domestic and international standards for this product.

Noble (Noble, 1984; Noble et al., 1984) and co-workers at the University of California-Davis have successfully used descriptive analysis alone and in conjunction with other sensory analysis methods to conduct research on wine. This group is responsible for development of the wine wheel (Noble et al., 1987) as well as considerable information about definitions, references, and specific varieties of wine (Noble and Shannon, 1987; Aiken and Noble, 1984; Ohkubo et al., 1987). As with the other generic methods discussed in this paper, the methodology used is reported in several papers. There appear to be several unique aspects of this research that merit mention. The panelists, who are actually called judges, are asked to smell each reference standard at the beginning of each session (Heymann and Noble, 1987), order of presentation is randomized by judge, and data are analyzed using univariate and multivariate techniques.

Variants

Occasionally, situations arise where a portion of the entire analysis, rather than the complete analysis, can be used for routine sensory testing. These are called variants. The distinguishing feature of a variant is that it evolves once the entire descriptive analysis has been completed and specific distinguishing attributes of the product are identified.

For example, a manufacturer of a formulated potato product was concerned that samples of product made at different plants had noticeably different textures. To further complicate the problem, the within-plant texture was not uniform. While perceptible differences were apparent, these differences could not be measured instrumentally or with specific attribute scales such as hardness, crispness, etc. Product differences seemed to be related to the way the product broke down upon mastication.

A descriptive analysis panel that focused on texture was selected and trained. About 30 perceptible texture attributes were originally identified by the panel. Eventually these were reduced to 21 unique attributes, which were evaluated over the following phases: appearance, surface properties, first bite, mastication, residual, and after swallow. Each attribute was defined and references were developed for each. Training samples representing a wide variety of textures were provided by research and development. These samples were used to make sure all attributes had been identified, to enable the panel to practice and gain confidence, and to monitor both individual and total panel performance.

Simultaneous with panel training, extensive consumer research gave indications as to what textural characteristics were desired by consumers. Finally, based on consumer input, an optimum product was identified. All this was actually part of phase I.

The objective of the second phase was to monitor plant production. First the amount of variation in the textural attributes of product produced at each plant was determined.

At the outset, the sensory description of the product was confused and unreliable. Using the descriptive analysis method the project team was able to better understand what happens as the product breaks down during chewing and apply this information to guide product improvement. This was accomplished in the following steps:

1. Identify the textural attributes.
2. Show how process changes affected texture.
3. Determine optimum texture (consumer).
4. Modify process to improve product and prevent product drift in the future.

With persistence, patience, and guidance provided by the descriptive analysis panel, the problem was solved. Adjustments in equipment speed resulted in a more uniform product. Currently production is controlled by monitoring just two attributes: piece firmness and mash. The complete sensory analysis using all the attributes is used to monitor the production panel to control product drift.

The resolution of this problem was made possible by the use of descriptive analysis testing. Textural properties were complex and hard to monitor. Consumer acceptability was dependent on the interrelationship of all the texture attributes but especially those during mastication. The complexity of this issue is described by Muñoz and Civille (1987), who reported that past studies have tended to focus on unidimensional reactions to discrete and simple textural properties and not on the consumer reaction to all the textural changes which occur during mastication. This phenomenon was verified in the research described in the formulated potato project.

One final example to demonstrate how a variant of descriptive analysis can be used for research guidance involves apple juice. Juice can be made by fresh pressing or, when fresh apples are not available, by combining concentrate and essence. Descriptive analysis has also been successfully used in this research to identify and document the flavor characteristics of apple juice. If juice made from concentrate and essence is going to taste like *fresh pressed* juice, improvements must be made in the processing and handling of essence. To provide direction for these improvements a series of juices was analyzed. In particular, the headspace aroma of several varieties of fresh pressed juices was characterized using descriptive analysis.

Graphs, sometimes called spider plots, depicting the aroma attributes for juice made from three varieties of apple are shown in Fig. 11.4. Each line represents a

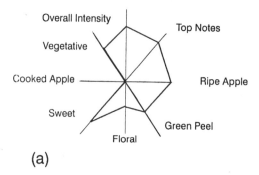

APPLE JUICE AROMA
Golden Delicious

(a)

FIG. 11.4 Intensity of each aroma attribute is shown by the distance from the center point.

unique attribute, and the distance from the midpoint along each line is a measure of the attribute intensity. These plots of fresh pressed juice are used as a target when blends of essence and concentrate are made in an attempt to match the aroma of freshly pressed juice as closely as possible. Cooked is included as an attribute, even though it is recorded as zero. This is done to verify that this note, which is not found in fresh juice but always found in processed juice, actually does not occur in fresh juice. Information such as has been described is very useful to provide direction for screening and blending essence/concentrate mixtures.

Applications

Panel operation and application is very dependent on the end use for which the panel has been trained. Dedicated panels are usually used to track one product in great detail. This type of panel is most often used for quality assurance. These panels usually have preprinted ballots, with carefully defined analysis instructions and detailed references. Other panels are used to provide guidance to R&D for new or reformulated product work. Still other panels are used to troubleshoot problem product or track competitive product changes. If carefully supervised and maintained, it is entirely possible for one panel to be used in all of the above applications.

Food material is not static. Every product, even shelf-stable, vacuum packed,

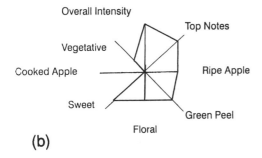

(b)

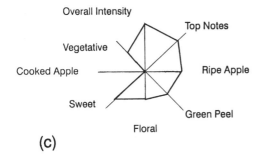

(c)

FIG. 11.4 *(cont.)*

or dry product, changes somewhat with time. Testing is destructive, retention samples are almost impossible to maintain. Many chemical and physical methods of analysis exist, but only human subjects can complete the total sample description in a way that reflects the actual sensory perception. Descriptive analysis is the best available means when a detailed analysis of the complex sensory perceptions is required.

By reviewing both descendants and variants of legitimate descriptive analysis systems, we have seen how this method can be applied to product development, process modification, and shelf-life testing. Other applications include package performance, competitive analysis, and quality assurance. Further applications of descriptive analysis are discussed in detail by Gillette (1984).

FUTURE ADVANCES *(The Next Generation)*

While there have been many refinements and modifications in descriptive analysis, there have been few major advances in quite some time. In addition to greater understanding of sensory perception in general, there are three areas where major advances could lead to substantial improvements in the descriptive analysis method. These areas include better understanding and application of time-related measurements, further development of universal references and definitions, and simpler and more flexible methods for data analysis.

Time-Related Measurements

In the descriptive analysis method, the order of occurrence of each attribute is very important. Some of the systems in use today (Meilgaard et al., 1987) attempt to measure this in a general way by identifying one or more attributes and monitoring over time. Use of a ballot or scorecard offers little provision for variation between products in regard to the order in which the attributes appear. This lack of flexibility is an inconvenience at least, but on occasion has been the cause of erroneous or misleading information. Research on time/intensity is being conducted, but for the most part this seems to pertain to individual attributes, their intensity, and their duration and does not measure simultaneous change with time of multiple attributes. Recently, ASTM Committee E-18 on the Sensory Evaluation of Products authorized the formation of a new task group that will study temporal or time intensity, or, as it is now called, time-related measurements. To completely document the sensory properties of the sample, we must learn how to identify, describe, and quantify each attribute at precisely the moment it is perceived and record this information in the correct perspective to all other perceptible attributes. Further understanding of perception, especially sensory interactions (Lawless, 1987), will also help in understanding and performing time-related measurements.

Universal Definitions and References

Much of the time spent in applying this method is consumed by developing and preparing the attribute definitions and references. Very little of this information has found its way into the literature since many companies consider this information proprietary. It would be much better if the operators of descriptive analysis panels considered the data, rather than the documentation of the method, were proprietary. However, this is seldom the case. Again, ASTM Committee E-18 is providing leadership in this area. Many member groups are working to develop a comprehensive lexicon that will serve as a basic primer of descriptive terms, references, and examples. Already this task group has served as a catalyst by connecting various interest groups, who are working together to collect hundreds of pieces of information on a wide variety of terms.

Data Analysis

The last area pertains to data analysis. Different end uses of the data have different needs. On one end, quality control and purchasing often claim to want everything reported as just one *simple* number, while researchers unraveling complex flavor issues tend to generate many, many numbers. Somehow a realistic balance must be found. Considerable advances in data analysis, especially multivariate analysis, have been made and reported (Powers, 1988, 1989). The universal availability of personal computers along with statistical software packages have enabled even the most unskilled to apply what was previously considered complicated statistics to routine data analysis. To correctly apply these methods is not simple. Furthermore, there still lacks universal agreement as to what these methods mean and how they should best be applied and interpreted.

CONCLUSION

Employing descriptive analysis is not a casual matter. While the merit of this method cannot be challenged, it is not a panacea for all sensory analysis problems. Its application should be used judiciously. Furthermore, this method is not for the faint. Experienced leadership, extensive training, careful maintenance, and fairly sophisticated data analysis are essential to assure a successful outcome. Success is measured not only by the results of an individual or series of panels but by the ability to compare data from different panel groups or to build a reliable and meaningful database over time. All this is possible using descriptive analysis. When there are reports that the method yields meaningless data or fails to be reproducible, it is usually because it has not been properly employed. As with any laboratory instument, the first period of use after its arrival results in some mistakes, some false conclusions, and inevitably, some skepticism. However, if one persists with repeated trials and patience, the instrument gradually becomes trusted and eventually relied upon as part of the laboratory team.

Descriptive analysis, when properly utilized, is an important method to conduct sensory analysis. Its weakness is that it is extremely time consuming, expensive, and dependent on both a highly trained panel and the skill of its leader. Its strength is that it provides a comprehensive qualitative and quantitative description, which results in reliable measurements and reproducible data using human subjects. Descriptive analysis provides valuable information about the sensory properties of the sample which are not obtainable from any other source.

REFERENCES

Aiken, J. W. and Noble, A. C. 1984. Comparison of the aromas of oak and glass-aged wines. *Am. J. Enol. Vitic.* 35: 196.

Ang, C. Y. W. and Lyon, B. G. 1990. Evaluations of warmed-over flavor during chill storage of cooked broiler breast, thigh and skin by chemical, instrumental and sensory methods. *J. Food Sci.* 55: 644.

ASTM. 1990. *Special Technical Publication Descriptive Analysis Testing.* American Society for Testing and Materials, (Committee E-18) Philadelphia. (in preparation)

Brandt, M. A., Skinner, E. A., and Coleman, J. A. 1963. Texture profile method. *J. Food Sci.* 28: 404.

Caul, J. F. 1957. The profile method of flavor analysis. *Adv. Food Res.* 7: 1.

Caul, J. F. 1959. *The Flavor Profile in Review. Flavor Research and Food Acceptance.* Sponsored by Arthur D. Little, Inc., pp. 65–74. Chapman & Hall, Ltd., London.

Civille, G. C. 1979. Sensory evaluation methods for the practicing food technologist. Descriptive analysis. Sponsored by Institute of Food Technologists. (M. R. Johnston, Ed.). Ch. 6.

Civille, G. V. and Lawless, H. T. 1986. The importance of langauge in describing perceptions. *J. Sensory Stud.* 1: 203.

Civille, G. C. and Liska, I. H. 1975. Modifications and applications to foods of the General Foods sensory texture profile technique. *J. Texture Stud.* 6: 19.

Civille, G. V. and Szczesniak, A. A. 1973. Guidelines to training a texture profile panel. *J. Texture Stud.* 4: 204.

Crippen, K. L., Spanier, A. M., Miller, J. A., and Boihem, L. L. 1988. Warmed-over flavor (WOF) in beef: The effect of internal cooking temperature and storage time on the descriptive flavor profile. Paper No. 65. Presented at the 48th Annual Meeting of the Institute of Food Technologists. New Orleans, LA, June 19–22.

Gillette, M. 1984. Applications of descriptive analysis. *J. Food Protection.* 47: 403.

Heymann, H. and Noble, A. C. 1987. Descriptive analysis of commercial Cabernet Sauvignon wines from California. *Am. J. Enol. Vitic.* 38: 41.

Johnsen, P. B. and Civille, G. V. 1986. A standard lexicon of meat WOF descriptors. *J. Sensory Stud.* 1: 99.

Johnsen, P. B., Civille, G. V., and Vercellotti, J. R. 1987. A lexicon of pond-raised catfish flavor descriptors. *J. Sensory Stud.* 2: 85.

Johnsen, P. B., Civille, G. V., Vercellotti, J. R., Sanders, T. H., and Dus, C. A. 1988. Development of a Lexicon for the description of peanut flavor. *J. Sensory Stud.* 3: 9.

Johnsen, P. B. and Kelly, C. A. 1990. A technique for the quantitative sensory evaluation of farm-raised catfish. *J. Sensory Stud.* 4: 189.

Lawless, H. 1987. Sensory interactions in mixtures. *J. Sensory Stud.* 1: 259.

Lynch, N. M., Kastner, C. L., Caul, J. F., and Kropf, D. H. 1986. Flavor profiles of vacuum packaged and polyvinyl chloride packaged ground beef: A comparison of cooked flavor changes occurring during product display. *J. Food Sci.* 51: 258.

Lyon, B. G. 1980. Sensory profiling of canned boned chicken: Sensory evaluation procedures and data analysis. *J. Food Sci.* 45: 1341.

Lyon, B. G. 1987. Development of chicken flavor descriptive attribute terms aided by multivariate statistical procedures. *J. Sensory Stud.* 2: 55.

Lyon, B. G., 1988. Descriptive profile analysis of cooked, stored, and reheated chicken patties. *J. Food Sci.* 53: 1086.

Lyon, B. G. and Lyon, C. E. 1989. Texture profile of broiler pectoralis major as influenced by postmortem deboning time and heat method. *Poultry Sci.* 69: 329.

Meilgaard M., Civille, G. V. and Carr, B. T. 1987. Descriptive analysis techniques. *Sensory Eval. Tech.* 2: 1.

Meilgaard, M. S., Reid, D. S., and Wurborski, K. A., 1982. Reference standards for beer flavor terminology stem. *J. Am. Society Brewing Chemists.* 40: 119.

Munoz, A. 1986. Development and application of texture reference scales. *J. Sensory Stud.* 1: 55.

Munoz, A. and Civille, G. 1987. Factors affecting perception and acceptance of food texture by American consumers. *Food Reviews International.* 3:285.

Noble, A. C. and Shannon, M. 1987. Profiling Zinfandel wines by sensory and chemical analyses. *Am. J. Enol. Vitic.* 38: 1.

Noble, A. C., Arnold, R. A., Buechsenstein, J., Leach, E. J., Schmidt, J. O., and Stern, P. M. 1987. Modification of a Standardized system of wine aroma terminology. *Am. J. Enol. Vitic.* 38: 143.

Noble, A. C., Williams, A. A., and Langron, S. P. 1984. Descriptive analysis and quality ratings of 1976 wines from Bordeaux communes. *J. Sci. Food Agric.* 35: 88.

Noble, A. C., 1984. Precision and communication: Descriptive analysis of wine. Wine Industry Technical Symposium, pp. 33–41.

Ohkubo, T., Noble, A. C., and Ouch, C. S. 1987. Evaluation of California Chardonnay wines by sensory and chemical analyses. *Sci. des Aliments*, 7(4): 573.

Powers, J. J. 1988. Current practices and applications of description methods. *Sensory Analysis of Foods*. (J. R. Piggott, Ed.). pp. 187–266. Elsevier, London.

Powers, J. J. 1989. Applying multivariate statistical methods to enhance information obtainable from flavor analysis results. *Flavor Chemistry of Lipids*. (D. B. Min and T. H. Sonse, Ed.), pp. 434–462. American Oil Chemists Society, Illinois.

Rainey, B. A. 1986. Importance of reference standards in training panalists. *J. Sensory Stud*. 1: 149.

Sanders, T. H., Vercellotti, J. R., Blakenship, P. D., Crippen, K. L., and Civille, G. V. 1989. Interaction of maturity and curing temperature on descriptive flavor of peanuts. *J. Food Sci*. 54: 1066.

Spanier, A. M., McMillin, K. W., and Miller, J. A. 1990. Enzyme activity levels in beef: Effect of postmortem aging and end-point cooking temperature. *J. Food Sci*. 55: 318, 326.

St. Angelo, A. J., Vercellotti, J. R., Legendre, M. G., Vinnett, C. H., Kuan, J. W., James, Jr., C., and Dupuy, H. P. 1987. Chemical and instrumental analyses of warmed-over flavor in beef. *J. Food Sci*. 52: 1163.

Stone, H. and Sidel, J. L. 1985. *Sensory Evaluation Practices*, pp. 194–226. Academic Press, Inc., Orlando, FL.

Stone, H. J., Oliver, S., Woolsey, A., and Singleton, R. C. 1974. Sensory evaluation by quantitative descriptive analysis. *Food Technol*. 28: (11): 24, 26, 28, 29, 32, 34.

Stone, H. S., Sidel, J. L., and Bloomquist, J. 1980. Quantitative descriptive analysis. *Cereal Foods World*. 25: 642.

Szczesniak, A. S., Brant, M. A., and Friedman, H. H. 1963. Development of standard rating scales for mechanical parameters of texture and correlation between the objective and the sensory methods of texture evaluation. *J. Food Sci*. 28: 397.

Szczesniak, A. S. 1963. Classification of textural characteristics. *J. Food Sci*. 28: 404.

12

Selection of Terms for Descriptive Analysis

John R. Piggott

University of Strathclyde
Glasgow, Scotland

INTRODUCTION

The simple, obvious function of quantitative descriptive sensory analysis is to describe samples of foods, beverages, or in a wider context any object that makes an impression on the human senses. This discussion will be directed at the use of quantitative descriptive analysis in the food and beverage industry, but there is no reason why the method should not be used in any industry where people's perceptions of a product or material are important. There are examples in the literature of psychologists investigating the sensory assessment of many products, such as textiles or paper, and most of the population must be familiar with the market researcher's desire for us to profile everything from airlines to zoos. The roots of current practices can be found in the semantic differential technique and in Kelly's Role Construct Repertory Test (Kelly, 1955), subsequently developed into the Repertory Grid Method (RGM), though the underlying theory might be different. The RGM has indeed been adapted to help in the provision of terms for free-choice profiling (FCP) by, for example, McEwan and Thomson (1988). Stripped of underlying theory, the various methods of quantitative descriptive analysis simply assume that assessors can meaningfully use terms to describe sensations, i.e., that a perception can be broken down into its components. There may be little theoretical justification for this, but in practice the method works and can be seen to produce meaningful and useful results with both internal and external validity.

The way in which the task is carried out, including the selection of terms used for description, depends on a number of factors. Selection of terms for a panel must take into account the purpose of the analysis, the samples, the assessors, definition of the terms, and to whom the results of the analysis must be reported.

FACTORS AFFECTING CHOICE OF TERMS

Purpose of Analysis

The function of the methods of descriptive analysis such as quantitative descriptive analysis or FCP is not as obvious as the names suggest; it may be simply to describe a sample, but more often it will be to find and identify differences between samples, and then to relate these differences to some other variables. A simple description of a sample, done in isolation, has very little meaning, just as a chemical or physical analysis such as a gas chromatogram means little on its own. Some comparison must always be made, either an explicit comparison with one or more other samples analyzed at the same time, or an implicit comparison with samples analyzed in the past or to be analyzed in the future. If the purpose of the analysis is simply to discriminate between samples, the ideal vocabulary is the shortest. If the samples vary in only one characteristic, then only one word is required. This will not normally be the case, and certainly it will be unusual if this can be known with confidence when the system is set up, so the vocabulary will usually be longer. The main criterion in this case is that the terms used are good discriminators.

Alternatively, the analysis may be required to provide a full description of the characteristics of the samples. These are different problems, and different sets of terms might be required in each case. In the second case, a complete vocabulary is required, even if the samples do not differ in a particular attribute.

Samples

For a single experiment, the full sample set will obviously be available, and so there should be no problem in ensuring that all possible characteristics of the samples are covered by the vocabulary. This is normally the case where FCP is being used, and it is important that all samples are used when the vocabulary is being developed. However, if the vocabulary is expected to be required for samples in the future, great care must be taken when the vocabulary is developed to ensure that as wide a range of samples is available as possible, so that a complete vocabulary is developed to describe all possible samples.

Assessors

The panel of assessors may be typically a single expert, a panel of 12 or so trained assessors in a laboratory, or a panel of up to several hundred consumers. Any of these panels may be used in different circumstances for different purposes. The single expert is traditionally used in industries where the product flavor (and other sensory characteristics) is under incomplete control, such as Scotch whisky or coffee, and thus must be examined for blending or classification. The single expert normally works in a small group within a company or an industry association, and tends to develop a vocabulary which may be difficult for an outsider to understand (see Lang, 1983). However, the main characteristic of the vocabularies developed by such experts seems to be that they are very detailed and contain a large number of words to attempt to describe precisely fine nuances of flavor or texture.

The laboratory panel, the conventional sensory analysis tool, tends to work with a shorter vocabulary, and while exact description of such fine detail may be lost, the properties of the samples should still be captured.

Consumer or untrained laboratory panels, used, for example, for modeling market behavior, seem to use the same kind of vocabulary (Guy et al., 1989) as trained laboratory panels, though with less precision (Clapperton and Piggott, 1979).

These assessors and their strengths and weaknesses must be considered when the terms to be used are chosen. Learning to recognize odors (i.e., learning to attach a label to a smell) is a slow process (Cain, 1977), and lengthy training can be reduced if the assessors are familiar with the terms. This is particularly important if the terms are to be used in different laboratories or different countries, and in this case some compromise may be required. It might be more effective to use a nonoptimal vocabulary for a single panel, if that vocabulary can then be used without change by a different panel. It is particularly important that the cultural and linguistic background of the assessors is taken into account when any terms are proposed.

Definition of Terms

To stabilize the panel's use of the vocabulary and to assist in the training of new recruits, it is helpful to have access to a set of reference standards which can be used to illustrate the meanings of the terms. This means that suitable materials must be available for use as standards. Different considerations apply if the standard is to represent a flavor note, a texture, or an appearance characteristic; foods are generally quiet, and so profiling the sounds made by a food has not been a common requirement.

Flavor   In the case of flavor notes, materials should be chosen so that, ideally (1) a standard contains a true representation of the flavor note as it appears in the product; (2) it contains the flavor note required and no other; and (3) it is readily available in the required purity. The standard may be a pure chemical, which has the advantage that it is absolutely stable across time and space, given constant purity. *Sweet* is, for example, very easy to define anywhere, in any language, by using sucrose as a standard. Alternatively, the standard may be a complex material which, while being a good representation of the flavor note, has the disadvantage that it might not be stable. The odor of cardboard, for example, depends on the manufacturer and the process used, and may be little use if it is to be used in a number of different countries. It might thus be more successful to choose a different descriptive term, for which an unequivocal defining standard is available, rather than to choose the "best" term for description or discrimination.

Clapperton (1973, 1976) and Clapperton et al. (1976) provide an early example of the derivation of a vocabulary and selection of reference standards. Meilgaard et al. (1979) summarized the principles upon which the system was based as follows:

Each separately identifiable flavor characteristic has its own name.

Similar flavors are placed together.

No duplication of terms for the same flavor characteristic.

The system is compatible with the "EBC Thesaurus for the Brewing Industry."

Subjective terms such as good/bad, young/mature, balanced/unbalanced are not included.

The meaning of each term is illustrated with readily available reference standards.

Compatibility with the "EBC Thesaurus" is obviously a special requirement of this case, but the desirability of compatibility with existing terminology is an issue which must be considered when formulating a vocabulary.

Texture   Standards for definition of texture terms have more often been used to define points on a well-understood scale, to reduce the variability in scaling, and as such should have little influence on the choice of terms.

Appearance Color is relatively easy to define by means of color chips or samples, and to measure in most foods by various spectroscopic methods. However, where a food is not a uniform color, sensory descriptions might be required, and there are other characteristics of foods, such as translucency, which render instrumental analysis less straightforward (MacDougall, 1989). Terms may need to be chosen carefully to ensure that representative standards are available.

Results

In order for the results of the analysis to have any utility, they must also be communicated. This might be explicit, in that the results will be incorporated in a report which another person will use to make a decision, or implicit in that the individual doing the analysis will use the results to make a decision. In this case, the experimenter will often need to make a report to justify the decision taken. The terms used must therefore make sense to the recipients of a report. This again might mean the use of a nonoptimal vocabulary, for the sake of a "meaningful" set of terms. For research workers a common form of report is a scientific journal, and in this case the reader of the report seems to be all too often forgotten. Specialized terms that are incomprehensible outside a company, an industry, or a country should not be used. Lawless (1985) raised the question of the "meaningfulness" of quantitative descriptive analysis results: Can the results be related back to the original samples? This can be regarded as the ultimate test of any quantitative descriptive analysis system.

SELECTION OF TERMS

The ideal vocabulary consists of terms which meet all the above conditions. That is, they accurately and precisely describe all required characteristics of all the samples, they are understood by the assessors who agree with each other about their meanings, they can be easily defined with reliable standards, and they are understood by the recipients of a report. However, in practice these demands may be contradictory, and some compromises may be required.

Whisky

This laboratory has extensive experience of descriptive analysis of whisky and other distilled alcoholic beverages. The development and use of the whisky vocabulary over about 12 years can be used as an example of the evolution of a successful system (Piggott and Canaway, 1981). An initial vocabulary (Piggott and Jardine, 1979) was compiled by searching the literature on Scotch whisky for descriptive terms already in use. The processes used were also examined to attempt to identify sources of flavor variation. Ambiguous and synonymous terms were deleted or condensed, and imprecise terms expanded and redefined, to yield a list of 35 terms. Eleven assessors used this to describe the flavors of 10 commercially available whiskies. After duplicate assessments had been made, the assessors were trained by familiarizing themselves with 21 reference materials. The other descriptors could not be defined in this way, so a verbal definition was sought through group discussion. After the training period, the whiskies were again presented to the assessors twice. In both pre- and post training analyses, broadly similar patterns of whiskies were found. This showed that the system had enabled a panel of inexperienced assessors to differentiate between a variety of whiskies. Training was shown to improve the reproducibility of the results.

The next stage involved improvement of the vocabulary, to eliminate synonymous terms, redefine ambiguous ones, and incorporate suggestions received from the industry. This gave a list of 30 terms. A panel of 15 untrained and inexperienced assessors profiled 64 whiskies. Analysis of these results showed that the method allowed the untrained assessors to distinguish well between all whiskies, both to separate individual whiskies and to group similar ones effectively. The ways in which the terms had been used were examined, and a revised list of 37 was developed. This was tested by running a similar experiment with a selection of 71 whiskies. Analysis of the results gave an improved separation of different whiskies and clustering of similar ones and a list of 26 terms. This was subsequently shortened by the omission of two terms (Canaway et al., 1984), which proved to be unnecessary for the description of the vast majority of samples (Table 1).

This vocabulary was designed to include all the terms regularly used by the assessors and those terms which described flavor notes not approximately covered by other items. Reference standards were subsequently identified for the majority of the terms in the list by examining chemical catalogs and discussions with other workers in the whisky industry and in sensory analysis. A number of standards was then proposed for each term and offered to the assessors for comment. The materials the assessors chose were then adopted as reference standards for the vocabulary (Table 1). In the case where the assessors did not approve of any of the proposed materials, the experimenters started again—we did not enforce standards on the assessors.

Format   In contrast to some other approaches (Meilgaard et al., 1979; Swan and Burtles, 1980), a concise single-tier scheme was adopted. The principal reason was that this was the form of vocabulary which the assessors' use of the terms generated. It was thought unnecessary to use a much larger vocabulary. If

TABLE 12.1 Whisky Descriptors and Reference Standards

Descriptor	Standard	Approximate concentration (μg/ml)
Pungent	Formic acid	10,000
Solvent	2-Methylpropan-1-ol	1,000
Spicy	Eugenol	1
Grainy		
Malty	Malted barley	
Mouldy	2, 4, 6-Trichloroanisole	1
Fruity (estery)	*iso*-Amyl acetate	10
Fruity (other)	Ethyl octanoate	10
Floral	2-Phenylethanol	100
Smooth		
Vanilla	Ethyl vanillin	10
Soapy	1-Decanol	100
Sour	Equimolar mixture of acetic, propionic, butyric, and valeric acids	100
Nutty		
Buttery	Diacetyl	1
Grassy	cis-3-Hexen-1-ol	1,000
Phenolic		
Oily	Heptanal	1
Woody		
Meaty		
Sulfury	Dimethyl sulfide	1
Catty	Mixture of sodium sulfide and mesityl oxide	100 each
Fishy	Trimethylamine	1
Sweet		

one small area of flavor is of interest, it is possible to expand the vocabulary in that area to assist the assessors, although this is not regarded as essential.

Time required The development of the whisky vocabulary was a long process, which could be justified in the earlier days of quantitative descriptive analysis by the relative novelty of the technique, but would be difficult to justify now. However, the vocabulary has been used since 1980 for routine description of whiskies from many countries, of grape and other fruit brandies, and of other light spirits such as vodkas and white rums; the time spent in development was time well spent.

Involvement of assessors At every stage after the initial trial run the assessors were consulted, both about terms and about reference standards. This has the advantages that the experimenters cannot impose "foreign" terms on the assessors, it ensures that the standards agree with what the majority of the assessors think the terms mean, and it helps to maintain panel motivation by demonstrating that the development and use of the vocabulary is a cooperative process.

Some of the terms, such as smooth, have proved impossible or difficult to define either verbally or with reference materials, but these have been retained at the assessors' insistence. When this term has been omitted experimentally, the assessors have put it back. It is useful during routine use of a quantitative descriptive analysis system to allow the assessors to make free-form comments; these are often of the form "yucch" or "yummy," but occasionally useful words are suggested and should be noted for possible inclusion in a later revision of the vocabulary. It may be observed in passing that systems for computerized sensory data collection might not allow assessors to input free-form comments, and so notepads and pencils must still be used!

It is also important that the experimenter exercises some judgment, and acts as a filter to screen out terms which are not helpful; while the assessors and their wishes are the most important components of the entire quantitative descriptive analysis system, they must occasionally be overruled by a sympathetic experimenter in the interests of common sense and rationality.

Flexibility   As noted above, the vocabulary has been changed once by the omission of two terms which subsequently proved to be unnecessary; the vocabulary is not regarded as permanent for all time, but good reason is required to change it.

Panel origin   The vocabulary discussed here was developed for use by a panel in the University, largely drawn from academic and technical staff of the department with a few graduate students. The assessors are mainly of British origin and English-speaking, and are laboratory trained and of considerably higher intelligence than average. As such they form a rather homogeneous group, and it has been relatively easy to develop a vocabulary which they could all use successfully. Problems can arise when such a vocabulary is transferred to other panels or other cultures, or vocabularies developed elsewhere are imposed on a panel.

Development of Vocabulary for Industrial Panel

A quantitative descriptive analysis procedure was developed to assist in the QC of rectified spirit by an industrial panel (Wilkin et al., 1983). The odor of rectified spirit is difficult to describe because the odor notes are weak and are masked by the strong odor of ethanol. An initial examination of six samples by the whisky panel in the University showed that the method had potential, and a subsequent experiment was carried out with 28 samples of spirit. This allowed the extraction of 11 terms, which best described and discriminated between the samples, from the vocabulary above, to form a specialized subset for the samples to be considered. Reference materials were identified for all the terms by following the procedure outlined above. A new panel of 12 assessors was recruited from the employees of the company, and the assessors were trained to recognize the odor notes selected (Piggott and Canaway, 1981). This procedure breaks the rules set out above, in that the panel was not consulted about the terms. However, given the practical constraints at the time, it was not thought to be realistic to attempt this.

Comparison of Vocabularies

Two other laboratories published vocabularies (Table 2) for descriptive sensory analysis of whiskies at about the same time as the work described above, and reported work based on the techniques (Shortreed et al., 1979; Jounela-Eriksson, 1981). To investigate the performance of the vocabularies in general use under laboratory conditions, each of the three was used by a panel of assessors to describe the odors of 24 whiskies (Canaway et al., 1984; Piggott et al., 1985).

Twelve assessors, with little or no experience of psychophysical experiments or of whisky, took part in the experiment. Twenty-four samples were chosen to provide examples of as many whisky types as possible, from new malt distillates to samples more than 12 years old. The assessors were divided into six pairs, and each pair used the three vocabularies in a different order, to avoid bias caused by a preference for, for example, the first vocabulary used. The samples were scored for each term in the same way as the original authors in each case.

Principal components analyses showed that all three vocabularies permitted the successful separation of the new distillates from the other samples, indicating that these groups of samples were clearly different. The two longer vocabularies enabled the separation of the grain whiskies, and the Strathclyde vocabulary permitted a clear distinction to be made between the matured malts and blended whiskies.

Of the three vocabularies, that published by Shortreed et al. (1979) suffered from the greatest redundancy; for example, 13 descriptors were used to describe the new distillate samples on component 1, whereas eight and five were used from the other vocabularies. This could be an advantage with a highly trained panel of assessors, who could use the descriptors with greater precision. However, until satisfactory reference standards are found, the training is likely to be difficult. The relatively poorer performance of the other vocabulary (Jounela-Eriksson, 1981) suggested that care should be taken when using such a short

TABLE 12.2 Two Other Vocabularies Used for Descriptive Sensory Analysis of Whiskies

Jounela-Eriksson (1981) (13 terms)	Shortreed et al. (1979) (43 terms)	
Typical whisky aroma	Pungent	Glycerin-like
Solvent-like	Prickle	Honey
Fusel oil	Nose warming	Vanilla
Estery (fruity)	Nose drying	New wood
Grain-like	Medicinal	Developed extract
Woody	Peaty	Defective cask
Spicy (peppery)	Kippery	Nutty
Tallowy	Leathery	Buttery
Phenolic (medicinal)	Tobacco	Fatty
Tarry	Sweaty	Rancid
Burnt	Stale fish	Sickly
Pungent	Cooked mash	Cheesy
Malty	Cooked vegetables	Vinegary
Aroma of sweet sap	Toasted	Cabbage water
	Malt extract	Rubbery
	Husky	Coal gassy
	Hay-like	Stagnant
	Leafy	Earthy
	Floral	Musty
	Fragrant	Blotting paper
	Fruity	Metallic
	Solvent	

vocabulary, unless it is known to be adequate for the intended use. Care must also be taken when a set of descriptive terms is translated from one language to another, or transferred from one culture to another. It is important that the assessors understand the terms in the vocabulary, for example, by carefully defining the terms with standards and training the assessors.

A vocabulary for descriptive sensory analysis should contain a sufficient number of terms to describe the range of samples likely to be encountered, but should not be so long as to be confusing and tedious in use. The results obtained showed that, under the conditions of this study, the Strathclyde laboratory's vocabulary was the most useful for describing and classifying the range of samples used. This conclusion supports our view that the best vocabulary is developed in house by a panel as close as possible to the panel which will be using it.

Incorporating Quantitative Descriptive Analysis into a Traditional System

Possibly the most difficult and most demanding problem in the development of a vocabulary occurs when the language of the report is specified, but is different from the language of the panel. Contrasts between American and British English can make this difficult even for laboratory panels reporting to a scientist, but a problem has now arisen where a whisky company wishes to use quantitative descriptive analysis to monitor quality of newly distilled product. The management and the traditionally trained product experts wish to use the traditional language of the industry for describing product flavor and defects, but the panel is to be drawn from laboratory and office staff with little knowledge of these specialized terms. What we hope to be able to do in this case is to compare expert and University panel assessments of samples and of reference standards to develop a translation table of equivalences of terms. The panel will then be trained with a set of suitable standards to use the terms required. This also breaks the rule of allowing the panel to choose the terms, but in this case they do not know them.

CONCLUSION

Many factors must be considered when terms are selected for use in a quantitative descriptive analysis method, including the purpose of the analysis, the samples, the assessors, definition of the terms, and to whom the results of the analysis must be reported. The most successful vocabulary is likely to be found when the assessors are carefully selected and then allowed to choose the terms they wish to use, guided by a sympathetic experimenter with a personal interest in and familiarity with the product and common sense.

REFERENCES

Cain, W. S. 1977. Physical and cognitive limitations on olfactory processing in human beings. In *Chemical Signals in Vertebrates* (D. Muller-Schwarze and M. M. Mozell, Ed.) pp. 287–302. Plenum, New York.

Canaway, P. R., Piggott, J. R., Sharp, R., and Carcy, R. G. 1984. Comparison of sensory and analytical data for cereal distillates. In *Flavour Research of Alcoholic Beverages* (L. Nykänen and P. Lehtonen, Ed.) pp. 301–311. Foundation for Biotechnical and Industrial Fermentation Research, Helsinki.

Clapperton, J. F. 1973. Derivation of a flavour profile method for sensory analysis of beer flavour. *Journal of the Institute of Brewing* 79: 495–508.

Clapperton, J. F. 1975. The development of a flavour library. *Proceedings of 15th Congress of the European Brewery Convention,* Nice p. 823. Elsevier, Amsterdam.

Clapperton, J. F. and Piggott, J. R. 1979. Flavour characterisation by trained and untrained assessors. *Journal of the Institute of Brewing* 85: 275–277.

Clapperton, J. F., Dalgleish, C. E., and Meilgaard, M. C. 1976. Progress towards an international system of beer flavour terminology. *Journal of the Institute of Brewing* 82: 7–13.

Guy, C., Piggott, J. R., and Marie, S. 1989. Consumer profiling of Scotch whisky. *Food Quality and Preference* 1: 69–73.

Jounela-Eriksson, P. 1981. Predictive value of sensory and analytical data for distilled beverages. In *Flavour '81* (P. Schreier, Ed.) pp. 145–164. Walter de Gruyter, Berlin.

Kelly, G. A. 1955. *The Psychology of Personal Constructs.* Norton, New York.

Lang, J. W. 1983. Blending for sale and consumption. In *Flavour of Distilled Beverages* (J. R. Piggott, Ed.), pp. 256–263. Ellis Horwood, Chichester.

Lawless, H. T. 1985. Psychological perspectives on winetasting and recognition of volatile flavours. In *Alcoholic Beverages* (G. G. Birch and M. G. Lindley, Ed.), pp. 97–113. Elsevier Applied Science, London.

McEwan, J. A. and Thomson, D. M. H. 1988. An investigation of the factors influencing consumer acceptance of chocolate confectionery using the repertory grid method. In *Food Acceptability* (D. M. H. Thomson, Ed.), pp. 347–361. Elsevier Applied Science, London.

MacDougall, D. B. 1989. Measurement of food and beverage colour. In *Distilled Beverage Flavour* (J. R. Piggott and A. Paterson, Ed.), pp. 85–96. Ellis Horwood, Chichester.

Meilgaard, M. C., Dalgleish, C. E., and Clapperton, J. F. 1979. Beer flavour terminology. *Journal of the Institute of Brewing* 85: 38–42.

Piggott, J. R. and Jardine, S. P. 1979. Descriptive sensory analysis of whiskey flavour. *Journal of the Institute of Brewing* 85: 82–85.

Piggott, J. R. and Canaway, P. R. 1981. Finding the word for it: Methods and uses of descriptive sensory analysis. In *Flavour '81* (P. Schreier, Ed.), pp. 33–46. Walter de Gruyter, Berlin.

Piggott, J. R., Carey, R. G., and Canaway, P. R. 1985. Sensory analysis and evaluation of whisky. In *Alcoholic Beverages* (G. G. Birch and M. G. Lindley, Ed.), pp. 69–84. Elsevier Applied Science, London.

Shortreed, G. W., Rickards, P., Swan, J. S., and Burtles, S. M. 1979. The flavour terminology of Scotch whisky. *Brewers Guardian* 108(11): 55–62.

Swan, J. S. and Burtles, S. M. 1980. Quality control of flavour by the use of integrated sensory analytical methods at various stages of Scotch whisky production. In *Olfaction and Taste VII* (H. van der Starre. Ed.), pp. 451–452. Information Retrieval, London.

Wilkin, G. D., Webber, M. A., and Lafferty, E. A. 1983. Appraisal of industrial continuous still products. In *Flavour of Distilled Beverages* (J. R. Piggott, Ed.), pp. 154–165. Ellis Horwood, Chichester.

13

Procrustes Analysis and Its Applications to Free-Choice and Other Sensory Profiling

Dolores C. Oreskovich, Barbara P. Klein, and John W. Sutherland

University of Illinois at Urbana-Champaign
Urbana, Illinois

INTRODUCTION

Descriptive analysis, also referred to as quantitative sensory (Powers, 1988), consensus (Tunaley et al., 1988), or fixed-choice (MacFie, 1987) profiling, is the method by which information about products has been obtained by sensory scientists for many years. Descriptive analysis provides a detailed description of the perceived quantitative and qualitative characteristics of a product. A number of descriptive analysis techniques have been developed, including flavor profiling (Cairncross and Sjöstrom, 1950; Caul, 1957), texture profiling (Brandt et al., 1963; Szczesniak, 1963; Szczesniak et al., 1963; Muñoz, 1986), Quantitative Descriptive Analysis™ (Stone et al., 1974; Stone and Sidel, 1985), and Spectrum™ (Meilgaard et al., 1988), and described extensively by these and other investigators (Piggott and Canaway, 1981; Powers, 1988). Each descriptive analysis method varies in the approach used for obtaining product information. A particular method is chosen based on the questions to be answered and the product being evaluated. All descriptive analysis methods are similar in that potential panelists are recruited, screened, selected, and trained. However, the number of panelists used, length of training, and type of products evaluated can vary. After training, panelists evaluate products (usually in a controlled environment), using established experimental procedures specific for each method. The data collected are then analyzed using statistical techniques.

Descriptive analysis is confined to the assessment of a product's attributes, usually considered independently of one another. Therefore, the only way a product is judged by a panelist is through individual assessment of each attribute. By collectively examining assessments of each attribute, a "profile" of the product can be provided, as examplified by the Quantitative Descriptive Analysis™ "spider web" (Stone et al., 1974). Although a profile of a product can provide information about each attribute, little consideration is given to the impression created when attributes interact when descriptive analysis is used. The flavor profile "amplitude" attempts to measure these interactions. In the past, descriptive analysis data have been analyzed using Analysis of Variance (ANOVA), a univariate statistical approach in which information is obtained about samples or products by pooling panelists' responses. Univariate statistics can provide information about the ability of panelists to collectively identify and quantify attributes, but disclose no information about relationships existing among attributes. This limitation makes it difficult to obtain information about the overall perception of a product.

More recently, multivariate analyses such as multidimensional scaling, factor analysis, and principal component analysis have been used to interpret data from descriptive analysis (Powers, 1984). The use of multivariate analysis is restricted by sample size. Meilgaard et al. (1988) suggested that the sample size be at least the square of the number of responses. Despite sample size restrictions, multivariate analysis takes into account the correlation between responses and therefore can provide additional information about the differences between products and panelists that univariate analysis of the same data may not.

Several descriptive analysis methods have been examined closely, improved upon, and ultimately matured into well-accepted and established techniques in sensory science. Although some of the methods are proprietary, attempts are being made to describe standardized procedures (see Chapter 12). Descriptive analysis and its applications have expanded and been improved upon as a result of extensive research and is now widely accepted as a standard methodology in the sensory science area (IFT, 1981; Stone and Sidel, 1985; Meilgaard et al., 1988).

Just as the benefits of descriptive analysis have been documented, so have the limitations. Descriptive analysis may require a considerable amount of time to recruit, screen, and train panelists to evaluate specific products. Once panelists are trained, the panel must be maintained over a lengthy period of time. Not only is panel training and maintenance time consuming, it can also be expensive. In addition, most descriptive analysis methods suggest that testing be done in environmentally controlled facilities, which can further escalate costs.

Since descriptive analysis relies on individuals to obtain information, it is limited by the ability of those individuals to perform the requested tasks, as well as to find words to adequately express their perception of the product. Once words are found, it may be difficult to obtain complete agreement among panelists on their interpretation and meaning. In addition, some individuals may perceive the same stimuli differently or may vary in their sensitivity to a particular attribute. The latter may be reflected in scoring differences in intensity values given by panelists. Panelists may also differ in the range of the scale they use. When product differences are large, the variance between products may be great enough to override the variance arising from these scoring and scaling differences (Powers, 1984). But when products are similar, these variances can no longer be ignored. Training helps panelists become consistent in performing repeated evaluations of the same product and thus promotes panel consistency. However, differences in the way a panelist evaluates a product from session to session or over replications can never be totally eliminated. Interpretation of the data from descriptive analysis requires the use of statistical techniques that can identify differences between samples in the presence of differences between individuals. In most cases the panelist-to-panelist differences are ignored and one or more of the panelists that differ are eliminated.

In this chapter, an alternative to the more conventional descriptive analysis procedures, free-choice profiling, is described and its usefulness is addressed. For analyzing free-choice profiling data, a multivariate statistical method, Procrustes analysis, has been rediscovered and applied specifically to sensory profiling data. This statistical technique will be discussed as a means of analyzing data obtained from free-choice profiling, as well as the more traditional descriptive analysis data. A historical and theoretical background of Procrustes analysis, as well as techniques and applications of Procrustes analysis to sensory data, are discussed below. A step-by-step example of the application of the basic principles of Procrustes analysis to descriptive sensory data is given.

FREE-CHOICE PROFILING IN SENSORY TESTING

Free-choice profiling, also referred to as free word association (Moskowitz and Howard, 1982; Lyon, 1987), is a relatively new technique used to obtain information about a product. Free-choice profiling differs from conventional descriptive profiling in that products are evaluated by members of a panel who describe perceived qualities of that product using their own individual list of terms, rather than a common scorecard. Terms are not shared or used collectively by panel members and may be mutually exclusive. The number of terms used by panelists may also vary, based on panelist experience and familiarity with the product. Terms must be defined and understood by their originator to ensure consistency in their use. Free-choice profiling is similar to traditional descriptive methods in that panelists must be able to detect differences between similar products, verbally describe specific attributes of products, and quantify those attributes. Moskowitz and Howard (1982) were able to use free word association successfully for market research purposes. Free-choice profiling, however, was first developed by Williams and coworkers (1981) and applied by Williams and Langron (1984) for the evaluation of commerical port wines. This technique has been further described by Arnold and Williams (1986).

Free-choice profiling has a number of procedures in common with descriptive analysis and, therefore, can be considered another type of descriptive analysis. Panelists are recruited, selected, and trained in scale usage, impartiality of judgments, and consistent term usage so that reproducible results may be obtained. However, the degree of training is less, because difficulties associated with achieving agreement among panelists do not have to be confronted.

Since free-choice profiling is fairly new, it has had only limited use and therefore procedural guidelines have not been established or standardized. Standardization of guidelines comes only from extensive testing of established methods obtained through sound research and, finally, acceptance by potential users as an adequate technique for obtaining product information.

Researchers who have used free-choice profiling thus far have varied in the characteristics evaluated, products tested, selection and screening procedures, training practices, scaling and sampling presentation, facilities, and testing conditions as well as analysis of the data. Different examples of how free-choice profiling has been applied to obtain sensory information about a product are discussed below.

Characteristics Evaluated

Similar to other descriptive analysis methods, free-choice profiling can be used to describe a product in terms of one or more characteristics. Free-choice profiling has been used to describe products in terms of a single characteristic by Williams and Arnold (1985) in the evaluation of the aroma of coffees, or Marshall and Kirby (1988) in the evaluation of texture of cheeses, similar to texture or flavor profiling. Free-choice profiling can be used to describe a product in terms of a number of characteristics, such as appearance, flavor, aroma, texture, or any combination of these, similar to the Spectrum™ or

Quantitative Descriptive Analysis™ methods. For example, several investigators have used free-choice profiling for the evaluation of appearance in addition to aroma, flavor, or texture (Williams and Langron, 1984; McEwan et al., 1989; Guy et al., 1989). The characteristics being judged may be restricted by the sensory analyst, but the number of terms used by each panelist is limited only by their perceptual and descriptive skills.

Products Tested

Although some traditional descriptive methods may be restricted to the evaluation of specific characteristics of a product, i.e., texture and flavor profiling, none is restricted to a specific class of products; neither is free-choice profiling. Free-choice profiling studies have dealt with a variety of products. Free-choice profiling has been used most extensively in the evaluation of beverages, such as wines (Williams and Langron, 1984), coffees (Williams and Arnold, 1985), whiskies (Guy et al., 1989), and meat products, such as chicken patties (Lyon, 1987), frankfurters (Oreskovich et al., 1990), and pork roasts (Oreskovich et al., in preparation). In addition to beverages and meats, other products evaluated include chocolates (McEwan et al., 1989), cheeses (Marshall and Kirby, 1988), and yogurts (Dijksterhuis and Punter, 1990).

Screening and Selection of Panelists

Traditionally, panelists for descriptive analysis are selected on a number of criteria depending on the descriptive method to be used. The crtiteria could include ability to discriminate between products, communicate perceived perceptions, product usage, task comprehension, availability, interest, health, attitude, level of confidence, etc. Lyon (1987) based selection of free-choice profile panelists on availability and willingness to participate, as well as ability to discriminate and verbally communicate perceptions. Marshall and Kirby (1988) used criteria described above, such as availability and health, and others to screen and select free-choice profiling panelists. Individuals were given a short questionnaire concerning likes, dislikes, and the extent of their texture vocabulary. Panelists were selected based on their responses. Oreskovich et al. (1990) used methods similar to those described for traditional descriptive analysis. They selected subjects based on responses to questions pertaining to food, interest, availability, and health. These subjects were then asked to complete a series of acuity tests including odor matching, basic taste identity, texture ranking, and a series of triangle tests. Panelists were further screened based on their performance. In other studies, however, a screening process was not used and selection was based on past experience of the panelists. Williams and Langron (1984) grouped panelists into expert and nonexpert wine tasters, using their familiarity with wine tasting as a criterion. Williams and Arnold (1985) selected panelists who had previous experience in descriptive profiling to evaluate coffee aroma. In contrast, McEwan et al. (1989) and Guy et al. (1990) selected subjects with no previous experience in sensory profiling.

Studies with free-choice profiling have routinely used from 8 to 20 panelists in evaluation of products. These numbers are comparable to conventional descriptive analysis techniques. Guy et al.(1989), on the other hand, recruited 100 subjects to study the usefulness of free-choice profiling by consumers. It is possible to develop panels of a larger size than used in other descriptive methods because free-choice profiling does not require extensive training. It has not been established that larger numbers are necessary or desirable because of the difficulty in interpreting the data.

Training of Panelists

The training of panelists is the major factor that distinguishes descriptive analysis from other sensory testing methods and is common to all conventional descriptive techniques. A training period of about 6 months is not uncommon for most descriptive methods, and the time is usually product dependent. A flavor profile panel, for example, may require approximately 60 hours of training and 100 hours of practice per panelist (ASTM, in preparation). About 3 months is required for training a Spectrum™ panel. It is suggested for training a Quantitative Descriptive Analysis™ panel that six to eight one-hour sessions are needed for the development of terms or attributes, as well as additional sessions to orient panelists to new products. For most descriptive analysis methods, the training procedures are well standardized. Standards or references are routinely used to acquaint panelists with product attributes and intensities and the ranges that might be encountered during testing. In comparison, free-choice profiling studies vary considerably in training protocols. In some studies, panelists received no training, while other investigators used extensive training and held numerous practice sessions. Reference samples were sometimes presented to train or familiarize free-choice profile panelists with products or ingredients.

Even investigators who frequently use free-choice profiling do not follow fixed training procedures. In the evaluation of wines by expert and nonexpert wine tasters, panelists were merely introduced to the concept of free-choice profiling and scaling with no mention of any training (Williams and Langron, 1984). Panelists were asked to identify different attributes of wine singularly or in combination with other attributes and to assign intensity values to them. Unlabeled samples were then introduced, and panelists were asked to assign their own descriptive terms. Terms used by each individual were defined and categorized into appearance, aroma, and flavor groups. A list of terms used by three expert and nonexpert tasters are given in Tables 13.1 and 13.2. These two groups

TABLE 13.1 Descriptive Terms Used by Three
Expert Wine Tasters

Expert	Appearance	Aroma	Flavor
Taster 1	Depth	Cleaness	Cleaness
	Fresh	Freshness	Fresh
	Brightness	Fruitiness	Body
		Richness	Grip
		Smoothness	Round
		Concentration	Aftertaste
Taster 2	Tawny	Clean	Body
	Ruby	Fruity	Firmness
	Purple	Green	Coarse
	Cloudy		Tannin
			Hard
			Crisp
			Sour
Taster 3	Red	Fruit	Tannin
	Blue	Esters	Acid
	Brown	Wood	Sweetness
	Intensity	Oloroso	Chocolate
		Spirit	Body
		Burnt	Green

Source: Williams and Langron, 1984.

used different terms to describe the same wines, based on their experience. Clearly, more terms were developed by the more experienced taster. Similarly, panelists had no formal training in the evaluation of the aromas of six varieties of coffee (Williams and Arnold, 1985) or in the appearance, flavor, and texture of five samples of milk chocolate (McEwan et al., 1989). However, in the evaluation of chocolates, eight samples, five of which were the real samples to be tested, were provided to help in the development of terms by individual panelists.

In a study reported by Marshall and Kirby (1988), panelists were trained in the free-choice profile technique to evaluate the texture of five unflavored processed cheese analogs over 10 30-minute sessions during which panelists, as a group, developed general guidelines to assess the texture of samples both manually and orally. Samples of cheese analogs that varied in moisture and fat content (similar to the model cheeses to be evaluated), as well as samples of different varieties of natural cheeses, were presented to encourage the individual development of texture terms. Panelists were also trained in the use of scales.

TABLE 13.2 Descriptive Terms Used by Three Inexperienced Wine Tasters

Nonexpert	Appearance	Aroma	Flavor
Taster 1	Clarity	Volatile acidity	Acid
	Intensity	Fruitiness	Sweetness
	Redness	Woodiness	Smoothness
	Yellowness	Alcoholic Strength	Body
		Overall intensity	Aftertaste
Taster 2	Intensity	Burnt	Astringent
	Red		Acidy
	Brown		
Taster 3	Color	Smell	Taste
	Red		Acidity
	Brown		Tannin
			Softness

Source: Williams and Langron, 1984.

After 6 days of training, panelists were asked to define their terms and to eliminate any that had similar meanings. Individual ballots were compiled following the developed guidelines, and terms were placed under two subheadings, manual and oral. The remaining four training sessions were used by panelists to refine their ballots.

More extensive training was provided to panelists evaluating meat products. Lyon (1987) used free word association to develop a list of terms to describe the taste and aroma of both fresh chicken patties and stored chicken patties that had been reheated. A 10-member panel was selected and trained on basic taste and odor recognition, as well as threshold perception. Training to help generate terms took place over an 8-week period in which panelists were presented with chicken patties that varied according to methods used for cooking and reheating.

Oreskovich et al. (1990) used a free-choice profile panel to evaluate the flavor and texture of two samples of frankfurters differing in fat content. Subjects selected for the free-choice profile panel practiced using a 15-cm continuous line scale during each training session. Panelists were offered several brands of commercial frankfurters differing in fat, meat type, and salt content, as well as frankfurters processed at the university meat science laboratory to help generate terms. Terms were then defined and incorporated into individual score sheets. Panelists were asked to generate values representing their perceived intensity for each attribute on a control frankfurter. Six one-hour sessions over 3 weeks were used to train panelists to evaluate frankfurters and to become familiar with their own intensity values given to the control sample.

In an ongoing study by Oreskovich et al. (in preparation), similar criteria were used to generate terms in the training of panelists for the evaluation of texture and flavor of whole muscle pork products. Panelists determined intensity values for a control sample of pork to use for the comparison with other products. After terms were generated and defined, panelists were trained in their use for 20 hours over a 5-week period.

In the majority of free-choice profiling studies, panelists develop terms individually with no suggestion of terms being added or eliminated by an outside mediator. Guy et al. (1989), however, added two terms, smoothness and maturity, to all consumer free-choice profiling ballots for whiskies. This was done to assess whether consumers could evaluate important attributes even if they did not generate the descriptors themselves. What the consumers characterized as maturity of whiskies, the trained panel characterized as smooth, sweet, vanilla, and malty. In addition, the consumers' perception of smooth, mellow, and sweet in the whiskies were characterized as estery and woody by the trained descriptive panel. This emphasizes the necessity for panelists to use their own terms, even if they are not as specific as the investigator would like.

Scale/Scorecard and Replications

Scales and scorecards used in descriptive analysis are usually very specific and are dependent upon the method used. For example, Quantitative Descriptive Analysis™ uses a 6-inch unstructered scale, while Spectrum™ scales are similar, being 15 cm in length. A 15-cm unstructured line scale is adopted in many descriptive techniques. Final scorecards are derived by consensus with guidance of a group leader. Flavor profiling has a unique scale that does not adapt easily to statistical analysis. In its earliest format, the scorecard was individualized with panelists reporting flavor notes as they appeared. Texture profiling depends on a standardized series of references to align panelists' judgments, and evaluations are made relative to the reference scales.

In free-choice profiling, no standardized scaling method or scorecard has emerged. The number of terms on an individual scorecard is idiosyncratic and reflects the subjects ability to describe the product attributes. Category and unstructured line scales (6.5–15 cm long) are commonly used. In the evaluation of wines (Williams and Langron, 1984) and coffees (Williams and Arnold, 1985), six-point category scales were incorporated in the score sheet. Marshall and Kirby (1988) used a 12-cm line scale. A 6.5-cm continuous line scale was used for the evaluation of chocolate samples (McEwan et al., 1989).

In Lyon's (1987) study of chicken patties, a nine-point category scale was used in one phase and a 10-cm line scale in another. A 15-cm continuous line scale was employed for the evaluation of frankfurters and pork roasts (Oreskovich et al., 1990; Oreskovich et al., in preparation). The diversity of scales and scorecards in free-choice profiling is evident.

Three or four replications by each judge for each treatment is normally recommended for any sensory testing to ensure reliable results (Larmond, 1977), but the actual number applied is the choice of the sensory analyst. Complex products with minor differences may require more replications. Similarly, the number of samples presented at one time is dependent on the product's characteristics and the judgment of the analyst.

The free-choice profiling studies reflect the same diversity in numbers of samples and replications seen in other testing methods. Since free-choice panelists are not extensively trained, the number of samples presented at one time is usually limited to two or three. Samples are often presented monadically (Williams and Langron, 1984; Williams and Arnold, 1985) or in pairs (Marshall and Kirby, 1988). Generally the number of replications was three or four (Williams and Arnold, 1985; Lyon, 1987; Marshall and Kirby, 1988; McEwan et al., 1989).

In studies by Oreskovich et al. (1990; in preparation), for each of four sessions, four coded samples of frankfurters (two treatment and two control) were randomly presented to panelists. Seven replications of eight different pork sample variations were presented in 14 sessions over 7 weeks. In addition to the scorecards, each panelist was given his or her intensity scores for the control product at the beginning of each session. This helped to maintain consistency in the panelist's evaluation.

Facilities and Testing Conditions

It is accepted practice in descriptive analysis to present samples to panelists using standard procedures under conditions in which lighting, temperature, humidity, odors, and sounds can be controlled to minimize distractions and potential biases. Similar practices have been implemented in many of the free-choice profiling studies. Testing of samples by panelists was done under controlled conditions utilizing individual sensory booths to ensure independent judgments. However, in the evaluation of whiskies by consumers trained in the free-choice profiling technique (Guy et al., 1989), samples were evaluated at their homes and ballots returned by mail.

Statistical Analysis

The procedures used in the aforementioned studies vary considerably, illustrating the lack of uniformity in the use of free-choice profiling at the present time. Unlike more conventional descriptive analysis techniques, the data collected cannot be analyzed and compared using univariate or traditional multivariate analysis because of the lack of uniformity in scorecards among panelists. Thus, the information must be examined differently.

Analysis of free-choice profiling data differs from that associated with descriptive analysis in that it is nicely interpreted using Procrustes analysis. Individual panelist's responses to a set of samples can be geometrically arranged into a single configuration where scores for each sample can be defined as a point in space, with each attribute defining a different dimension for each attribute in the space. Configurations of individual panelist's responses to each sample can then be matched and compared using Procrustes analysis. Procrustes analysis can also be used to analyze descriptive analysis data, but was not considered as an alternative statistical method until Harries and MacFie (1976) used it to examine the textural characteristics of beef roasts. It should be noted that Procrustes analysis can provide sample, panelist, as well as attribute information.

PROCRUSTES ANALYSIS

The advantages and disadvantages of using free-choice profiling or more conventional descriptive analysis techniques have been suggested above. Since its inception, free-choice profiling has been thought to have some advantages over conventional descriptive profiling. Free-choice profiling is less time consuming, since panelists require less training. Panelists may feel more at ease with their judgments because it is not necessary for them to agree upon a common list of descriptors. In addition, the frustration associated with trying to force agreement among panelists is thought to be alleviated. Similar to descriptive analysis, free-choice profiling is not exempt from dealing with potentially large variations that can be associated with any method that uses individuals as analytical tools. However, by using Procrustes analysis to interpret free-choice profiling data, many of these panelist-to-panelist variations can be reconciled. As with any statistical tool, there are certain limitations and disadvantages to Procrustes analysis.

Background

The name "Procrustes" was first used in 1962 by Hurley and Cattell to describe the matching of configurations. "Procrustes," a Greek term meaning "to beat out," described a Greek innkeeper in Attica who seized travelers and tied them to the iron framework of his beds where he stretched the legs of short men and cut the legs of tall men so that they would all fit his beds. The name

"Procrustes" conceptually describes, to some extent, what is done with data in a Procrustes analysis.

In conventional descriptive analysis as well as free-choice profiling, panelists evaluate samples by assigning to them intensity values for various attributes. The values can be placed in a matrix in which samples and attributes are represented by rows and columns, respectively. Each element in the matrix is the intensity value perceived for that sample and attribute. A single matrix is created for each panelist as seen in Fig. 13.1. The matrices for several panelists can be represented geometrically, with each sample defining a point in space, and each attribute corresponding to a different dimension. A panelist's responses to a set of samples then defines a configuration of points in that space. Procrustes analysis is a multivariate statistical tool in which the configurations of individual panelists can be matched and compared. Procrustes analysis uses geometrical transformations, whereby the configurations are adjusted to match each other as closely as possible. Transformations allow for reexpression of the configurations. The closeness of the matrices to one another can be characterized by summing the squared distances between the points (matrices) and a common centroid (target or consensus) matrix. This sum of squares is referred to as the Procrustes statistic.

Procrustes transformation procedures are aimed at optimizing a particular criterion. Criteria are considered to be indirect measures of the similarities between the transformed and target matrix and are usually defined or selected by the experimenter. Three criteria commonly used with Procrustes analysis are least-squares, inner product, and a consensus criterion. The Procrustes statistic described earlier is used as a statistical check to determine if the designated criterion is being met.

	Attribute 1	Attribute 2	Attribute 3
Sample 1	5.9	7.1	6.9
Sample 2	4.2	3.1	7.0
Sample 3	6.0	7.0	2.9
Sample 4	3.9	2.9	3.1

Intensity Values

FIG. 13.1 Data matrix. Matrix configuration of simulated data set for panelist 1. Rows represent samples and columns represent attributes. The numbers within each column and row represent the intensity values assigned by a single panelist.

Initially, Procrustes analysis was used to match one configuration to a target configuration, which has been defined as the pair-wise approach. Moiser (1939) and Green (1952) first used the analysis for the comparison of different types of statistical analyses. Green (1952) described the matching of configurations as a transformation in which one matrix is rotated to a target matrix under specified constraints. Specifically, Green (1952) restricted the matrices to having the same number of columns and to be of "full column rank." The "column rank of a matrix" is the number of linearly independent columns in the matrix. If the column rank is equal to the number of columns in the matrix, the matrix is said to have "full column rank."

Green (1952) used a least-squares criterion to develop three analytical methods for obtaining transformed matrices. The least-squares criterion minimizes the distances between corresponding points in the configurations (Procrustes-PC, 1989). Hurley and Cattell (1962), who were the first to coin the term "Procrustes," used a rotation matrix to transform one matrix of responses to match a designated target matrix for hypothesis testing.

Cliff (1966) transformed matrices for matching and for fitting to a target matrix, as described above. However, he restricted these to orthogonal transformations. Schönemann (1966) derived his own solution for matching matrices and referred to it as the "orthogonal Procrustes problem." This solution was a more generalized approach in that it was not restricted by Green's (1952) restraints in which matrices had to be of "full column rank." The matching of one configuration to a target configuration as described by Cliff (1966) and Schönemann (1966) was later referred to as the "two-sided" orthogonal Procrustes problem by Schönemann (1968).

Schönemann and Carroll (1970) found the matching of matrices helpful when comparing different methods of multidimensional scaling. Schönemann and Carroll (1970) presented a least-squares technique designed to be programmable for use with computers for fitting matrices by rotation, translation, and central dilation. A scaling factor for central dilation that allowed for the expansion or contraction of data points, taking differences in spread between configurations into account, was also included.

Instead of rotating one matrix to fit another, it became possible to rotate many matrices to fit a common centroid matrix: generalized Procrustes analysis. This centroid matrix was first introduced by Kristof and Wingersky (1971) and is a key element in an iterative procedure in which the least-squares criterion is met without any consideration to translation and scaling. Gower in 1975 described the centroid matrix as representing the average or consensus configuration. Unlike Kristof and Wingersky (1971), Gower included translation and scaling constraints and provided an elaborate starting procedure in which matrices were initially standardized prior to rotation. In addition, he provided a computational technique by which results from Procrustes analysis can be summarized in an analysis of variance format.

TenBerge (1977) made modifications, specifically the rotation and scaling steps, of Gower's method. TenBerge and Kroll (1984) derived transformations for multiple matrices having different numbers of columns and proposed an inner product criterion by which angles between corresponding point vectors were minimized.

Peay (1988) used a different criterion: consensus. The consensus criterion maximizes the total variance of the common centroid configuration. Peay (1988) was able to combine and expand features of techniques derived and discussed by other investigators that incorporated scaling and matching multiple asymmetric matrices (unequal column numbers). One computer program that is available for Procrustes analysis uses the consensus criterion (Procrustes-PC, 1989).

It is often suggested that principal component analysis be used in place of Procrustes analysis and vice versa. Principal component analysis is also a multivariate technique. Data for analysis is arranged in a matrix as described in the Procrustes procedure. However, the purpose of applying principal component analysis to a set of data is to reduce the number of dimensions (attributes) in which samples may be found, with minimal loss of information. Principal component analysis determines the orientation or directionality of a single set of data, which can indicate relationships between attributes and differences between samples. This differs from Procrustes analysis which matches a minimum of two sets of data. As explained by Piggott and Sharman (1986), principal component analysis searches for a sequence of linear combination of attributes that can account for the maximum variation associated with the data. The number of axes or principal components needed to characterize the majority of variability for a single data set is dependent upon how correlated the original set of variables are.

Principal component analysis is applied to only one data set at a time. This data set may be a very large set in which raw data are entered. Alternatively, it is most often applied to a data set consisting of sample means averaged across panelists and over replications for each attribute (Powers, 1984). Use of a single large data set is not recommended (MacFie, 1987) because the resulting configuration is very crowded and may be difficult to interpret.

Whether panelist effects are identified by using principal component analysis or a univariate analysis method, variations associated with panelists should be accounted for prior to any analysis to determine differences between samples. Because generalized Procrustes analysis and principal component analysis provide the experimenter with very different information, it would not be fair to suggest that one be used over the other. In fact, it has been suggested that principal component analysis be used in conjunction with Procrustes analysis. By applying principal component analysis to the final configuration obtained by Procrustes analysis, a visual representation of the configuration can be constructed that allows for easier interpretation.

Early investigators (Gower, 1966) mentioned using principal coordinate analysis following Procrustes analysis. Principal coordinate analysis is similar to principal component analysis in that the number of dimensions present in the final configuration obtained by applying Procrustes analysis to the data can be reduced. These two analyses differ in that principal coordinate analysis starts with a matrix of distances and principal component analysis with a covariance or correlation matrix (Piggott and Sharman, 1986). Gower (1966) suggested that principal coordinate analysis be used as a matter of computational convenience especially when there are more attributes than samples. This was suggested at a time when computer analysis was not as advanced as it is today. Because of this advancement either principal coordinate or principal component analysis may be used on data following Procrustes analysis, each providing similar information.

Approaches

Pair-wise In the pair-wise Procrustes analysis approach, pairs of configurations are compared. These configurations may represent panelists, samples, or attributes. For simplicity, discussion will be restricted to panelist configurations. As illustrated in Fig. 13.2, samples are represented by points and the attrib-

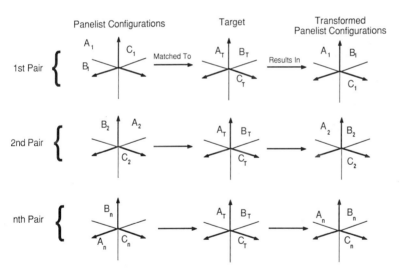

FIG. 13.2 Pair-wise Procrustes analysis. Pairs of panelist configurations are transformed; for the first pair, panelist 1 is matched to a target configuration. As a result of the analysis, a transformed panelist 1 configuration was computed. Letters (A, B, C) represent three samples. Subscripts (1, 2, n) represent panelists, and subscript T represents the target configuration. Bold axes within each configuration represent the dimensions in which samples could be found.

utes represent the dimensions. In the pair-wise approach, one panelist configuration is usually designated as a "target" to which other panelist configurations are compared. The target is a matrix of values for each attribute that is set by the investigator. The target matrix can be based on theoretical expectations, previous experience or obtained by some mathematical computation of existing values for identical attributes (Cliff, 1987). The configuration (matrix) representing a panelist is mathematically transformed in Procrustes analysis to form a new transformed panelist configuration, which more closely matches the target configuration. Through Procrustes analysis, the distance between a pair of configurations, referred to as the Procrustes statistic, is obtained. In the pair-wise approach, a Procrustes statistic can be calculated between the newly transformed panelist configuration and the target configuration.

A Procrustes statistic is computed by summing the squared distances between the corresponding points of two configurations, whether it be between a pair of configurations representing panelists or between a configuration and a target. A Procrustes statistic computed for a pair of panelists can be illustrated in the form of a lower triangle matrix (Table 13.3), as in data taken from Banfield and Harries (1975). A low Procrustes statistic, as obtained from expert panelists 1 and 2 (0.513), for example, depicts how similar these two panelists were in their perceptions of the differences between samples. A high value such as that

TABLE 13.3 Procrustes Statistics for Each Pair of Expert and Inexperienced Panelists

	Expert					Inexperienced				
	1	2	3	4	5	6	7	8	9	10
Expert										
1	0.000									
2	0.513	0.000								
3	0.566	0.522	0.000							
4	0.591	0.569	0.514	0.000						
5	0.502	0.390	0.624	0.546	0.000					
Inexperienced										
6	0.747	0.764	0.810	0.776	0.752	0.000				
7	1.149	1.085	1.088	1.104	1.161	1.153	0.000			
8	1.412	1.449	1.483	1.517	1.513	1.589	1.608	0.000		
9	1.184	1.169	1.253	1.296	1.173	1.318	1.536	1.504	0.000	
10	1.078	0.987	1.041	1.065	1.065	1.113	1.289	1.486	1.569	0.000

Source: Banfield and Harries, 1975.

between inexperienced panelists 7 and 8 (1.608) shows a large disagreement between this pair of panelists in determining differences between samples.

After Procrustes analysis has been applied to the data, principal coordinate or principal component analysis can be used to provide a visual image. This is important since Procrustes analysis provides information in multidimensional space, which makes it difficult, if not impossible, to visualize. Although either analysis is appropriate to use following Procrustes analysis, in that each can provide a visualization in two dimensions, principal coordinate analysis has been used most often. Interpretation of the Procrustes data is then easier because the number of dimensions in which the points can be found can be reduced without sacrificing any valuable information. Principal coordinate analysis can be used to produce a reduced set of axes from each configuration, while attempting to keep the distances between panelists and between attributes fixed or constant. Newly constructed axes for panelists can then be plotted to give a graphical display of their relative positions allowing for easier interpretation of the original configuration.

Generalized The generalized approach is based on the pair-wise concept, but differs in that all configurations are compared as a group (Fig. 13.3) to a target configuration and then a new consensus configuration is derived. The target configuration can vary depending upon the experimenter and is derived by computing the mean of individual configurations or it can be the configuration

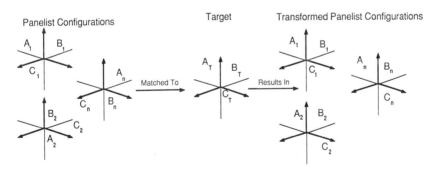

FIG. 13.3 Generalized Procrustes analysis. Several panelist configurations are transformed to more closely match a target configuration. As a result of the analysis, transformed panelist's configurations were computed. Letters (A, B, C) represent three samples. Subscripts (1, 2, n) represent panelists, and subscript T represents the target configuration. Bold axes within each configuration represent the dimensions in which samples could be found.

for a particular individual. The consensus configuration, with samples represented as points, is most often derived from individual configurations by the following process. The dimensions of the consensus configuration are equal to the number of attributes used for the evaluation of samples. When generalized Procrustes analysis is applied to conventional descriptive analysis data, the number of attributes is equal. However, in free-choice profiling the number of attributes may be different for each panelist. To obtain the consensus configuration for free-choice profiling data, columns of zeros are added to individual matrices so that all panelist configurations will have equal dimensionality. Individual configurations are then mathematically matched to a target configuration, similar to the pair-wise approach. Matching is achieved through a series of iterative steps, by transforming individual configurations to the target configuration in such a way that intersample distances for individual configurations are maintained (Fig. 13.4). The consensus configuration is then computed as the mean of all transformed individual configurations. The consensus configuration is now the new "target" to which transformed individual configurations are matched. This iterative process continues until the distance between the newly transformed configurations and the consensus configuration is minimized as measured by the Procrustes statistic. The Procrustes statistic can be determined at any transformation point and is calculated for a pair of configurations as in the

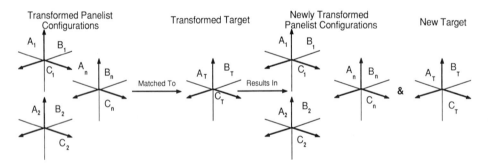

FIG. 13.4 Iterative procedure for generalized Procrustes analysis. In the generalized Procrustes approach, transformed configurations for each panelist goes through a series of iterative steps in which configurations are continually matched and compared to newly computed target configurations. The iterative process continues until the distance between the newly transformed configurations and the target (consensus) configuration is minimized as measured by the Procrustes statistic to a point where continued transformations result in a minimal change meeting some predetermined tolerance. Letters (A, B, C) represent three samples. Subscripts (1, 2, n) represent panelists, and subscript T represents the target configuration. Bold axes within each configuration represent the dimensions in which samples could be found.

pair-wise approach, such as between a panelist and a consensus configuration. The distance will be minimized to a point where continued transformations result in a minimal change in the Procrustes statistic meeting some predetermined tolerance. A final consensus configuration can then be computed (Fig. 13.5).

Similar to the pair-wise approach, principal coordinate or principal component analysis can be applied to transformed individual configurations, as well as the final consensus configuration to obtain those attributes or factors that can clearly differentiate between samples. These attributes or dimensions can account for decreasing proportions of total variance. Just as in the pair-wise approach, a graphical two-dimensional display of the consensus configuration can be obtained by plotting sample scores on the first two axes. These new axes can be interpreted in terms of each panelist's own initial attributes list in one of two ways: by calculating correlations of the original attributes for each panelist with the newly transformed attributes of the consensus or by obtaining a rotation matrix for each panelist, which transforms his or her original centered configuration to the consensus (Arnold and Williams, 1986).

When generalized Procrustes analysis is applied to free-choice or conventional descriptive analysis data, panelists' configurations are matched in such a way that the relative positions between samples for each individual are unchanged. As a result, a consensus configuration is computed that can be used in place of untransformed sample means for comparing other analytical measurements or sensory data (Langron et al., 1984).

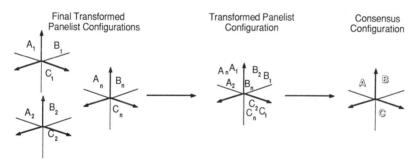

FIG. 13.5 Final transformed panelist and consensus configuration. Transformed panelist configurations can be superimposed to form one configuration. In addition, a final consensus configuration can be computed as the mean of all finally transformed panelists' configurations. Letters (A, B, C) represent three samples. Subscripts (1, 2, n) represent panelists, and hollow letters represent the final configuration for each sample known as the consensus configuration. Bold axes within each configuration represent the dimensions in which samples could be found.

ILLUSTRATION OF PROCRUSTES ANALYSIS

Transformation steps of Procrustes analysis can include initialization, rotation/ reflection, and scaling. In the various Procrustes applications that are to be presented later in this chapter (see Applications), the exact steps and the order in which they have been applied has varied greatly. Although the authors of this chapter feel that all the steps are important, the experimenter must decide the relevance of each transformation step for the particular data to be analyzed. In addition, the way in which each transformation is computed can also vary. Each transformation attempts to reconcile differences between the panelist and target configurations. To demonstrate this, a simulated data set will be used to illustrate each transformation step.

Three panelists used conventional descriptive analysis to evaluate three texture attributes (hardness, cohesiveness, and springiness) of four different whole wheat bread formulations. All attributes were evaluated using a 15-cm continuous line scale. A response for each panelist was constructed, where the rows represent different breads (A, B, C, D) and columns represent the three texture attributes (Table 13.4). Table 13.4 shows the intensity values of each attribute given by panelist 1 in the evaluation of the four breads. Each row of the matrix is the coordinates for a point representing each bread sample, with each attribute representing a dimension in the space in which a particular sample can be found. The different shapes in Fig. 13.6 represent the different bread formulations: sphere-bread 1, cube-bread 2, cone-bread 3, and cylinder-bread 4. The different shading of these shapes as well as labels designated as P1–P3 represent panelists, where light gray is panelist 1 (P1), middle gray is panelist 2 (P2), and black is panelist 3 (P3). All panelist configurations have been plotted in three-dimensional space in which the x axis represents hardness, the y axis cohesiveness, and the z axis springiness. The size of the symbol indicates its position,

TABLE 13.4 Simulated Data Set for Panelist 1 in the Evaluation of Whole Wheat Bread Formulations

Samples	Attributes		
	Hardness	Cohesiveness	Springiness
A	5.9	7.1	6.9
B	4.2	3.1	7.0
C	6.0	7.0	2.9
D	3.9	2.9	3.1

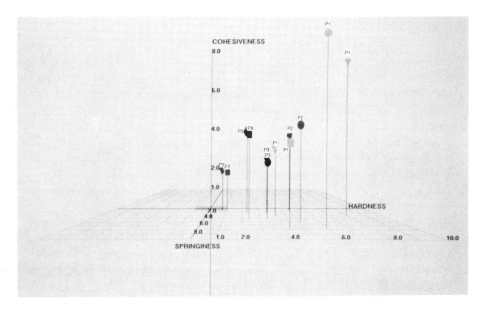

FIG. 13.6 Original configuration. The original configuration of simulated data set presented in three-dimensional space. Shapes represent breads 1–4. Sphere-bread, 1; cube-bread, 2; cone-bread, 3; and cylinder-bread, 4. The different shading of these shapes as well as labels P1–P3 represent panelists, where light gray is panelist 1 (P1), middle gray is panelist 2 (P2), and black is panelist 3 (P3). Axes represent attributes 1–3: X-hardness, Y-cohesiveness, and Z-springiness. The size of the symbol indicates its position with symbols that appear larger being closer to the viewer.

with symbols that appear larger being closer to the viewer. For panelist 1, the scores appear to be more spread out while scores for panelist 3 appear more condensed. It is difficult to determine differences between the treatments; however, breads 1 and 3 appear to be similar. Procrustes statistics and scaling factors for each panelist at each transformation step are found in Table 13.5. Procrustes statistics were calculated based on original data between each panelist and the initial consensus configuration. The initial consensus configuration was calculated as the mean of the three panelists' configurations. As previously mentioned, a large Procrustes statistic can indicate disagreement between a panelist and the target or consensus. In Table 13.5, a large Procrustes statistic for panelist 1 (35.002) compared to values for panelists 2 and 3 indicates that the configuration for panelist 1 is much further away from the initial consensus configuration. Panelist 3, however, appears to be the closest to the initial consensus configuration as suggested by a low Procrustes statistic (14.346).

TABLE 13.5 Procrustes Statistics Calculated Before and After Each Transformation and Scaling Factors for Iteration 1

Panelist	Original	Initialization without normalization	Rotation and reflection	Scaling	Scaling factors
1	35.002	8.968	3.202	1.573	0.865
2	17.766	6.948	4.179	3.069	0.869
3	14.346	10.344	4.980	1.939	1.306
Total	67.113	26.260	12.362	6.581	

Initialization

Translation is the first transformation step in Procrustes analysis. It has been suggested that translated matrices be initially standardized prior to rotation and reflection. This standardization process was also referred to as "initial scaling" and later referred to as "normalization" by Peay (1988). In the development and explanation of Procrustes analysis, most researchers refer to both these steps, translation and standardization, as simply "translation." To minimize any possible confusion, we will refer to this two-step process as "initialization" with the first step referred to as "translation," and the second "normalization." Note that the normalization step should not be confused with "normalizing the data" by converting sets of scores so that they are normally distributed, and that the term "scaling" is reserved for a later step in the analysis.

Translation   The translation step involves the moving of individual configurations about the origin (zero). Translated matrices are computed by subtracting from each individual matrix its matrix of column means. Each element of the translated matrix is then the deviation from its mean. The resulting matrix can also be referred to as a "deviation" or "shifted" matrix. The translation step can account for the variation due to different levels of the scale being used by different panelists. It can remove the effect of individual panelists who consistently under- or overscore a particular attribute. These scoring differences may reflect an individual's response to the perception of different stimuli or a difference in an individual's sensitivity to a particular attribute. In the initial data (Fig. 13.6) all the configurations were positive and grouped together. After translation, the configurations are centered about the origin. It appears that scores for panelist 2 (Fig. 13.7) are much more spread out than they appeared in the initial configuration (Fig. 13.6). Panelist 1, however, appears to have shifted the most after translation, and panelist 3 the least. The consensus was calculated as the mean of all panelist's translated configurations. The Procrustes

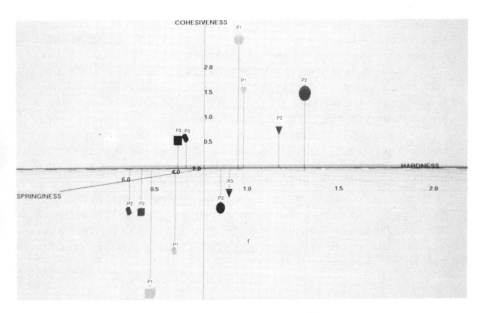

FIG. 13.7 Initialized configuration without normalization. Initialized configuration without normalization of simulated data set presented in three-dimensional space. Shapes represent breads 1–4. Sphere bread, 1; cube-bread, 2; cone-bread, 3; and cylinder-bread, 4. The different shading of these shapes as well as labels P1–P3 represent panelists, where light gray is panelist 1 (P1), middle gray is panelist 2 (P2), and black is panelist 3 (P3). Axes represent attributes 1–3: X-hardness, Y-cohesiveness, and Z-springiness. The size of the symbol indicates its position with symbols that appear larger being closer to the viewer.

statistic was calculated between each panelist's translated configuration and the new consensus. The Procrustes statistic (Table 13.5) for panelist 1 confirms what can be seen in Fig. 13.7. The difference between the Procrustes statistic for panelist 1 initially (35.002) and after translation (8.968) is 26.034, which is much greater than for panelist 2 with a difference of 10.818 and panelist 3 of 4.002. This large difference confirms that panelist 1 was more affected by the translation step than panelist 2 or 3. It can therefore be inferred that panelist 1 overscored or consistently assigned higher intensity values to all the bread samples in comparison to panelists 2 and 3.

Normalization   Normalization is a computational procedure that adjusts a matrix so that its elements are of comparable magnitude to other normalized matrices. The matrices are then expressed on a common basis so that no one measurement, because of its magnitude, can unduly influence the comparison of samples. After normalization, elements of translated matrices for each panelist are converted so that all values fall between −1 and 1. The application of the normalization step is optional but usually follows the translation step. Gower (1975) and Peay (1988) suggested that normalization be applied to translated matrices that were unusually disproportionate in magnitude prior to transformation by rotation and reflection. Such incommensurable translated matrices may present themselves when comparing panel matrices derived from two different sensory methodologies or when comparing matrices obtained from sensory assessment to matrices derived from analytical measurements. Peay (1988) was able to derive a more generalized version of Gower's normalization step, in that it could be used for sets of translated matrices that did not necessarily have identical column order. When the normalization process is applied, an adjustment must be made to the final transformed matrices if they are to be compared relative to the units of the initial matrices. Because scores for panelists in our simulated data set were of similar magnitude, normalization was not applied to the panelist configurations.

Rotation/Reflection

The next step in the analysis is to optimally fit or match initialized configurations to the target configuration by using the rotation step. The target configuration to be matched in this illustration was calculated as the mean of the translated panelist configurations. This step is accomplished by rotating panelist configurations to most closely match the target configuration. Rotation matrices are calculated and then multiplied by the initialized panelist configurations to obtain transformed configurations. Rotation procedures may differ based, for example, on whether they allow a particular configuration to be reflected. When generalized Procrustes analysis is applied to free-choice profiling data, the rotation/reflection transformation procedure typically has a very dramatic effect. This is so, because this is the first step in comparing individual configurations that may have different dimensions or attributes associated with them.

Rotation and reflection can account for variation associated with individual panelists' different interpretations of the same term. This type of variation can come about when a particular descriptor that may be difficult to define or interpret is used. For instance, in this example, the terms cohesiveness and hardness in the evaluation of bread may be confused by a panelist. In the evaluation of tea, the terms bitter and astringency have been commonly confused. This variation is reflected in the panelist's scores resulting in a changed orientation of the configuration compared to other panelists. After rotation/reflection the configurations appear to be better matched (Fig. 13.8). The

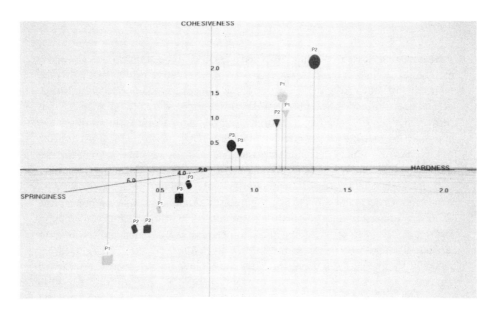

FIG. 13.8 Initialized and rotated configuration. Initialized, rotated, and reflected configuration of simulated data set presented in three-dimensional space. Shapes represent breads 1–4. Sphere-bread, 1; cube-bread, 2; cone-bread 3; and cylinder-bread, 4. The different shading of these shapes as well as labels P1–P3 represent panelists, where light gray is panelist 1 (P1), middle gray is panelist 2 (P2), and black is panelist 3 (P3). Axes represent attributes 1–3: X-hardness, Y-cohesiveness, and Z-springiness. The size of the symbol indicates its position with symbols that appear larger being closer to the viewer.

different breads seem to be much more differentiated after rotation than after translation. Bread 1 is characterized as being much harder yet as springy as bread 2. Bread 2 appears to be much softer than bread 1 and 3. Bread 3 appears to be more cohesive than the other breads. Bread 4 was given low intensity values for all three attributes. The Procrustes statistics for the panelists were calculated between each panelist configuration and the new consensus. The new consensus was calculated as the mean of newly rotated panelist configurations. The Procrustes statistic (Table 13.5) for panelist 2 confirms what is seen in Fig. 13.8. The difference between the Procrustes statistic for panelist 2 after translation (6.948) and after rotation (4.179) is 2.769 in comparison to 5.748 and 5.364 for panelists 1 and 3, respectively. The small difference in the Procrustes statistic for panelist 2 in comparison to the other panelists confirms that panelist 2 was the least affected by the rotation/reflection step. These results suggest that panelist 2 was the most consistent within himself in his use of the scorecard.

Scaling

Scaling is the last step on the Procrustes analysis procedure, and again, there are many alternative procedures that can be used to "scale" the data. One alternative is not to scale the data at all (Schönemann and Carroll, 1970; Gower, 1975; TenBerge, 1977; Langron, 1983). If it is decided to use the scaling step, then scaling can occur in two directions. The points in a particular configuration can either be expanded or contracted. Dilation is another term used to describe this expansion. When the expansion or stretching occurs about some center point, it is referred to as "central dilation." Scaling is optimized by computing scaling factors that meet some predetermined criteria and are then applied to the translated, rotated, and reflected configurations. Scaling factors can indicate the dispersion of a set of data points for a particular panelist. One panelist may use a very small range of the scale, while another may use a much larger range to express differences between samples on a particular attribute. A large scaling factor will indicate that a panelist used a small range of the scale while a small number will indicate the opposite. When scaling is applied, this variation associated with different scale usage by panelists can be addressed. If the distance between points is small, for instance, that distance can be enlarged; however, if the value is large, it can be decreased depending upon how the criteria can best be met. If the scaling factor is greater than 1, the distance of the points from the origin are increased (stretched) and if they are less than 1, they are decreased (shrunk) to match the target configuration. The configurations appear as shown in Fig. 13.9. Panelist 2 appears to have moved very little relative to the amount of movement by panelists 1 and 3. The Procrustes statistic for each panelist to a new consensus was calculated. The new consensus was calculated as the mean of the newly scaled panelist's configurations. The Procrustes statistics (Table 13.5) for the panelists confirm what is seen in Fig. 13.9. The difference between the Procrustes statistic for panelist 3 after rotation (4.980) and after scaling (1.939) is 3.041 in comparison to a value of 1.629 and 1.110 for panelists 1 and 2, respectively. This confirms that panelist 2 was least affected by scaling and panelist 3 was the most affected. In addition to the Procrustes statistic, scaling factors have been calculated for each of the panelists. Panelist 3 had a scaling factor greater than 1 (1.306) indicating that a limited range of the scale was used, whereas panelists 1 and 2 with scaling factors less than 1 (0.865 and 0.869, respectively) used a much larger range of the scale.

Iterative Procedure

The iterative procedure begins at the point where individual panelists' configurations have been normalized and a consensus configuration has been com-

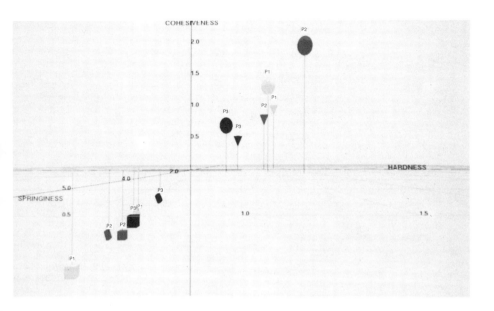

FIG. 13.9 Initialized, rotated, and scaled configuration. Initialized, rotated, reflected, and scaled configuration (iteration 1) of simulated data set presented in three-dimensional space. Shapes represent breads 1–4. Sphere-bread, 1; cube-bread, 2; cone-bread, 3; and cylinder-bread 4. The different shading of these shapes as well as labels P1–P3 represent panelists, where light gray is panelist 1 (P1), middle gray is panelist 2 (P2), and black is panelist 3 (P3). Axes represent attributes 1–3: X-hardness, Y-cohesiveness, and Z-springiness. The size of the symbol indicates its position with symbols that appear larger being closer to the viewer.

puted. The consensus configuration is the mean of all newly normalized panelists' configurations. When all normalized configurations have been rotated and scaled (referred to as newly transformed configurations), a single iteration is complete. After the first iteration, a new consensus configuration is derived as the mean of all newly transformed panelists' configurations. It is this new consensus configuration or target that newly transformed configurations computed from iteration one are rotated and scaled to match. For our simulated data, five iterations were computed, and panelist configurations are shown in Fig. 13.10. When examining Fig. 13.10, the data appear to be closer to the origin (zero) in general. In addition, there seems to be greater variation between panelists in evaluating bread 4. Breads 1 and 2 had similar scores for springiness, although they differed in hardness. Hardness scores for bread 3 were comparable to that of bread 1; however, cohesive scores were higher. Table 13.6 shows the Procrustes statistic after each iterative step. Results of each iterative step show a

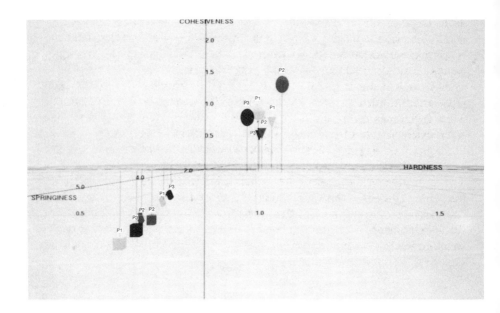

FIG. 13.10 Panelist configurations after five iterations. Final configuration after five iterations of simulated data set presented in three-dimensional space. Shapes represent breads 1–4. Sphere-bread, 1; cube-bread, 2; cone-bread, 3; and cylinder-bread, 4. The different shading of these shapes as well as labels P1–P3 represent panelists, where light gray is panelist 1 (P1), middle gray is panelist 2 (P2), and black is panelist 3 (P3). Axes represent attributes 1–3: X-hardness, Y-cohesiveness, and Z-springiness. The size of the symbol indicates its position with symbols that appear larger being closer to the viewer.

TABLE 13.6 Procrustes Statistics Calculated Before and After Iterations 1–5 and Final Scaling Factors

Panelist	Iteration 1	Iteration 2	Iteration 3	Iteration 4	Iteration 5	Scaling Factors
1	1.573	0.960	0.714	0.593	0.495	0.656
2	3.069	2.579	2.319	2.130	1.969	0.648
3	1.939	1.113	0.897	0.820	0.788	1.560
Total	6.581	4.653	3.930	3.543	3.252	

decrease in the sum of the Procrustes statistic over all panelists from 4.653 to 3.252. The point at which a change in the Procrustes statistic between iterations is less than a predetermined tolerance is when the iterative process is terminated. In this example, the difference in the total Procrustes statistic after iteration 4 (3.543) to iteration 5 (3.252) was 0.291. This difference is less than our predetermined tolerance of 0.300. Gower (1975) stopped the iterative process when differences in the sum of the Procrustes statistics were less than 0.0001 when examining data from three panelists' evaluations of nine beef carcasses. Scaling factors computed after the fifth iteration (Table 13.6) show similar trends as scaling factors calculated after iteration 1 (Table 13.5).

The final consensus configuration is shown in Fig. 13.11 in which all transformed panelists' configurations have now been combined to give a relative comparison of the samples. The final consensus configuration is the mean of all finally transformed configurations and can be used in place of the untransformed panel mean. The consensus configuration shows similar sample arrangement as Fig. 13.9.

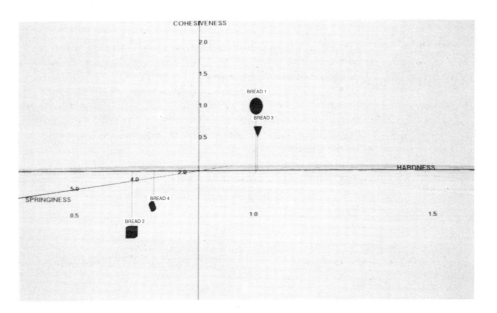

FIG. 13.11 Final consensus configuration. Final consensus configuration of simulated data set presented in three-dimensional space. Shapes represent breads 1–4. Sphere-bread, 1; cube-bread, 2; cone-bread, 3; and cylinder-bread, 4. Axes represent attributes 1–3: X-hardness, Y-cohesiveness, and Z-springiness. The size of the symbol indicates its position with symbols that appear larger being closer to the viewer.

Table 13.7 gives the final rotation matrix for each panelist, which when multiplied to the initial panelist matrices, best matches the final consensus configuration. By examining the final rotation matrices, information about how each panelist used the terms to determine sample differences can be obtained. Those values for each attribute closest to 1 indicate good agreement with the transformed attributes. For example, the first transformed attribute was best described by panelists 2 and 3 as hardness because values of 0.984 for panelist 2 and 0.981 for panelist 3 were obtained. Panelist 1 however, used a combination of both hardness and cohesiveness to differentiate between the samples as suggested by a lower value of 0.862 for hardness and a higher absolute value for cohesiveness (–0.504). The second transformed attribute was cohesiveness in which panelists 1 and 2 had similar meaning with values of 0.857 and 0.985, respectively. Panelist 3 however may have perceived this attribute much differently or reversed the scale for this attribute as noted by the negative value (–0.973). The last transformed attribute was springiness. Springiness of the samples appears to be evaluated similarly by all the panelists as noted by the closeness of the values (0.988, 0.991, and 0.992, respectively) for the three panelists. If the panelists had used different, rather than common terms for the evaluations as in free-choice profiling, Procrustes analysis would allow one to determine which descriptors were used similarly.

Procrustes Statistic

The Procrustes statistic is not considered a transformation step, but when it is calculated, it can give an indication of the relative orientation of configurations from a target or consensus configuration as shown in our illustration. The Procrustes statistic can be calculated at any point in the transformation procedure: initially, after initialization, rotation/reflection, scaling or after each iterative procedure. As in the simulated data, calculating the Procrustes statistic

TABLE 13.7 Final Rotation Matrices for Individual Panelists

	Transformed attributes[a]								
	First			Second			Third		
Panelist	H	C	S	H	C	S	H	C	S
1	0.862	–0.504	0.044	0.493	0.857	0.151	–0.113	–0.108	0.988
2	0.984	0.143	–0.108	–0.151	0.985	–0.077	0.095	0.092	0.991
3	0.981	–0.194	0.011	–0.191	–0.973	–0.126	–0.035	–0.122	0.992

[a]Transformed attributes include hardness (H), cohesiveness (C), and springiness (S).

for each panelist after each step of the transformation can indicate the degree to which a particular transformation has affected each individual's configuration. The Procrustes statistic calculated before and after translation can give an indication as to how panelists used the scale. For instance in our simulated data, a large difference between the Procrustes statistic value for panelist 1 compared to the other panelists indicated that a different level of the scale was used by panelist 1 than by panelist 2 or 3. In addition, after the last iteration, a large Procrustes statistic can indicate that a panelist is perceiving the samples differently than the other panelists.

The Procrustes statistic can be calculated not only for panelists as described above but also for samples. The Procrustes statistic for samples is calculated as the sum of squared distances for all panelists from the consensus positioning to an individual panelist's final positioning for each sample. A small Procrustes statistic value for samples indicates a greater agreement among panelists with respect to the positioning of a given sample.

APPLICATIONS OF PROCRUSTES ANALYSIS TO SENSORY DATA

Procrustes analysis, either pair-wise or generalized, can be an effective way of obtaining information about samples, their attributes, and the panelists assessing them. Although Procrustes analysis has most often been used for free-choice profiling data, it has other applications as well. In most of the studies described here, Procrustes analysis was used for multiple reasons: (1) to assess panelists' performance; (2) to describe sample differences; (3) to relate instrumental data with sensory; (4) to compare methodologies; and (5) for acceptance and preference testing. These various applications will be described here.

Panelist Performance

Procrustes analysis has been used to provide performance information about individuals, which can be helpful in determining proper panel selection and training. Panelist performance information allows sensory scientists to measure a panelist's consistency over replications and sessions. Researchers have found Procrustes analysis useful in determining panelist performance, as well as identifying discrepancies among panelists in their interpretation of terms and in their use of scales. Several examples of how Procrustes analysis has been used to obtain such information are given below.

Banfield and Harries (1975) compared the visual assessments of beef carcasses made by experts, trainees, and inexperienced panelists. Pair-wise Procrustes analysis followed by principal coordinate analysis was used to provide information about choosing and training panelists. The Procrustes statistics for comparison of each pair of panelists (see Table 13.3) estimates the agreement within or among the groups. In this case, the large values for the five inexperienced panelists (6–10) show that they were less consistent among themselves and with the other group. The size of the Procrustes statistic can indicate whether a potential panelist approaches the ability of a trained judge. Visualizing panelists' scores, using principal component analysis, while they are being trained can also be helpful in determining training progress.

When sensory profiling is used to describe product characteristics, the results are usually expressed as panel means. However, in free-choice profiling the panelists use different vocabularies, so Procrustes analysis provides a consensus configuration. The steps used in achieving the consensus (e.g., translation, rotation/reflection, and scaling) enables the experimenter to identify panelists' differences in vocabulary usage, and to determine if panelists are in agreement. Through replications, panelists who are not consistent in the use of their scales can be identified (Langron, 1983).

Williams and Langron (1984) confirmed the effectiveness of generalized Procrustes analysis in interpretation of free-choice profiling data when port wines were evaluated by experts and nonexperts. The Procrustes statistics, in addition to principal coordinate analysis, showed that, overall, color was used more consistently by both expert and nonexpert tasters than aroma or flavor in differentiating between samples. When dimensions for individual panelist configurations were examined, however, it appeared that several panelists used all three attributes (color, aroma, and flavor) to differentiate between samples. Thus, a multidimensional configuration, such as that achieved with a Procrustes consensus configuration, indicates how panelists make their evaluations and distinguish between samples.

Danzart (1988) reported that Procrustes analysis was useful in detecting those descriptive analysis panelists with similar perceptions of different products. Danzart (1988) was able to group panelists who had similar perceptions in their evaluations of leg and breast of chickens from 12 farms for juiciness, tenderness, and flavor.

Replication of sample evaluations by free-choice profiling panelist provides a means of measuring consistency of responses (Marshall and Kirby, 1988). Comparison of the consensus configuration for panelists or samples can be visualized by principal coordinate analysis, and the Procrustes statistic for each individual panelist can identify anyone who is different. By examining differences in replications with Procrustes analysis, variations in panelist performance can be identified and accounted for (Harries and MacFie, 1976; Williams et al., 1981; Williams et al., 1984; Williams and Arnold, 1985; MacFie, 1987; Tunaley et al., 1988; McEwan et al., 1989).

Sample Information

One of the most important applications of Procrustes analysis is its use in distinguishing those attributes by which samples can be most clearly differentiated. This is in addition to monitoring panelist performance in evaluating the same samples.

Harries and MacFie (1976) used a subset of data from the Banfield and Harries (1975) study of 180 beef carcasses cited above to determine the interrelationship between attributes. Ten attributes were used by expert and inexperienced evaluators to visually assess the carcasses. Generalized Procrustes analysis of the data followed by principal coordinate analysis revealed that several of the attributes (points) were in close proximity. Close association of attributes suggests that some attributes can be combined or eliminated without affecting the evaluator's ability to discriminate between the carcasses. For example, of the 10 attributes used by expert evaluators, attributes representing conformation of the rump and buttock were grouped similarly, as were the conformation of the loin and forerib. Harries and MacFie (1976) suggested that two of these four attributes be eliminated. For the inexperienced panelist a single attribute, overall conformation, was used in place of three other attributes (proportion of muscle and conformation of the buttock and rump) to discriminate between carcasses. By using Procrustes analysis, attributes for which panelists are most discriminating can be identified.

Differentiation between samples using generalized Procrustes analysis and principal coordinate analysis can be obtained from sensory profiling data of any type (texture, descriptive, flavor, free-choice). It is particularly useful when multiple attributes are evaluated and panelists use them differently to discriminate between samples. Examples can be found in a number of studies of products, including chocolate (McEwan and Thomson, 1988), sweeteners (Tunaley et al., 1988), and wines (Williams and Langron, 1984).

Sensory Data Versus Instrumental Measurements

Generalized Procrustes analysis can be applied to either instrumental data and/or sensory data to obtain a consensus configuration to which the other is compared. The comparisons between instrumental and sensory measurements provide a means of testing the reliability of instruments to provide sensory information about samples.

Williams and Langron (1983) were able to relate analytical color measurements to the appearance evaluation of eight commercial wines by free-choice profile panelists by using multiple regression. Vectors obtained from the correlations of the tristimulus color measurements (L, a, b) and the derived variables (saturation and hue angle) with the transformed attributes were superimposed on the consensus sample plot (Arnold and Williams, 1986). By examining the length of these vectors and the direction in which they were pointing, information about the relationship of the samples with the tristimulus color measurements could be examined. Even though panelists used different words to describe the appearance of the ports, they were still able to evaluate similar color properties of the samples.

Daget et al. (1983) compared the texture properties of gel systems by both quantitative descriptive analysis and by mechanical measurements. Data from sensory and mechanical measurements were examined using canonical analysis and principal component analysis and then compared using Procrustes analysis. By applying Procrustes analysis to the data, it was determined to what extent each method was reproducible. In addition, information about the sensitivity of each method was provided, as well as how related terms used by panelists described the physical properties of the gels.

Marshall and Kirby (1988) examined the composition (fat and moisture) of unflavored cheese analogs and related these variables to the texture data obtained by free-choice profiling. Calculated consensus scores for each cheese for the first two principal component axes of the consensus configuration were derived from the generalized Procrustes analysis for each replicate. These scores were then regressed separately against the percent of fat and moisture content of nonfat solids. The generalized Procrustes analyses of each individual's calculated scores gave information about each panelist's sensitivity to changes in composition. The calculated consensus scores for each replicate were related to the known moisture content of nonfat solids and fat content of cheeses, and suggested that panelists were most sensitive to changes in the moisture content of nonfat solids.

Sensory Methodology Comparisons

Consensus configurations for samples evaluated by different sensory methods can be compared through the use of the Procrustes analysis. This includes comparisons of descriptive analytical techniques, as well as different mathematical treatments of the same data. Procrustes analysis can be used to validate results obtained by different methods.

Langron et al. (1984) showed that the consensus configuration from generalized Procrustes analysis was more meaningful than untransformed panel means from sensory profiling of an apple cultivar during storage. This further suggests that descriptive analysis data could be subjected to general Procrustes analysis.

In a study of three methods of evaluating coffees, Williams and Arnold (1984) noted that generalized Procrustes analysis was helpful when comparing conventional descriptive profiling with free-choice profiling. Grouping of samples was similar with both methods, and free-choice profile panelists seemed to be in greater agreement than descriptive panelists. However, this was not clearly demonstrated.

McEwan et al. (1989) suggested that "conventional" free-choice profiling of chocolates was preferred to a structured free-choice technique based on Kelly's repertory grid method (Kelly, 1955). The conclusion was based on the similarity of results obtained by the two methods and the greater simplicity of the conventional method.

Procedures for conducting sensory tests have also been compared using Procrustes analysis. For example, in a study by Williams et al. (1984), Procrustes analysis was used to determine how the assessment of aroma and flavor of 10 wines by 10 panelists differed when the ability to see the product was varied by presenting samples in red or clear glass. Generalized Procrustes analysis was applied to scores averaged over replications for each panelist and for each attribute. Sample and panelists consensus configurations were obtained for each of five evaluations (appearance in red glass; aroma and flavor in red and clear glass). Pair-wise Procrustes analysis was then applied to the five sample configurations in which appearance was compared with aroma and flavor evaluated in clear and red glass. The Procrustes statistic was calculated and principal coordinate analysis applied to provide a visual representation of the relationship between the five evaluations. Results showed that the panel aroma and flavor scores differed depending upon whether the appearance of the wines could be evaluated. Scores from individuals varied, indicating how each panelist may have been influenced differently.

Acceptance and Preference Testing

The ultimate goal of all sensory evaluations is to relate panel evaluations of a product and its attributes to the way consumers perceive and finally accept that product. Once this relationship is found, the important product attributes may be optimized.

Laslett and Bremner (1979) determined the relationship between scores for six attributes (aroma, off-aroma, flavor, off-flavor, toughness, and moistness) of fish minces and acceptability as evaluated by the same panel. Generalized Procrustes analysis was used to determine if the panelists were evaluating the fish minces in similar ways. The consensus configuration for the panelists indicated that the panelists were similar so the mean attribute scores could be used to derermine the relationship between acceptance and sensory scores.

Williams et al. (1988) addressed some of the problems encountered when trying to relate sensory information with acceptance data. They found that by using Procrustes analysis, a sample consensus configuration could be derived in which dimensions that relate to acceptability could be superimposed. In addition, subgroups of panelists who perceive samples similarly could also be determined. The importance of bridging the gap between objective information and hedonic information in determining what aspects of a product are significant to consumers when they select products and how such characteristics may be optimized was reemphasized.

Benedict et al. (1988) discussed a new procedure using generalized Procrustes analysis integrated with a technique called "natural grouping" in which individual subjects divided a set of products based on their perceived similarities until no further separations were possible. After each partitioning each individual described the attributes which separated each group of products. By collecting data in this manner a completely individualized perceptual configuration was obtained for each individual for the same set of products. These configurations are referred to as "perceptual maps" and can provide some insight into consumers' perceptions of food products. Generalized Procrustes analysis was used to construct a consensus of all the individual configurations which allowed for easier interpretation of the data that took into account individual variation.

CONCLUSIONS

Procrustes analysis is a multivariate statistical tool that allows individual panelists configurations, such as those obtained from free-choice profiling, to be matched and compared by initialization, rotation/reflection, and scaling. The final transformed configurations obtained from this analysis can be combined to form a consensus configuration that can be used in place of sample means that are traditionally averaged across panelists and replications.

Free-choice profiling differs from conventional descriptive analysis in the use of terms, in the development of individual scorecards, and the analysis of the data. Free-choice profiling is newer than other more conventional descriptive analysis techniques, so procedures have not been standardized. However, researchers have reported shorter training times than that required by other profiling methods. In addition, because panelists are evaluating samples on an individual basis, variations associated with forcing conformity in term and scale use by panel members are alleviated.

By using free-choice profiling and analyzing the resulting data using Procrustes analysis, sample information, as well as panelist information can be provided. Panelists responses can then be evaluated and outliers identified. One note of caution in using free-choice profiling and applying Procrustes analysis to data is that a false sense of security may be obtained from its use. The analyst must still examine the data and judge whether it is meaningful within the context of the test. Sensory professionals must not be misled by the power of this statistical tool. Training is indeed very important in getting accurate and reliable information about a product. Free-choice profile panelists still must be trained in the use of scales, and to be consistent.

Procrustes analysis is a powerful technique that has not only been used to analyze free-choice profiling data, but descriptive analysis data as well. Other statistical methods used in conjunction with Procrustes analysis, such as principal component and principal coordinate analysis, can help intepret results obtained from applying Procrustes analysis to sensory data. From its application, information about the performance of panelists and differences and similarities between samples can be identified. In addition, it has been used to compare different sensory methodologies. Recently, Procrustes analysis has been helpful in providing insight to consumer acceptance and preferences when compared to trained panel responses. Procrustes analysis can be a valuable tool to the sensory scientist who understands not only the way in which it can be applied to sensory data, but also its limitations.

ACKNOWLEDGMENTS

Special appreciation goes to Donald J. Colby for construction of the three-dimensional data plots with the cooperation of the National Center for Supercomputing Applications and the Beckman Institute on the University of Illinois Urbana-Champaign campus. Dolores Oreskovich gratefully acknowledges the support of the Ford Foundation, sponsor of the Ford Foundation Doctoral Fellowship, the National Pork Council Women, sponsor of The American Home Economics National Pork Fellowship, and the National Hispanic Scholarship Fund. This work was part of Project No. 60-326 of the Agricultural Experiment Station, College of Agriculture, University of Illinois at Urbana-Champaign.

REFERENCES

Arnold, G. M. and Williams, A. A. 1986. The use of generalized Procrustes techniques in sensory analysis. In *Statistical Procedures in Food Research* (J. R. Piggott, Ed.), pp. 233–253. Elsevier Applied Science, London.

ASTM draft. In preparation. Descriptive analysis testing. Special technical publication. E-18.06.04.

Banfield, C. F. and Harries, J. M. 1975. A technique for comparing judges' performance in sensory tests. *J. Food Technol.* 10: 1.

Benedict, J., Steenkamp, E. M., and Van Trijp, H. C. M. 1988. Free-choice profiling in cognitive food acceptance research. In *Food Acceptability* (D. M. H. Thomson, Ed.), pp. 363–376. Elsevier Applied Science, London.

Brandt, M. A., Skinner, E. Z., and Coleman, J. A. 1963. Texture profile method. *J. Food Sci.* 28: 404.

Cairncross, S. E. and Sjöstrom, L. B. 1950. Flavor profiles-a new approach to flavor problems. *Food Technol.* 4: 308.

Caul, J. F. 1957. The profile method of flavor analysis. *Adv. Food Res.* 7: 1.

Cliff, N. 1966. Orthogonal rotation to congruence. *Psychometrika* 31: 33.

Cliff, N. 1987. Transformation of component loadings and computation of component scores. In *Analyzing Multivariate Data* (N. Cliff, Ed.), pp. 319–347. Harcourt Brace Jovanovich, Orlando, FL.

Daget, N., Collyer, P., and Vuatez, L. 1983. Profile analysis of food products: Comparisons between sensory and physical measurements of gel systems. In *Food Research and Data Analysis* (H. Martens and H. Russwurm Jr., Ed.), p. 421. Applied Science, England.

Danzart, M. 1983. Evaluation of the performance of panel judges. In *Food Research and Data Analysis* (H. Martens and H. Russwurm Jr., Ed.), pp. 305–343. Applied Science, London.

Dijksterhuis, G. B. and Punter, P. H. 1990. Generalized Procrustes analysis of conventional profiling data. Poster presented at Weurman Flavour Research Symposium 90, Geneva.

Dijksterhuis, G. B. and Punter, P. H. Interpreting generalized Procrustes analysis 'analysis of variation' tables. Submitted to *Food Quality and Preference*.

Gower, J. C. 1966. Some distance properties of latent root and vector methods used in multivariate analysis. *Biometrika* 53: 325.

Gower, J. C. 1975. Generalized Procrustes analysis. *Psychometrika* 40: 33.

Green, B. F. 1952. The orthogonal approximation of an oblique structure in factor analysis. *Psychometrika* 17: 429.

Guy, C., Piggott, J. R., and Marie, S. 1989. Consumer profiling of Scotch whisky. *Food Quality and Preference* 1: 69.

Harries, J. M. and MacFie, H. J. H. 1976. The use of rotational fitting technique in the interpretation of sensory scores for different characteristics. *J. Food Technol.* 11: 449.

Hurley, J. R. and Catell, R. B. 1962. The Procrustes program: Producing direct rotation to test a hypothesized factor structure. *Behav. Sci.* 7: 258.

IFT. 1981. Sensory evaluation guide for testing food and beverage products. *Food Technol.* 35(11): 50.

Kelly, G. A. 1955. The repertory test. In *The Psychology of Personal Constructs: A Theory of Personality* (G. A. Kelly, Ed), p. 219. W. W. Norton & Co., New York.

Kristof, W. and Wingersky, B. 1971. Generalization of the orthogonal Procrustes rotation procedure to more than two matrices. *Proceedings,* 79th Annual Convention, American Psychological Association 89–90.

Langron, S. P. 1983. The application of Procrustes statistics to sensory profiling. In *Sensory Quality in Foods & Beverages: Definition, Measurement & Control* (A. A. Williams and R. K. Atkin, Ed.), pp. 89–95. Ellis Horwood Ltd., Chichester.

Langron, S. P., Williams, A. A., and Collins, A. J. 1984. A comparison of the consensus configuration from a generalized Procrustes analysis with the untransformed panel mean in sensory profile analysis. *Lebensmittel-Wissen. Technol.* 17: 296.

Larmond, E. 1977. *Laboratory Methods for Sensory Evaluation of Food.* Minister of Supply and Services, Ottawa, Canada.

Laslett, G. M. and Bremner, A. H. 1979. Evaluating acceptability of fish minces and fish fingers fron sensory variables. *J. Food Technol.* 14: 389.

Lyon, B. G. 1987. Development of chicken flavor descriptive attribute terms aided by multivariate statistical procedures. *J. Sens. Studies* 2: 55.

MacFie, H. J. H. 1987. Data analysis in flavour research: Achievements, needs and perspectives. In *Flavour Science and Technology* (M. Martens, G. A. Dalen, and J. H. Russwurm, Ed.), pp. 423–437. John Wiley and Son Ltd, New York.

Marshall, R. J. and Kirby, S. P. J. 1988. Sensory measurement of food texture by free-choice profiling. *J. Sens. Studies* 3: 63.

McEwan, J. A., Colwill, J. S., and Thomson, D. M. H. 1989. The application of two free-choice profile methods to investigate the sensory characteristics of chocolate. *J. Sens. Studies* 3: 271.

McEwan, J. A. and Thomson, D. M. H. 1988. An investigation of the factors influencing consumer acceptance of chocolate using the repertory grid method. In *Food Acceptability* (D. M. H. Thomson, Ed.) pp. 347–362. Elsevier Applied Science, London.

Meilgaard, M. C., Civille, G. V., and Carr, T. C. 1988. *Sensory Evaluation Techniques.* CRC Press, Boca Raton, FL.

Mosier, C. I. 1939. Determining a simple structure when loadings for certain tests are known. *Psychometrika* 4: 149.

Moskowitz, H. R. and Howard, R. 1982. Sensory analysis of "thickness." *Cosmetics and Toiletries* 97: 34.

Muñoz, A. 1986. Development and application of texture reference scales. *J. Sens. Studies* 1: 55.

Oreskovich, D. C., Klein, B. P., and Sutherland, J. W. 1990. Variances associated with descriptive analysis and free-choice profiling of frankfurters. Annual Meeting of the Institute of Food Technologists, Anaheim, CA. *Abstracts*.

Oreskovich, D. C., Klein, B. P., and Sutherland, J. W. In preparation. Development of a standardized language for evaluating the flavor and texture of pork by descriptive analysis and free-choice profiling.

Peay, E. R. 1988. Multidimensional rotation and scaling of configurations to optimal agreement. *Psychometrika* 53: 199.

Piggott, J. R. and Canaway, P. R. 1981. Finding the word for it—Methods and uses of descriptive sensory analysis. In *Flavour '81* (P. Schreier, Ed.), pp. 33–46. Walter de Gruyter & Co., Berlin.

Piggott, J. R. and Sharman, K. 1986. Methods to aid interpretation of multidimensional data. In *Statistical Procedures in Food Research* (J. R. Piggott, Ed.), pp. 181–232. Elsevier Applied Science, London.

Powers, J. J. 1984. Using general statistical programs to evaluate sensory data. *Food Technol.* 38(6): 74.

Powers, J. J. 1988. Current practices and application of descriptive methods. In *Sensory Analysis of Foods* (J. R. Piggott, Ed.), pp. 187–266. Elsevier Applied Science, London.

Procrustes-PC V2.0. Computer software. Oliemans Punter and Partners, Inc. SG Utrecht, The Netherlands.

Schönemann, P. H. 1966. A generalized solution of the orthogonal Procrustes problem. *Psychometrika* 31: 1.

Schönemann, P. H. 1968. On a two-sided orthogonal Procrustes problems. *Psychometrika* 33: 19.

Schönemann, P. H. and Carroll, R. M. 1970. Fitting one matrix to another under choice of a central dilation and a rigid motion. *Psychometrika* 35: 245.

Stone, H. and Sidel, J. L. 1985. *Sensory Evaluation Practices*. Academic Press, Orlando, FL.

Stone, H., Sidel, J., Oliver, S., Woolsey, A., and Singleton, R. C. 1974. Sensory evaluation by quantitative descriptive analysis. *Food Technol.* 28(11): 24.

Szczesniak, A. S. 1963. Classification of textural characteristics. *J. Food Sci.* 28: 385.

Szczesniak, A. S., Brandt, M. A., and Friedman, H. H. 1963. Development of standard rating scales for mechanical parameters of texture and correlation between the objective and the sensory methods of texture evaluation. *J. Food Sci.* 28: 397.

TenBerge, J. M. F. 1977. Orthogonal Procrustes rotation for two or more matrices. *Psychometrika* 42: 267.

TenBerge, J. M. F. and Kroll, D. L. 1984. Orthogonal rotations to maximal agreement for two or more matrices of different column orders. *Psychometrika* 49: 49.

Tunaley, A., McEwan, J. A., and Thomson, D. M. H. 1988. An investigation of the relationship between preference and sensory characteristics of nine sweeteners. In *Food Acceptability* (D. M. H. Thomson, Ed.) pp. 387–400.

Williams, A. A. and Arnold, G. M. 1984. A new approach to the sensory analysis of foods and beverages. In *Progress in Flavour Research 1984, Proceedings of the 4th Weurman Flavour Research Symposium* (J. Adda, Ed.), pp. 35–49. Elsevier Science, Amsterdam.

Williams, A. A. and Arnold, G. M. 1985. A comparison of the aromas of six coffees characterised by conventional profiling, free choice profiling and similarity scaling methods. *J. Sci. Food Agric.* 36: 204.

Williams, A. A., Baines, C. B., Langron, S. P., and Collins, A. J. 1981. Evaluating tasters' performance in the profiling of foods and beverages. In *Flavour '81* (P. Schreier, Ed.), pp. 83–92. Walter de Gruyter, Berlin.

Williams, A. A. and Carter, C. S. 1977. A language and procedure for the sensory assessment of Cox's Orange Pippin apples. *J. Sci. Food Agric.* 28: 1090.

Williams, A. A. and Langron, S. P. 1983. A new approach to sensory profile analysis. In *Flavour of Distilled Beverages: Origin & Development* (J. R. Piggott, Ed.), pp. 219–224. Ellis Horwood Ltd, Chichester.

Williams, A. A. and Langron, S. P. 1984. The use of free-choice profiling for the evaluation of commercial ports. *J. Sci. Food Agric.* 35: 558.

Williams, A. A., Langron, S. P., Timberlake, C. F., and Bakker, J. 1984. Effect of colour in commercial port blends. *J. Food Technol.* 19: 659.

Williams, A. A., Rogers, C. A., and Collins, A. J. 1988. Relating chemical/physical and sensory data in food acceptance studies. *Food Quality and Preference* 1: 25.

14

The Uses of Qualitative Research in Product Research and Development

**Edgar Chambers IV and
Elizabeth A. Smith**

Kansas State University
Manhattan, Kansas

WHAT IS QUALITATIVE RESEARCH?

The United States is a nation of self-reported dieters, yet its people chronically overeat. Consumers want fruits and vegetables that are pure and natural, yet they also want low cost and only will buy foods that look perfect. Purchasers demand quality, yet they buy convenience. There appears to be a large difference between what consumers say, think, and actually do. Chay (1989) asks "How do we unlock the consumer's mind? It is clearly evident that there is an enormous gap between attitudes and behavior." For the sensory professional this presents tremendous problems because response to product attributes is linked to expectations and biases.

Definition

Qualitative research provides detailed information about people's attitudes, opinions, perceptions, behaviors, habits, and practices. Qualitative research is just that, *qualitative*. It is vocabulary, meanings, behaviors—not intensities or frequencies. All of these factors drive consumer behavior and, ultimately, product selection and use. Qualitative research answers the questions what and why but not how much or how often. It is concerned with describing and understanding products and ideas rather than measuring them. Quantitative research methods provide the measure of intensity or amount.

Neither qualitative nor quantitative research alone can provide all of the information a research project needs. Qualitative testing usually needs to precede quantitative testing to help establish criteria for data collection and to follow quantitative testing to aid in the explanation of the quantitative data. Using both general classes of methods in conjunction with each other enhances and reinforces the use and validity of both methods.

The integration of qualitative and quantitative studies should be part of programmatic research generated out of previous studies and build on the base of previous findings. That enables each research component to build on the strengths and compensate for the weaknesses of prior research.

Differences and Similarities Between Qualitative and Quantitative Research

Table 14.1 shows some differences and similarities between qualitative and quantitative data. The table is not an absolute guide, but represents the authors' general idea about the methods as commonly used. Qualitative data generally is not numerical, it is verbal-descriptive, and ultimately dependent on the skill of the interviewer. Because of the small (usually) numbers of respondents involved, it is not likely that they represent everyone in a population, although general ideas and trends certainly become apparent. Qualitative studies are used to generate ideas, probe for reasons, and cause respondents to think about a product or topic in depth. Conversely, quantitative data are numerically based in counts or scale values and usually are dependent on the skill with which a questionnaire has been designed. Quantitative studies are best for research where amount or intensity are critical, i.e., respondents are asked to measure specified characteristics.

Both quantitative and qualitative studies require that questions be asked. Quantitative research allows the researcher to ask many more questions, whereas qualitative research allows a more in-depth study of fewer questions. Both methods can be misleading if respondents do not give carefully considered, truthful answers. Qualitative methods have an advantage over quantitative methods in helping to determine if the attitudes and behaviors consumers describe are likely to be accurate. In theory, qualitative research requires detailed answers with reasons, and consumers may be more likely to reveal their true

TABLE 14.1 Differences and Similarities of Qualitative and Quantitative Research[a]

Categories	Qualitative Group	Qualitative Individual	Quantitative
Number of respondents	8–12 in a group 2–4 groups	< 50	100–1000 +
Demographics	Screened	Screened	Screened - Assume to represent larger population
Procedures	Verbal dynamic interactive	Verbal dynamic interactive	Written consistent
Responses	Group dependent	Independent (except for interviewer bias)	Independent
Ability to:			
1. Generate ideas	Excellent	Fair	Poor
2. Probe	Good	Excellent	Poor
3. Numerically measure	Poor	Poor	Good
4. Compare replications	Fair	Poor	Excellent
Analysis	No statistics	Statistics limited	Full range of statistics

[a]Represents common practice—complexity and purpose of study may change numbers or procedures.

feelings when careful, thorough consideration is required. Also, with qualitative research the same question often can be asked in slightly different ways requiring slightly different descriptions of the same phenomenon. Respondent answers can be challenged with additional questions by the interviewer or by other respondents. In practice, the ability to discover true beliefs is largely dependent on the interviewer's ability to ask the right question at the right time.

COMMON TYPES OF QUALITATIVE RESEARCH

Focus groups, focus panels, and in-depth or one-on-one interviews are the common types of qualitative research. Although other methods such as gaming, psychodrawing, collage, image response, association, analogy, personalization, and psychodrama are available and have been suggested to provide greater ability to uncover true beliefs, they are used much less frequently than the common interviewer methods (Cooper, 1987).

Here a distinction needs to be made. Qualitative research is a group of methods; it is not marketing research, although the methods typically are used for that segment of testing. Qualitative methods are *methods;* they can be applied to many types of research including sensory research and product development.

Focus Groups

A focus group usually consists of 8 to 10 respondents, recruited according to predefined specifications. The respondents engage in a discussion led by a qualified group moderator (Wu, 1989). The moderator (1) directs the flow of the discussion; (2) recognizes important points and encourages the group to explore and elaborate on them; (3) observes the nonverbal communication among respondents, between respondents and moderator, and between respondents and the subject matter; (4) creates an atmosphere that allows respondents to relax and lower some of their defenses; (5) synthesizes the understanding gained about the objectives; and (6) tests ideas generated by the information gained (Gordon and Langmaid, 1988). The group discussion often is observed by researchers and other interested individuals in an adjacent room or through a video-conference. Observers of the focus groups later may constitute a focus group as a formal de-briefing is held using qualitative methods to ensure that the observers have a common experience for making future decisions.

Focus groups are taped for a permanent record. Audio tapes are far easier to transcribe than video tapes, but video tapes are needed if action is involved or body language is likely to be important. Often a combination of audio and video tapes are best. A transcription of the tapes is made to provide detail, and viewing the actual tapes may be an important step when a critical issue is what the respondents do. In reality, many people involved with the project simply read the report written by the moderator that summarizes and links each group's discussion with that of other groups in the study.

Properly structured focus groups can be used to obtain almost any kind of qualitative information from almost any type of respondents (Templeton, 1987). It is the most versatile of qualitative research methods and is used more often than all other qualitative methods.

Focus Panels

Typically, respondents for focus groups are consumers who participate infrequently in consumer research and have not discussed the topic previously for other research. Alternatively, focus panels are focus groups that return one or more times to explore areas that have been discussed previously and to have follow-up discussions (Bellenger et al., 1976). For example, a focus panel may discuss a concept, take the product home to use, and return to discuss the product's attributes. Other uses of focus panels include iterative testing and discussion of prototypes and competitors or long-term testing of product consumption and use.

Both focus groups and focus panels can be conducted while respondents are doing something. Sequentially preparing food mixes according to varying package directions over several days in a central test kitchen is a common example of a focus panel situation where action is required. The discussion then can focus on the unprepared product, the preparation, and/or the finished product.

In-depth Interviews

Some ideas cannot be explored most effectively in a group situation. Respondents may feel uncomfortable discussing certain subjects in group settings, such as flatulence production by different food products. Additionally, some projects need in-depth exploration of each respondents motivations or feelings. That is especially true when products are poorly defined or when small modifications of a product's characteristics are needed and considerable time is needed to probe for specifics that may be hard for the consumer to describe.

Typically, in a focus group each respondent has about 5 to 8 minutes of "talking time"; a one-on-one interview may provide 45 minutes of "talking time," (Seymour, 1988). A general outline of questions is followed, but ample time is provided to probe the respondent's answers and reasons. As with group situations, the original questions are carefully established to be sure the objectives of the study will be met.

INFORMATION NEEDS OF PRODUCT RESEARCH AND DEVELOPMENT

In a broad sense, research and development needs two types of information to provide the best products: (1) general perceptions about products or product categories, and (2) how consumers feel about current and future product choices.

Information About Consumer Perceptions

Product scientists must understand general perceptions about products, e.g., the role of aroma in home-baked bread, the value of texture in candy bars, the

importance of variations in flavor among sauce mixes, and the ease of preparation for microwavable products. The words, responses, and emotions that arise
when people talk about a product or product category are essential to understanding (Thompson, 1989). Those words could be technical jargon, sensory descriptors, or consumer vocabulary—all those words must be understood in order
to communicate ideas from all segments of testing.

Similarly, product users are the most logical people to tell researchers what
they think could improve products or what they want on their market shelves.
The problem is that consumers often say things that "sound good," but their
actions do not always back up their words. For example, a consumer might
respond on a questionnaire that brown bread is "good for you" and "I like it."
However, that same person continues to buy soft white bread and furtively enjoy
it. Qualitative research methods are used to probe that response to determine
what the individual truly wants and needs.

Even in qualitative research the consumer may originally state that "brown
bread is good." The ability to probe that response is the value of qualitative
research. The interviewer has a wide variety of techniques available to help
determine if that is the respondent's true belief and practice. The interviewer may
ask questions that require the respondent to discuss when various breads would
be eaten and by whom. For example, questions may be asked that (1) require the
respondent to describe people who would eat white bread instead of brown
bread, (2) determine when the respondent thinks it would be appropriate to eat
white bread, or (3) elicit examples of things to make with white bread that could
not be made with brown bread, i.e., the respondent may believe certain sandwich
spreads do not go with brown bread. If the moderator has made the respondents
feel comfortable during the introduction and warm-up phase, then probing is
made easier because the respondents may feel that they are talking to an impartial
friend. Techniques such as the "false close" when the moderator uses an excuse
to leave the room for a few minutes near the end of the discussion sometimes is
effective for focus groups or panels because the respondents say things to each
other that they may not say to the moderator. Of course those comments are
overheard and recorded.

Information About Products or Product Categories

Researchers need to know the attributes and characteristics consumers expect
in a product or category and need to understand and translate those characteristics
into technically viable options that improve the product. Qualitative research
methods are used to help generate ideas and understand terminology. The use of
qualitative methods with technical staff often is valuable to obtain input into how
ideas can be developed into actual products.

QUALITATIVE RESEARCH IN PRODUCT DEVELOPMENT

In general terms Baker et al. (1988) suggested that five main stages in product development are apparent:

1. Idea generation
2. Initial development
3. Preliminary evaluation
4. Final development
5. Final testing and marketing

Each of those stages requires research answers from more than one type of test. It is critical to determine what and of whom questions need to be asked. It is important to remember that any data are irrelevant if the wrong hypothesis is tested, the wrong attributes are measured, or the wrong people are studied.

Idea Generation

Idea generation often is best done with qualitative research using focus-type groups. Groups may be conducted using consumers and technical or research personnel. Common idea generation and problem-solving techniques that use group dynamics often use the focus group format with technical personnel. Certainly consumers are used to voice opinions about products they would like to have or changes they would like in existing products. Occasionally, R & D personnel who have been the observers of a consumer focus group become participants in a follow-up focus group to discuss and explore the ideas they have heard in the consumer groups. The ideas generated by those groups may be listed on a ballot to be ranked or rated with quantitative study using a larger population sample.

Initial Development

Research questions for this stage usually are answered by quantitative re-search—often with a trained descriptive sensory panel. Descriptive analysis of similar products on the market helps establish profile goals that meet the needs identified in earlier idea generation. Sensory studies of ingredients are conducted to determine appropriate sources for the desired profile goal. However, quali-tative aspects of this stage do exist. Primarily qualitative work is used at this point to determine attributes for measurement. Although the determination of attributes is necessary for every sensory descriptive project, most sensory professionals do not think of attribute determination as qualitative, *It is*. Stone and Sidel (1985) suggest that vocabulary determination is like a focus group. Verbal, nonnumeric attributes are considered, discussed in relation to the products, opinions are expressed, and the sensory panel or professional makes attribute decisions based on that input. Often those important decisions are made with informal decision making rather than formal qualitative research.

A procedure we used for a large consumer products company was to conduct focus groups with (1) consumers, (2) previously trained sensory panelists, (3) research staff, and (4) technical sales/quality assurance staff to determine what words and vocabulary each group might use to describe various test samples. After compiling all the possible vocabulary (over 200 terms in this instance), the trained panel then examined the list for words that actually described specific charateristics. Redundant terms, hedonic terminology, and terms that described a sensory phenomenon in terms of a cause were eliminated automatically. The trained panel attempted to use terms that could be defined as single concepts rather than integrated, multifaceted concepts. Several additional words were added at this stage. A preliminary ballot was designed and tested with a range of products in the product class. Words that the panel found to be difficult to use because of lack of clarity in the definition were redefined or eliminated, if appropriate. Several terms were divided into additional concepts and a final descriptive sensory ballot was developed. Although time consuming, the formal qualitative sessions allowed input from all facets of the development process and improved the usefulness of subsequent sensory testing.

Preliminary Evaluation of Products

The research questions for this stage should be answered with both qualitative and quantitative research. Generally this is a lengthy stage in the product development process and includes work conducted on prototypes through initial pilot plant products.

Qualitative research is needed to help determine:

1. Which attributes consumers believe are important or relevant in the products they have tested and how those attributes might relate to "liking" or "preference" of the product or product category
2. What the various words used to describe attributes may mean to consumers
3. How quantitative questionnaires can be designed for ease of use and reduction of misunderstanding

Input from one-on-one interviews where respondents are given products to consume may help to determine appropriate test methods and conditions for quantitative testing that are effective if certain attributes are measured. It is critical to spend time with consumers early in the development process discussing in detail how they used the product they tested; how it was prepared, served, and consumed; and what potential abuses the product may be subjected to. Subsequent success with quantitative studies may depend on understanding those testing issues.

The use of qualitative testing can help researchers to develop user-friendly questionnaires for testing their product variations—IF users have given input. Researchers who believe that they and consumers think alike probably will err in the questionnaire design.

Qualitative research, both group and individual, with consumers and technical personnel can help determine question content. Is the question really needed? Is this question sufficient to generate the needed information? Will the data be actionable? Can and will the respondent answer the question correctly? Are there external events that might bias the response to the question? Certainly, such formal procedures as focus groups need not be conducted prior to each study in order to answer those questions. However, occasional use of a trained moderator in a focus group situation could be helpful.

Question phrasing is an important issue that can be aided with qualitative methods. Do the words have the same meaning to all the respondents? Are the words or phrases ones that could lead the consumer to answer in a biased manner? Are there implied alternatives in the question? Are there unstated assumptions related to the question? Will respondents approach the question from the frame of reference desired by the researcher?

Quantitative questionnaire layout is an important point of investigation for qualitative studies. Are the questions in a sequence that avoids introducing errors? Is the questionnaire designed in a manner to avoid confusion and minimize recording errors (Seymour, 1988)?

Group or individual interviews after a quantitative study can help pinpoint aspects of a product that were not clearly delineated in a quantitative ballot. After home use tests, interviews or focus groups may be valuable in learning if problems with the product were encountered, if the questionnaire adequately allowed the participants to describe their feelings about the product, to gain additional information on open-ended responses, or to probe inconsistencies found in the data.

A focus group after a central location test may help the researcher understand certain responses to the tested product. For example, a researcher asked a quantitative question about the liking for intensity of spice in barbecue sauce. The researcher thought she was asking the consumer to rate how much he liked the "intensity level" of the "spiciness." In most cases consumers would respond to the question in precisely that manner. However, a focus group after the test found that the character of the "spiciness" was different among the samples tested and one of the products had a strong atypical spice flavor. Consumers rated the intensity low in the quantitative study. In reality consumers were attempting to tell researchers that the "typical" spice character was low; the researcher misunderstood the data until it had been explored qualitatively.

The relationship between consumer terminology, descriptive panel words, and technical jargon needs to be established to help product developers understand sensory responses. While much of that relies on quantitative model building through regression or other multivariate methods, qualitative data may help researchers to understand the larger issues and often is the only way to examine nuance in language.

Final Development

In general, this stage needs quantitative answers. Trained sensory panels continue to evaluate changes made by developers. Panelists screened for sensitivity might use difference testing to determine the effect of ingredient changes, preparation variations, packaging options, and cost savings before final product testing is conducted.

Final Product Testing

This stage needs to include both quantitative and qualitative methods. The final quantitative consumer and descriptive testing would be conducted. Where questions about the data are found, qualitative data may be needed to understand the data and its implications. Packaging, preparation instruction, and other important issues for consumers often need to be studied with qualitative studies. Quantitative, as well as trained sensory panel would be used establish the product's gold standard or fingerprint.

REQUIREMENTS FOR GOOD QUALITATIVE RESEARCH

There are at least nine requirements for a good qualitative research study:

    Clear objectives and statement of purpose
    Good moderator/interviewer
    Qualified respondents
    Appropriate group size
    Adequate number of groups
    Appropriate testing location, physical environment, and equipment
    Prepared observers

Review of past research
Appropriate analysis of data

Clear Objectives and Statement of Purpose

The objectives must be clearly defined and agreed upon by all personnel related to the study, i.e., the flavor chemists, the sensory analyst, the product developer, and management. Different people will have different ideas concerning the important aspects of the project. All divergent opinions need to be resolved. Who the respondents should be must be clearly stated (Axelrod, 1975). The key issues need to be delineated and examined in relation to the purpose of the study. For each area, ask the following questions: (1) Is it necessary? (2) How does this relate to the purpose of the project? and (3) What role will the answer to the question play in the decision-making process?

Good Moderator/Interviewer

A well-qualified moderator/interviewer is THE essential component of effective qualitative research—training is mandatory. A moderator must:

1. Use group dynamics skillfully

2. Listen

3. Be flexible, both in demeanor and being able to change course in midstream

4. Be animated and sponteneous, have a sense of humor, be sensitive, admit biases, and express thoughts clearly

The moderator will submit an outline for the study that follows the format described in Fig. 14.1 and will prepare the moderator's guide (Fig. 14.2 is one example). The moderator's guide is the outline for the discussion and covers the general issues that will be explored in the session. The guide may be changed as the study progresses if unexpected issues arise that need to be examined. Changes are made only if there is consensus among all parties involved and should not substantially change the study. It is important to use the same moderator or team of interviewers for an entire study to ensure continuity. Large projects may require several moderators or teams of interviewers, but caution must be exercised in their selection so that the fewest variables possible are introduced by changing moderators.

The decision to use either a company employee or a consulting moderator is

I. Background
   A. Purpose of the Study
   B. Expected Outcomes

II. Methodology
   A. How the research will be conducted and the respondents recruited
   B. Location, time, specifications for recruiting

III. Key Elements of the Moderator's/Interviewer's Guide
   A. Major issues to be covered
   B. Related activities (e.g., copy test, taste test, view pilot, etc.).

IV. Time Frame
   A. From start date to report date
   B. Schedule of activities

V. Assumptions
   A. Number of observers
   B. Travel requirements
   C. Areas of responsibility (e.g., who manages the recruiting, payment to respondents, etc.).

VI. Cost Estimate
   A. Line item cost estimate
   B. Invoice for deposit

VII. Moderator's/Interviewer's qualification
   A. Résumé/related experience
   B. Client reference list
   C. Brochure

FIG. 14.1 Proposal format for qualitative research projects.

crucial. Employees of the company have product and market knowledge that cannot be conveniently forgotten while the interviewing is proceeding. That knowledge may be a problem. When an employee is used to conduct focus groups on one's own products, the results are usually a self-fulfilling prophecy. Also, different projects require different moderators; one cannot assume that a moderator who can deal with the egos and technical background of product developers can also relate to a group of 6-year-old children. It may be possible to

1) Introduction of moderator
   a) State purpose
   b) Guidelines for the group
   c) "Muffins have been furnished to eat while we are working, also beverages, feel free to get up for more drinks"

2) Respondents introduce themselves to group
   a) Warm-up questions
   b) "How many have had muffin in last 3 months?" (check of recruiting)

3) General questions about muffins
   a) Times muffins are eaten - "Tell me a time when a person would eat a muffin."
      Expect: quick breakfast/coffee break/with salad/soup at lunch/afternoon snack "Would you want a different type of muffin at these different times?" Is there a type/time relation?
   b) Reasons to choose a muffin over other options - "Thinking specifically that you are in a restaurant, could you give me a reason you might choose a muffin over a piece of toast, or a roll, or a pastry, or a biscuit?"
      Expect: healthy/better for you/easy to eat/new, not the same old thing/the "in" thing/tasty - discuss responses - ask appropriate questions
   c) Reasons not to choose a muffin - "Still in a restaurant, can you think of a reason that you might decide not to have a muffin?"
      Expect: price/size/too big/too sweet/too many calories/do not like the additions
   d) Attitude toward additions such as raisins nuts/bits of fruit

4) Specific questions about oat bran muffin samples
   a) "Could you tell me what you liked best about these muffins?" (indicate a platter) -
      Expect: size/spice level/sweet level/lack of inclusions/texture. Discuss fully:
         - What makes the size/spice/sweetness/inclusions/texture good?
         - What do you mean by that?
         - What else makes you want to eat this muffin?
   b) "Could you tell me what you dislike or could be improved?"
      Expect: spice type/sweet level/lack of raisins or nuts/texture. Discuss fully:
         - What makes you not like the spice type/sweetness/texture?

FIG. 14.2 Condensed version of a moderator's guide for oat bran muffins.

           - What kind of spices/nuts/fruit do you think should be in muf-
             fins?
           - If you made these muffins how would you change the recipe
             next time?
  c) Continue questions with all three samples

5) Summary—linking—ordering

6) False close—will gather questions from observers

7) Close and thank you

FIG. 14.2 *(continued)*

use an employee when technical knowledge is essential or where the topic and
the employee have little relationship (i.e., an employee in one division *may* be
able to moderate focus groups for an unrelated division). In all cases the
moderators and interviews should have prior training; experience is necessary,
but not a substitute for training (ASTM, 1990).

Qualified Respondents

The desired composition of the groups is dependent on the objectives and
should be clearly delineated in the objectives statement. A screener is developed
that specifies the qualifications needed by the respondents selected for the study.
The screener may be as simple as "serves on a sensory descriptive panel" or may
be more detailed for specific types of consumer tests.

Strict attention must be paid to the objectives—a group for an idea generation
study may have a different composition from a group for a terminology study.
For example, Baker et al. (1988) indicates that the consumers most likely to try
new products and ways to prepare foods will fit into one or more of the following
categories:

Lives in an urban area
Is located in the far West or Northeast rather than in the Midwest or South
Has had at least one year of college
Is an employed woman rather than a full-time homemaker
Is in the medium high- and upper-income bracket
Is the wife of a white-collar worker rather than a blue-collar worker
Is aged between 25 and 40 years

That person may not be the target for qualitative study of a general-use product and may be of no help in trying to select appropriate terminology with qualitative studies.

Appropriate Group Size

For in-depth interviews, the size of the group typically is one respondent. The size for a focus group usually is between 8 and 12, although in special circumstances it may have as few as 4 or 5 respondents or as many as 20. The size is dependent on (1) types of group participants, (2) the subject matter, and (3) complexity of problem and length of moderator's guide. Large groups (greater than 12) may work for brainstorming and idea generation, but they are especially difficult to moderate and present special problems with quiet or timid responders. Children usually must be part of small groups. In other cases small groups should be used only with extreme caution because they often turn into interviews rather than discussions. Other factors may play a part such as moderator preference and the size of the interview area, but those should be secondary considerations.

Number of Groups

Three main issues determine the number of interviews or groups required for a good study.

1. Market segmentation by region, age, sex, social class, user/nonuser require at least two or three focus groups or 15–25 interviews for each grouping.
2. Realistically, the driving force behind some decisions is lack of time or money to pay for additional groups. However, short time or low funds must never compromise the integrity of the study. Too few groups or interviews may give the researchers a false sense of security. Each group is different and will give different responses, thus more groups give a broader picture.
3. Marginal utility: The first and second groups or sets of interviews in an unfamiliar area yields a great deal of new information. The third group or interview set may yield slightly less data. As more interviews occur it becomes obvious that little new information is to be gained from doing "one more group."

One should not attempt to apply "quantitative thinking" to a qualitative technique by increasing the number of groups in order to estimate a population attitude. Conduct quantitative research when quantitative answers are important.

Testing Location, Physical Environment, and Equipment

The choice of a testing location is dependent on the objectives of the study and the type of respondents needed. A decision must be made to determine if testing can be conducted locally or needs to be done in a market segment location. A neutral location usually is best, although they can be expensive; on-site testing usually is only viable when the biases inherent in a company's identity can be eliminated. The environment should, in fact, be similar to other types of sensory studies: comfortable, distraction free, suitably lit, neutral.

In addition to the environment, high quality, reliable equipment is critical for the recording of qualitative research. The recordings are proof of what the data are. The testing facility usually supplies the necessary equipment. It should be examined carefully, and extra equipment should be available in case of failure during the interviews.

Prepared Observers

Viewers (often lab personnel, management, sensory analysts) must come to groups or interviews with an open mind and not with the expectation of having opinions confirmed. They must concentrate on what is being said, not what they think is said, and pay attention to the nonverbal messages that are being sent. They should not be judgmental during the process. Observers should be quiet so that everyone can hear everything that is said and in most cases should not be visible to the respondents. They should have a copy of the moderator's guide to follow to facilitate taking notes. Often observers will have a chance to speak with the moderator before the close of the group in order to have the moderator ask the group to clarify or expand on ideas.

Review of Past Research

As with any research project a review of past work should be conducted. That review may be from previous experience, company files or from outside sources such as journals, popular magazine articles, or research service reports.

Analysis of Data

Qualitative data are descriptive; there are no statistical analyses. The moderator prepares a report for the researcher. A descriptive report would give a summation of all of the opinions and ideas heard and seen. Many reports stop at that point and the researcher draws conclusions based on the descriptions given.

The moderator may prepare an analytical report that would include the moderator's evaluation of the data and would include recommendations. The moderator's skill, expertise, and knowledge in an area is critical if an analytical report with recommendations is needed.

Of great importance in either report is that the moderator attempt to make objective evaluations of the descriptive data. If any information is to be gained from qualitative research then it is important that the experiment not "overinterpret" data by attempting to derive a level of understanding or meaning that does not exist. Similarly, the interviewer should not overprobe a topic to make it seem more important that it is.

SUMMARY

For product developers, qualitative research may explain how consumers feel about current food choices and what they want for future choices. Qualitative research can assist in studying the words, responses, or emotions that arise when consumers talk about a product. It may help researchers understand the characteristics that are expected in products and the vocabulary used to describe products and attributes. Qualitative research methods are applicable to sensory studies, but they do not substitute for quantitative data. Rather, when used together, quantitative and qualitative studies may provide more and better information than was previously gained.

REFERENCES

ASTM. 1990. Private communication on Committee E-18 document in progress.

Axelrod, C. J. 1975. 10 essentials for good qualitative research, *J. of Ad. Research*. 12: 27.

Baker, R. C., Hahn, P. W. and Robbins, K. R. 1988. *Fundamentals of New Food Product Development*. Elsevier, New York.

Bellenger, D. N., Bernhardt, K. L., and Goldstucker, J. L. 1976. *Qualitative Research in Marketing*. Amer. Marketing Assoc., Chicago, IL.

Chay, R. F. 1989. Discovering unrecognized needs with consumer research. *Res. Technol. Management*. 32(2): 36.

Cooper, P. 1987. Research is alive and well—and different—in Europe, *Marketing News*. (Aug 28): 58.

Gordon, W. and Langmaid, R. 1988. *Qualitative Market Research: A Practitioner's and Buyer's Guide*. Gower Publishing, Hants, England.

Stone, H. and Sidel, J. L. 1985. *Sensory Evaluation Practices,* p. 205. Academic Press, Orlando, FL.

Seymour, D. T. 1988. *Marketing Research: Qualitative Methods for the Marketing Professional.* Probus Publishing, Chicago, IL.

Templeton, J. F. 1987. *Focus Groups: A Guide for Marketing and Advertising Professionals.* Probus Publishing, Chicago IL.

Thompson, C. J. 1989. Putting the consumer experience back into consumer research: the philosophy and method of existential-phenomenology *J. Cons Res.* 16: 133.

Wu, L. S., 1989. *Product Testing with Consumers for Research Guidance* L. S. Wu (Ed.). ASTM, Philadelphia, PA.

15

Claim Substantiation for Sensory Equivalence and Superiority

Maximo Gacula, Jr.

The Dial Corporation Technical Center
Scottsdale, Arizona

INTRODUCTION

In general, claim substantiation is a statement of facts supported by evidence. The evidence is usually quantitative in nature and capable of being replicated under specified conditions. It is therefore evident that statistics and experimental design play a major role in furnishing evidence for establishing claims.

Suppose an optimal product formula is to be compared against a competitor or against a product class, the result of which will be used for advertising purposes. This situation is an example of claim substantiation, where there are no concrete guidelines on the type of evidence to provide to advertising agencies for claim support. As a result, claim substantiation data are generally evaluated on their own merits. Substantiation submitted to advertising agencies can be challenged through the National Advertising Division (NAD) of the Council of Better Business Bureaus (CBBB), a self-regulatory system for the industry. Their evaluation of claims is based on its own review and evaluation of advertiser's substantiation and, when necessary, on consultation with technical experts

(NAD, 1990). Claim substantiation cases and their resolutions are published in *NAD Case Report*, a CBBB publication. This publication is highly recommended reading material for sensory professionals involved in claim substantiation. Another source is the work of Stone and Sidel (1985), which discussed the issues of clinical and perceived efficacies for claim purposes. An example of a substantiation challenge involving taste test data between Diet Pepsi and diet Coke (NAD, 1990) is given in Fig. 15.1.

In this chapter, we present a statistical design and analysis to support claim substantiation.

TESTING OF CLAIMS HYPOTHESIS

The conventional approach for testing of hypotheses in scientific work has been to formulate a null hypothesis and an alternative hypothesis. For example, when studying the effects of a certain flavoring additive on sensory characteristics, one may use symbols μ_1 and μ_2 to denote, respectively, the mean of scores without and with the flavoring additive to formulate a null hypothesis

$$H_0: \mu_1 - \mu_2 = 0$$

The H_0 states that there is no effect of flavoring additive. An alternative hypothesis is

$$H_a: \mu_1 - \mu_2 \neq 0$$

which states that the difference between mean scores could be positive or negative. A process for reaching a decision about the validity of H_0 or H_a is called "testing of hypothesis." In reaching a decision, one may commit the so-called Type I and Type II errors due to sampling variations. The rejection of the null hypothesis when in fact it is true results in a Type I error, the probability of which is denoted by α. For example, a probability of $\alpha = 0.05$ indicates that on the average the test would wrongly reject the null hypothesis five times in 100 cases. The value of α is called the significance level of the statistical test. On the other hand, a Type II error results when the statistical test accepts the null hypothesis when in fact it is false, the probability of which is denoted by β.

Note that in the conventional approach, one does not prove the null hypothesis, rather one disproves or fails to disprove it. To prove a null hypothesis is irrelevant as it refers to a claim which the investigator suspects is false. Hence, on the basis of statistical evidence, the null hypothesis is either proved false or has insufficient evidence to disprove it. In his 1935 classic book, *The Design of Experiments*, Fisher wrote the following:

Basis of Inquiry: On April 16, 1990, NAD reported the resolution of a challenge of a multi-media campaign employing variations of the claim: "Diet Pepsi—the taste that beats diet Coke." The challenge was received from The Coca-Cola Company in February and was accompanied by a series of central location tests involving several thousand diet soft drink users. The challenger maintained that the overall results, individual test results and analysis of subgroups did not support the claim. In response, PepsiCo informed NAD of its previous decision, based on marketing considerations, to modify the comparative claim and, in March, the advertiser advised NAD that the claim would be discontinued prior to publication of the resolution. Consistent with established practice in NAD proceedings, the advertiser chose not to submit substantiation for review by NAD or the competitor.

Immediately prior to distribution of the April *NAD Case Report,* The Coca-Cola Company initiated a new challenge based on two concerns, (1) the comparative taste claim was featured in the March 28 issue of a weekly magazine (2) two television commercials, "Billy Crystal" and "Ray Charles," although modified, retained an implied message of the unsubstantiated taste preference claim. To support its challenge, The Coca-Cola Company resubmitted its original taste test data and details of a consumer perception study conducted on the "Billy Crystal" commercial in the original and modified forms. The challenger maintained the results indicated that the modified commercial continued to convey a taste superiority message for Diet Pepsi. NAD referred the new challenge to PepsiCo and requested substantiation for the taste claim and responses to the challenger's concerns. The inquiry was conducted under *NAD/NARB Procedures,* effective April 1, 1990.

Decision: Three issues were addressed.

1. The challenger expressed concern that the print advertisement published in March represented a resumption of use of the comparative taste claim, thereby triggering the need for PepsiCo to submit its substantiation. The advertiser explained that the magazine went to press in mid-February and the advertisement represented a single and inadvertent oversight. Publication preceded by three weeks the announcement of the resolution of the controversy in the *NAD Case Report.* Thus, the advertiser maintained its decision not to submit the requested substantiation for the taste preference claim, which has been discontinued.

NAD accepted the advertiser's explanation and agreed that the newly adopted procedures still permit an advertiser to choose not to submit substantiation or otherwise defend a claim that is being discontinued at the time of the inquiry. The benefits of this practice—providing an incentive for the advertiser to take the initiative in resolving controversies and saving the parties from time-consuming debate—outbalance concerns resulting from an incomplete public record.

FIG. 15.1 NAD case report on the claim: "Diet Pepsi—the taste that beats diet Coke." *(From NAD Case Report 20(6): 21–22, Aug. 20, 1990).*

NAD noted the case record contains no substantiation for the comparative claim featured in PepsiCo's advertising during the period January–March, 1990. However, NAD agreed the advertiser's confirmation that the comparative claim has been discontinued resolved the issue.

2. The challenger maintained its research indicated the modified "Billy Crystal" commercial retained an implied message of an unsubstantiated taste preference claim. The advertiser stated the modified commercial has been withdrawn from use at expiration of the contract. In comments on the challenger's data, the advertiser maintained that the modified commercial did not communicate taste superiority.

NAD reviewed details of the communication study, which employed a series of open-ended questions, and noted the challenger's interpretation of the modified commercial was based on treating responses such as "Diet Pepsi is better tasting than diet Coke," "He prefers Diet Pepsi over diet Coke," "Diet Pepsi beats diet Coke" and "Better than diet Coke" as referring to a taste preference message. Following previously established practice, NAD excluded general superiority mentions and concluded the residual percentage of panelists recalling a message that Diet Pepsi is better tasting than diet Coke was too small to indicate the existence of a taste preference claim.

NAD agreed the modified advertising contained no unsubstantiated implied claim.

3. The challenger maintained the modified "Ray Charles" commercial retained the implied message of an unsubstantiated taste preference claim. The advertiser presented details of a communication study for NAD's review and made a redacted version available to the challenger. The study concluded that the modification to the super reduced the taste superiority message, although that message was minimal even in the original form.

NAD agreed the research supported the conclusion that the modified advertising contained no unsubstantiated implied claim.

Advertiser's Statement: Pepsi-Cola Company accepts NAD's decision and appreciates NAD's thoughtful analysis and resolution of this dispute. We would also like the case record to reflect our views of the studies presented by the challenger to support its position. Those studies were seriously flawed by a number of methodological problems that, in our judgment, rendered the results invalid. In view of our earlier decision to modify the claim for creative reasons, however, we did not feel it was an appropriate use of NAD's or the parties' resources to pursue issues that had become moot. (#2813, closed 7/3/90)

FIG. 15.1 *(continued).*

In relation to any experiments we may speak of this hypothesis as the null hypothesis and it should be noted that the null hypothesis is never proved or established, but is possibly disproved, in the course of experimentation. Every experiment may be said to exist only in order to give the facts a chance of disproving the null hypothesis.

If we fail to reject the null hypothesis at a specified significance level, one concludes that there is no real difference between treatment means or that the observations possibly came from the same population. Note that one cannot conclude that the mean values are equal. In practice, failure to reject the null hypothesis is generally given less attention, and at times the results are shelved, particularly when it is desirable to show a difference, e.g., "improved product." On the other hand, the rejection of the null hypothesis attracts further inquiry and a search for a plausible explanation of the outcome is pursued that may lead to further experimentation. This is the general course of action encountered in many research studies.

In other experimental situations, the "acceptance" of the null hypothesis is the desired result, for example, ingredient substitution for cost saving. This situation leads to a logical conflict with the traditional approach of disproving the null. Furthermore, the experimental design can be faulty, resulting in erroneous acceptance of the hypothesis. Therefore, the formulation of the null hypothesis and its analysis must be revised to accommodate the desired result. Such revisions have been the subject of several investigations in clinical trials (Westlake, 1972, 1976; Metzler, 1974; Dunnett and Gent, 1977; Spriet and Beiler 1979; Blackwelder, 1982) and in sensory evaluation (Gacula, 1990). These revisions are as follows:

1. Use of the confidence interval method instead of test of significance.
2. Formulation of the null hypothesis with a specified difference.
3. Use of the power of the test.
4. Use of a control chart that sets limits based on a specified number of standard deviations.

These revisions provide one or more procedures to support claims for parity, superiority, and inferiority. These procedures are discussed and illustrated by examples in the following sections. First, let us define an experimental design and its role in providing claims support.

EXPERIMENTAL DESIGN AND CLAIMS SUPPORT

The key to successful claims substantiation is the use of a correct experimental design to support the claim hypothesis. As a review, an experimental design is a structured plan conceived before the experiment is to begin; this plan includes clear statement of purpose of the study, how treatments/interventions are to be applied, the number of experimental units to be used, i.e., number of panelists/ judges, accountability in conducting the experiment, the unit of measurement to be used to generate the data, development of the questionnaire, and the determination of the population to be sampled (Gacula, 1987). All aspects of this plan are important and must be thoroughly understood by the members of the research team.

An example of a claim is: "You can't beat the taste of diet Soda PopX." This is a superiority claim, and the product (Soda PopX) must be compared to all products in its class or category. The sensory attributes to define "taste" in the claim must be fully stated in the design. To maintain the same value of Type I error in all product comparisons, the paired comparison design is recommended as opposed to comparing all products in a randomized complete block design. The reason for this is that in a series of independent paired comparisons, the confidence level of each comparison is always $1.0 -$ Type I error, thus all product comparisons obviously have the same confidence level. In this claim, all product differences must be statistically significant at the α level of significance to support the superiority claim.

Another example is the implied superiority claim in the context of Wyckham (1987), such as "No other cereal you can buy has more natural food fibre than Super-Fibre Cereal." This claim entails very extensive chemical testing of all breakfast cereals in the market. The design to use is a group comparison consisting of obtaining random samples of each brand for chemical analysis. To support this claim, the differences in fiber content between Super-Fibre Cereal and the other brands must be statistically significant at the α level by the independent t-test or other appropriate test statistics.

The easiest to substantiate are claims that do not involve the competitors directly, such as "Brand X fights cavities effectively" or "Brand X is an antibacterial agent." To support this claim, one must only show that the product is significantly effective at the α level when compared to a placebo. A simple group comparison design is used for this type of claim, and the data are analyzed by an independent t-test or other appropriate nonparametric tests.

On the other hand, a parity or similarity claim such as "SweetenerX tastes just like sugar" involves comprehensive chemical testing and sensory evaluation. Although no product brands are compared, one has to provide data that show the active ingredient of SweetenerX to have a close identity with natural sugars on both taste and biochemical aspects. Note that the desired result of this claim is the acceptance of the null hypothesis. The statistical procedure to support this claim is given in the next section. See Gacula and Singh (1984) and other statistical references for a discussion of two-sample paired design, group comparison design, and the randomized complete block design.

TEST FOR EQUIVALENCE AND SUPERIORITY

For two treatments to be compared under the conventional null hypothesis, the Student's t statistic is usually used as the test statistic for the rejection or acceptance of the null. Rejection of the null hypothesis does not indicate the practicality of the size of difference between treatments, rather it only indicates that the difference is not equal to zero or there is evidence of the presence of real effects.

In other situations, such as in clinical bioequivalence or bioavailability studies, the interest is in the acceptance of the null hypothesis, i.e., the new drug is as effective as the standard drug. The use of conventional hypothesis testing in this situation leads to difficulties of interpretation because the desired result is the acceptance of the null, hence the popular test of significance procedure no longer applies. In this respect, it has been suggested that the confidence interval test is more appropriate (Westlake, 1972; Metzler, 1974; Shirley, 1976; Blackwelder, 1982). As stated by Cochran (1983), a confidence interval relates to the question "How large is the difference?" In this section, the procedure used to validate the acceptance of the null will be called a test for equivalence or parity.

Associated with the confidence interval of a parameter is the degree of confidence that the length of the interval will include the true value of the parameter. If the interest is in the difference between mean values of two populations, and if the hypothesized difference lies within this interval, then the mean values are said to be equivalent at a confidence level of $1 - \alpha$, where α is the level of significance determined in advance of the experiment. Suppose we wish to test the hypothesis, H_0: $\mu_1 - \mu_2 = \mu_D$. The confidence interval for the difference μ_D between population means μ_1 and μ_2 is known to be

$$[d - t_{\alpha/2} (SE), d + t_{\alpha/2} (SE)]$$

or simply

$$d \pm t_{\alpha/2}(SE) \tag{1}$$

where d is the estimate of μ_D, the difference between sample means from the two populations, t is a value obtained from the t distribution at α level of significance and appropriate degrees of freedom, and SE is the standard error of d (SE = $S/\sqrt{N}$, where S = standard deviation of d). For large N, the t distribution approaches the standard normal distribution of Z with mean zero and variance 1, for example, for $\alpha = 0.05$, the value of Z = 1.96 (see Table 15.1). Thus formula (1) can be written as

TABLE 15.1 Probabilities Associated with the Standard Normal Distribution; $Z = (X - \mu)/\sigma$.

Z	Decimal fraction of Z									
	.00	.01	.02	.03	.04	.05	.06	.07	.08	.09
0.0	0.0000	0.0040	0.0080	0.0120	0.0160	0.0199	0.0239	0.0279	0.0319	0.0359
0.1	0.0398	0.0438	0.0478	0.0517	0.0557	0.0596	0.0636	0.0675	0.0714	0.0753
0.2	0.0793	0.0832	0.0871	0.0910	0.0948	0.0987	0.1026	0.1064	0.1103	0.1141
0.3	0.1179	0.1217	0.1255	0.1293	0.1331	0.1368	0.1406	0.1443	0.1480	0.1517
0.4	0.1554	0.1591	0.1628	0.1664	0.1700	0.1736	0.1772	0.1808	0.1844	0.1879
0.5	0.1915	0.1950	0.1985	0.2019	0.2054	0.2088	0.2123	0.2157	0.2190	0.2224
0.6	0.2257	0.2291	0.2324	0.2357	0.2389	0.2422	0.2454	0.2486	0.2517	0.2549
0.7	0.2580	0.2611	0.2642	0.2673	0.2704	0.2734	0.2764	0.2794	0.2823	0.2852
0.8	0.2881	0.2910	0.2939	0.2967	0.2995	0.3023	0.3051	0.3078	0.3106	0.3133
0.9	0.3159	0.3186	0.3212	0.3238	0.3264	0.3289	0.3315	0.3340	0.3365	0.3389
1.0	0.3413	0.3438	0.3461	0.3485	0.3508	0.3531	0.3554	0.3577	0.3599	0.3621
1.1	0.3643	0.3665	0.3686	0.3708	0.3729	0.3749	0.3770	0.3790	0.3810	0.3830
1.2	0.3849	0.3869	0.3888	0.3907	0.3925	0.3944	0.3962	0.3980	0.3997	0.4015
1.3	0.4032	0.4049	0.4066	0.4082	0.4099	0.4115	0.4131	0.4147	0.4162	0.4177
1.4	0.4192	0.4207	0.4222	0.4236	0.4251	0.4265	0.4279	0.4292	0.4306	0.4319
1.5	0.4332	0.4345	0.4357	0.4370	0.4382	0.4394	0.4406	0.4418	0.4429	0.4441
1.6	0.4452	0.4463	0.4474	0.4484	0.4495	0.4505	0.4515	0.4525	0.4535	0.4545
1.7	0.4554	0.4564	0.4573	0.4582	0.4591	0.4599	0.4608	0.4616	0.4625	0.4633
1.8	0.4641	0.4649	0.4656	0.4664	0.4671	0.4678	0.4686	0.4693	0.4699	0.4706
1.9	0.4713	0.4719	0.4726	0.4732	0.4738	0.4744	0.4750	0.4756	0.4761	0.4767
2.0	0.4772	0.4778	0.4783	0.4788	0.4793	0.4798	0.4803	0.4808	0.4812	0.4817
2.1	0.4821	0.4826	0.4830	0.4834	0.4838	0.4842	0.4846	0.4850	0.4854	0.4857
2.2	0.4861	0.4864	0.4868	0.4871	0.4875	0.4878	0.4881	0.4884	0.4887	0.4890
2.3	0.4893	0.4896	0.4898	0.4901	0.4904	0.4906	0.4909	0.4911	0.4913	0.4916
2.4	0.4918	0.4920	0.4922	0.4925	0.4927	0.4929	0.4931	0.4932	0.4934	0.4936
2.5	0.4938	0.4940	0.4941	0.4943	0.4945	0.4946	0.4948	0.4949	0.4951	0.4952
2.6	0.4953	0.4955	0.4956	0.4957	0.4959	0.4960	0.4961	0.4962	0.4963	0.4964
2.7	0.4965	0.4966	0.4967	0.4968	0.4969	0.4970	0.4971	0.4972	0.4973	0.4974
2.8	0.4974	0.4975	0.4976	0.4977	0.4977	0.4978	0.4979	0.4979	0.4980	0.4981
2.9	0.4981	0.4982	0.4982	0.4983	0.4984	0.4984	0.4985	0.4985	0.4986	0.4986
3.0	0.4987	0.4987	0.4987	0.4988	0.4988	0.4989	0.4989	0.4989	0.4990	0.4990

$$d \pm Z_{\alpha/2}(SE) \tag{2}$$

In order for the confidence interval method to be valid, the samples must be taken independently from each population. Moreover, the samples must be "sufficiently" large to compensate for lack of normality and other distributional assumptions. The decision rule for test of equivalence and superiority is stated below.

Decision Rule for Equivalence

If the interval includes μ_D, one concludes that the two treatment means differ by μ_D with a confidence level of $(1 - \alpha)$ 100%. If in particular $\mu_D = 0$, then the means are considered nearly equal.

Although the interval may include μ_D, there are other important factors that should be considered for supporting an equivalency position. The first factor is the power of the test commonly defined in statistical books (Dixon and Massey, 1957; Quenouille, 1965; Lehmann, 1986) as the probability of rejecting the null hypothesis when it is false, or conversely, the probability of rejecting the null hypothesis when the alternative is true. Since in a parity position one desires the acceptance of the null, one needs a powerful test to reject the null hypothesis when it is false. Probabilities associated with the power of the test range from 0.0 to 1.0, with higher values indicating a powerful test. The power of a test can be calculated at any point in the alternative hypothesis and can be plotted for ease of examination. Since power refers to the probability of rejecting the null hypothesis when it is wrong, the greater the power of a test, the better it is.

The second factor is the experimental design, which specifies the sample size, randomization, replication, control of variation, and experimental execution. With a good experimental design, one can feel reasonably certain that a statistically nonsignificant result is evidence of parity.

The use of a plot similar to a quality control chart can also be used to support parity by plotting individual repetition of the experiment. Recent references on quality control charts are Schilling (1982) and Ryan (1989). Three standard deviations are recommended as the parity limits because we assumed under the null hypothesis that we are dealing with one population. Note that a three-standard deviation limit includes about 99% of observations in the population. If a majority of the differences fall within the parity limits, then there is evidence of equivalence. The basic rationale of this application is sampling. If each repetition represents a random sample from the same population, then most of the differences should fall within the specified parity limits. Figure 15.2 shows a typical control chart for substantiating a parity position. Formula (1) or (2) can be used to obtain the width of the parity limits, such that a 95% confidence limit is d $\pm$ 2.0 (SE) and a 99% confidence limit is d $\pm$ 3.0 (SE). In industrial practice, it is common to use a value of 2.0 instead of 1.960 for the 95% confidence limit and 3.0 instead of 2.576 for the 99% confidence limit. Since the standard error of the mean (SE) is used instead of S (the standard deviation of observations) in the above parity limits, the chart is called a $\bar{d}$ control chart to indicate that mean differences are plotted.

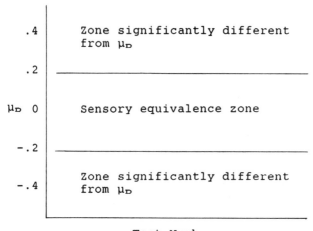

FIG. 15.2 A chart to support the position of sensory equivalence (parity) or superiority/inferiority. μ_D = population difference estimated by the sample difference d.

Decision Rule for Superiority Claim

If the lower limit of the confidence interval of the difference is greater than zero, the treatment or product with the larger mean is declared superior at the $(1 - \alpha)$ 100% level of confidence.

Note that the result of the confidence interval analysis can also be interpreted as a test of significance at the α level (Natrella, 1960; Barr, 1969; Jones and Karson, 1972; Cochran, 1983). In particular, Gacula and Singh (1984) applied the confidence interval significance test in a paired comparison of scale values obtained by the Thurstone-Mosteller model.

Calculation of Power of the Test

As stated earlier, although a nonsignificant result may indicate parity, we need to know the power of the test. The power of the test gives the probability of rejecting the null hypothesis when it is false. In substantiating a parity position, it is desirable that this probability should be high for the alternative hypothesis. Unfortunately, the power of the test cannot be adequately estimated before the study because of the necessity of knowing the estimate of variance in the data, and this estimate comes only from the study itself. A power of 0.50 may be a reasonable value as the lower limit on the following basis. If we plot the power function on the vertical axis and the critical value $[Z_{\alpha/2}(SE)]$ on the horizontal axis, the power at the critical value is 0.50 regardless of α level and sample size. However, one must aim at higher power so that the study can withstand scientific and legal challenges.

In clinical trials involving drug safety and efficacy, a power as high as 0.90 has been reported (Freiman et al., 1978). Because of the high risk involved in clinical trials for making a wrong decision, the sample size of the study is determined in advance of the experiment by calculating the sample size based on the variance of historical data. The aim in this calculation is to obtain a power close to 1.0.

It is known in the statistical literature that three factors can increase the power of the test. These factors are as follows:

(1) Type I error α. A test with large Type I error (i.e., 0.10, 0.20) will have more power. Thus, it is a trade-off between the significance level and the power. For a fixed sample size test, it is not possible to have a high significance level and expect to increase the power. The relationship between the power of the test and significance level α (Type I error) can be seen below:

Type I error	Confidence level	Power of test
0.01	0.99	Low
0.05	0.95	
0.10	0.90	↓
0.20	0.80	High

As stated above, it is evident that a compromise should be made between significance level and power.

(2) Sample size. In sensory and consumer testing, sample size refers to the number of panelists or judges, computed to provide an adequate protection from the risks of α and β errors. The larger the sample size, the smaller the Type I (α) and Type II (β) risks. Once the compromise between the significance level and the power is made, the power of the test can be further increased with larger sample size (N).

(3) Setting a reasonable difference between the value of the null and the alternative hypotheses. This factor is simply formulating a hypothesis with a specified difference for testing parity. In the absence of a method for specifying the difference, one resorts to subjected determination based on historical data. For example, one may use the estimate of random error as the specified difference, or one can also use a fraction of the standard error of the mean or difference.

The power of the test can be computed by

$$\text{Power} = 1 - \beta \qquad (3)$$

where β is the probability (P) of committing the Type II error. Assuming that we are sampling from a normally distributed population and that we desire to test

$$H_0: \mu_1 - \mu_2 \leq \mu_0 \text{ vs. } H_a: \mu_1 - \mu_2 > \mu_0$$

then

$$\beta = P \text{ (accepting } H_0 \text{ when } \mu_1 - \mu_2 \neq \mu_0)$$

An understanding of both the Type I and II errors can be obtained from graphical illustration in Fig. 15.3. The first step is to find the critical value Z_c of the test statistic $Z = (\bar{X}_1 - \bar{X}_2 - \mu_0)/SE$ that divides the distribution into the so-called acceptance and rejection regions (Fig. 15.3a). If the value of Z lies in the rejection region, we reject the null hypothesis (H_0); conversely, if it lies in the acceptance region we accept it.

The rejection region is specified once the level of significance α of the test is known, usually comprising 5 or 1% of the area under the normal curve. In Fig.

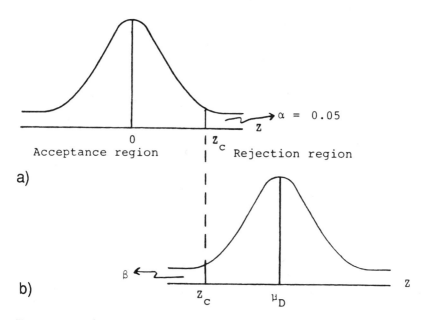

FIG. 15.3 Standard normal distribution of random variable Z with mean 0 and variance 1 illustrating the area corresponding to the α and β errors when the alternative hypothesis is actually true with mean μ_D.

15.3a, for $\alpha = 0.05$, the distance from 0 to Z_c is 1.645 (obtained from Table 15.1) by the following reasoning. The probability (α) of making a Type I error is 0.05; therefore, the probability of correctly accepting the null is 0.95. The area under the curve below 0 is 0.500, and between 0 and Z_c it is 0.450. This probability (0.450) (from Table 15.1) is equivalent to a Z_c (critical value) value of 1.645. Thus, any Z value greater than 1.645 is in the rejection region where $\alpha \leq 0.05$.

But suppose that the alternative hypothesis (H_a) is true, i.e., $\mu_1 - \mu_2 = \mu_D > \mu_0$. Then the distribution looks like that given in Fig. 15.3b. The area to the left of Z_c is the Type II error β, and the area to the right is the power of the test, hence power $= 1 - \beta$. In the form of a formula,

$$\text{Power} = P (Z > Z_c \text{ given } \mu_1 - \mu_2 = \mu_D)$$
$$= 1 - \Phi[(Z_c - \mu_D)/\text{SE}] \qquad (4)$$

where Φ denotes the cumulative distribution function of the standard normal random variable Z with mean zero and variance 1. The areas associated with the standard normal distribution are given in Table 15.1. Examples for calculating the Type II error and consequently the power of the test will be given in the following sections.

Sensory Equivalence

Many claim substantiations in the personal care, household, and food industries use human beings as the measuring instruments; specifically, the human basic senses of taste, sight, and smell. In this section, sensory equivalence is proposed to denote equivalence or parity in one or more of the basic tastes between two products. Obviously, sensory equivalence is used to support product parity. One unique application is the specification of a certain score on a rating scale that should be satisfied in a monadic evaluation for product acceptance.

During product development and reformulation, prototypes are monadically evaluated. It is common to establish a base point on the scale, such as a 6.0 on the 9-point hedonic scale, to be the value of the null hypothesis:

$$H_0: \mu_0 = 6.0$$
$$H_a: \mu_0 \neq 6.0$$

Although the above formulation is two-sided, one may formulate the one-sided, $H_a: \mu_0 < 6.0$ or $H_a: \mu_0 > 6.0$, depending on the experimental problem.

Using the confidence interval test, if the interval of μ_0 includes 6.0, then one may conclude sensory equivalence between the product and the perceived value of the null hypothesis. As discussed earlier, if more statistical evidence is required, the power of the test can be computed or the number of tests (repetition) increased and the result of each test plotted on a control chart (Fig. 15.2). The use of the control chart is useful because of the dynamic nature of consumer responses.

Example 1   Marketing research requires that the prototype of a product recently developed by R&D (Research and Development) have an average score of at least 6.0 on the 9-point hedonic scale before further consumer testing can proceed. A monadic test was conducted with 100 panelists with the following result for "overall liking" of the product:

Mean, $\overline{X}$	5.2
Standard deviation, S	1.4

In this example, H_0: $\mu_0 = 6.0$ and H_a: $\mu_0 \neq 6.0$. For large sample size ($N > 30$) one may use $Z_{\alpha/2}$ to approximate $t_{\alpha/2}$ in formula (1) for any α values. At the 95% confidence level, the value of $Z_{\alpha/2}$ is 1.960 (see Table 15.1). That is, the area $\alpha/2 = 0.5000 - 0.4750 = 0.025$ corresponds to $Z = 1.960$. Thus the 95% confidence interval is, from formula (2)

$$5.2 \pm 1.960 \ (1.4/\sqrt{100}) = 5.2 \pm 0.27$$

resulting in an interval of 4.9 to 5.5, which can also be written as (4.9, 5.5). The interval does not include 6.0, therefore, one concludes that the prototype is not at parity with the desired value of the null hypothesis. In fact, it is significantly inferior from the desired value of the null at the $1.0 - 0.95 = 0.05$ significance level. Note that only in the confidence interval test can one perform this operation to obtain the level of significance. This operation should not be used in other test procedures involving more than two means, such as in a multiple comparison test, because the significance level varies with the number of means.

Let us calculate the power of the test when the observed result $\overline{X}$ is actually true, that is, we reject the null and accept the alternative hypothesis. First we calculate the critical value $\overline{X}_c$ on the $\overline{X}$ scale corresponding the cut-off point -1.960 on the Z scale. Since $Z = [(\overline{X} - 6.0)]/(1.4/\sqrt{100})$, we find $\overline{X}_c$ to be $6.0 - 1.960(0.27) = 5.47$. Then, using Table 15.1 to obtain the probability of Type II error (in this case 0.159), we have

$$
\begin{aligned}
\text{Power} &= 1 - \Phi[(5.47 - 5.20)/0.27] \\
&= 1 - \Phi(1.00) = 1 - (0.5000 - 0.3413) \\
&= 1 - 0.159 \\
&= 0.841
\end{aligned}
$$

This result is reasonable because the difference between the null value (6.0) and the alternative (5.2) is quite large, hence the power should be relatively large suggesting the falsity of the null. See Fig. 15.4 for the graphical illustration of the results.

Suppose that the observed mean $\overline{X}$ is 6.10 (Fig. 15.5b) instead of 5.20 (Fig. 15.4b). What is the power of the test? Note that the value of H_a is shifted to the right of H_0. Corresponding to the cut-off point 1.960 on the Z scale, the cut-off point on the $\overline{X}$ scale is

$$\overline{X}_c = 6.0 + 1.96(0.27) = 6.53$$

and

$$
\begin{aligned}
\text{Power} &= 1 - \{1 - \Phi[(6.53 - 6.10)/0.27]\} \\
&= 1 - [1 - \Phi(1.593)] \\
&= 1 - [1 - (0.5000 - 0.4441)] \\
&= 1 - 0.944 \\
&= 0.056
\end{aligned}
$$

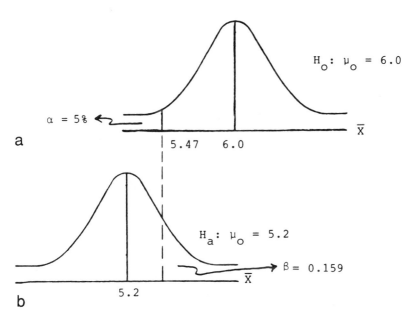

FIG. 15.4 Normal distribution curves for data from Example 1 showing the area of Type II error β when the alternative hypothesis is true.

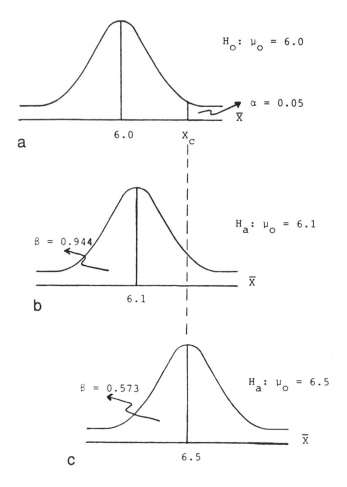

FIG. 15.5 Normal distribution curves for data from Example 1 showing the area of Type II error β when the value of the alternative hypothesis is close to the null hypothesis.

Again this result is reasonable because the observed value (6.1) is close to the null (6.0) and the power should be low (Fig. 15.5b) suggesting that the null is true. When the alternative is 6.50, the power is 0.457 (Fig. 15.5c). If we further increase the value of the alternative to 6.7, the power of the test becomes 0.764. This example clearly illustrates that as the value of H_a moves away farther from H_0 the power of the test approaches 1.0.

Example 2   In a consumer study, product A was compared to the leading brand product B. Using the paired comparison design the mean values for "overall liking" were:

	A	B
No. of panelists, N	92	92
Mean, $\overline{X}$	6.1	5.9

The standard deviation of the difference was computed to be 1.10. Thus the standard error is SE $= 1.10/\sqrt{92} = 0.11$. The 95% confidence interval of the difference is

$$(6.1 - 5.9) \pm 1.960(0.11) = 0.20 \pm 0.22$$

or an interval of $(-0.02, 0.42)$. Since the interval includes zero, one concludes at the 95% confidence level that products A and B are sensorially equivalent. However, one must compute the power of the test in order to ascertain the confidence of the conclusion of equivalency. In this example it is apparent that the value of the null is zero, hence the critical value $\overline{X}_c$ is $0.0 + 1.960(0.11) = 0.22$. Note that $\overline{X}_c$ is to the right of μ_0 (Fig. 15.6), therefore

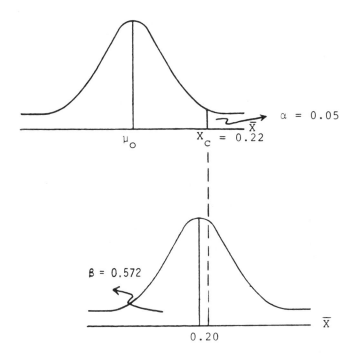

FIG. 15.6 Illustration of paired comparison test (Example 2) showing the area of Type II error β when power of test is low.

$$\begin{aligned}
\text{Power} &= 1 - [1 - \Phi((0.22 - 0.20)/0.11)] \\
&= 1 - [1 - \Phi(0.182)] \\
&= 1 - [1 - (0.5000 - 0.0714)] \\
&= 1 - (1 - 0.428) \\
&= 0.428
\end{aligned}$$

The power is less than 0.50, thus this result is a suspect to support parity. The conclusion should be suspended until additional consumer test data are available.

Example 3   The results of four central location tests (CLT) and one conducted by R&D are given in Table 15.2. The question is: Is product A sensorially at parity with product B? The standard error of difference from the paired comparison analysis was not available for this study, however, close approximation can be obtained by pooling the standard deviation (S) of the mean values. The pooling is computed as follows (see Gacula and Singh, 1984):

$$\begin{aligned}
S^2 = [&95(1.9)^2 + 95(1.8)^2 + 100(2.0)^2 + 100(1.6)^2 \\
&+ 89(2.0)^2 + 89(2.1)^2 + 81(2.1)^2 + 81(1.9)^2 \\
&+ 99(1.7)^2 + 99(1.8)^2]/464 = 7.137
\end{aligned}$$

and SE $= \sqrt{7.137/469} = 0.123$. The 99% confidence interval of the sensory equivalence zone is $3.0(0.123) = \pm 0.37$. This interval defines the entire width of the zone on the $\bar{d}$ chart.

Using this interval, the plot of the difference A − B is shown in Fig. 15.7. It is seen that 4/5 of the mean differences lie in the zone of sensory equivalence, and for practical purposes this is evidence of product parity. Perhaps the data for CLT2 may be examined to see if this result is due to location differences in product preference (assignable cause) or strictly sampling variations (unassignable causes).

TABLE 15.2   Mean Hedonic Scores (± standard deviation) for Products A and B in Central Location and Research and Development Tests to Illustrate the Position of Sensory Equivalence Using the $\bar{d}$ Chart

Product	CLT1	CLT2	CLT3	CLT4	R&D
A	5.7 ± 1.9	5.5 ± 2.0	5.9 ± 2.0	5.6 ± 2.1	5.4 ± 1.7
B	5.8 ± 1.8	5.9 ± 1.6	5.7 ± 2.1	5.4 ± 1.9	5.4 ± 1.8
A–B	−0.1	−0.4	0.2	0.2	0.0
N	96	101	90	82	100

CLT1–CLT4: Central Location Tests with consumers; R&D: Research and development personnel. Hedonic score ranged from 1 (disliked extremely) to 9 (like extremely).

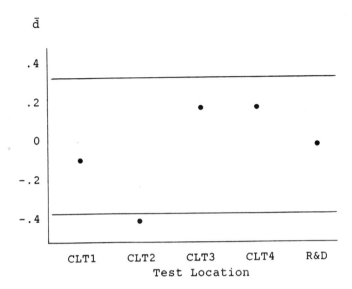

FIG. 15.7 The $\bar{d}$ control chart to substantiate sensory equivalence between two products at the 99% confidence level. The data plotted are differences between means as shown in Table 15.2. The width of the zone is $\pm 3\sigma$, where σ is estimated by the standard error of difference SE.

Effects of Sample Size on Power

As stated earlier, the number of observations N plays an important role in the power of the test by lowering the standard error of the mean. Recall that SE = $S/\sqrt{N}$, where S is the standard deviation of the observations. The formula for computing the power of the test can also be written as

$$\text{Power} = 1 - \Phi[((\mu_1 - \mu_2)/\text{SE}) + Z_\alpha] \tag{5}$$

where Z_α is the standard normal deviation corresponding to the area of the normal curve at α. In the form of (5), one can vary N and fix α at some given level in computing the power.

Let us use the statistics of Example 2 to illustrate the effect of N on power of the test. Using formula (5) the power was calculated for N = 41, 92, 184 with Type I error α equal to 0.05 and its corresponding $Z_\alpha = 1.960$ (Table 15.1). The results are shown in Table 15.3. The power of the test for rejecting a false null hypothesis increases as the value of the alternative hypothesis moves away from the null regardless of the sample size. However, the power approaches 1.0 faster

TABLE 15.3 Effect of Sample Size on the
Power of the Test with Probability of Type
I Error Fixed at 5% (Example 2)

		Power	
$\mu_A - \mu_B$	N = 41	N = 92	N = 184
0.00	0.025	0.025	0.025
0.05	0.048	0.064	0.090
0.10	0.084	0.138	0.234
0.15	0.138	0.256	0.457
0.20	0.213	0.413	0.695
0.25	0.306	0.586	0.870
0.30	0.414	0.743	0.959
0.35	0.530	0.861	0.991
0.40	0.643	0.936	0.999
0.45	0.744	0.975	1.000
0.50	0.828	0.992	1.000

This test is one-sided; $\alpha/2 = 0.05/2 = 0.025$; $Z_{\alpha/2} = 1.960$; standard deviation of difference $= 1.10$; $\mu_A - \mu_B = \mu_D$.

with larger sample sizes. For example, if $\mu_D = 0.45$ the power for N = 41, 92, 184 are, respectively, 0.744, 0.975, and 1.0. Note in this table that when the null hypothesis is true, $\mu_D = 0$, both the Type I error ($\alpha/2$ being one-sided) and the power of the test have the same value equal to 0.025 regardless of N. This result is always true because the significance level (Type I error) of the test is the maximum value of the power function when H_0 is true (Hogg and Craig, 1970).

NULL HYPOTHESIS WITH SPECIFIED DIFFERENCE

One method of improving the support for parity is by specifying a difference in the formulation of the null hypothesis (Blackwelder, 1982; Blackwelder and Chang, 1984). For example, FDA (Federal Drug Administration) requires a 1-log reduction by the CADE handwashing procedure (Cade, 1951) for antimicrobial efficacy. The null hypothesis may be written as

$$H_0: \mu_0 - \mu_1 \geq 1 \text{ log}$$

and the alternative

$$H_a: \mu_0 - \mu_1 < 1 \text{ log}$$

Here μ_0 is the average microbial population at the beginning of the test (no treatment applied) and μ_1 is the average count after the treatment has been applied. The decision rule using the confidence interval test for the above hypotheses is as follows:

If the lower limit of the interval is equal to or greater than 1 log, one concludes that the required reduction for antimicrobial efficacy has been satisfied. On the other hand, if the interval includes a value less than 1 log, then the difference does not meet the 1-log reduction.

The choice of the specified difference depends on the experimental problem. The specification of the difference minimizes doubts in a claim that an inappropriate experimental design was used to support the acceptance of the null hypothesis. This specification also provides the appropriate sample size to obtain the desired power of the test.

Example 4 This example illustrates the computational procedure for testing the null hypothesis with a specified difference to assure product equivalence or superiority. The approach here is to decide how close the difference between products would be before one can guarantee that the products are sensorially equivalent. The sensory analyst will proceed to do a full central location test only if the small-scale consumer test will result in a sensory equivalence with a specified difference $\mu_D = 0.25$ on the 9-point hedonic scale. In practice, this difference is close to 1 standard error generally observed in sensory evaluation work. For this example, let us use the partial data in Table 15.4 with 10 panelists.

The null and alternative hypotheses for this study are as follows:

$$H_0: \mu_1 \geq \mu_2 + 0.25$$
$$H_a: \mu_1 < \mu_2 + 0.25$$

Here, μ_1 is estimated by $\overline{X}_1$, the mean value for product 1, and μ_2 is estimated by $\overline{X}_2$, the mean value for product 2. The estimated average difference between products under the null is $d = 6.60 - 5.70 - 0.25 = 0.650$ with a standard error of difference (SE) of 0.314. Using formula (1), the 95% confidence interval is

$$0.650 \pm 2.262(0.314) = 0.650 \pm 0.710$$

or an interval of -0.06 to 1.36. Since this interval includes zero, the two products are sensorially equivalent. The sensory analyst had strong evidence that

TABLE 15.4 Sensory Data to Illustrate
the Computations for Testing the Null
Hypothesis with a Specified Difference
(Example 4)

Panelist	Product 1	Product 2	d
1	7	5	1.75
2	5	5	−0.25
3	8	6	1.75
4	6	7	−1.25
5	7	6	0.75
6	6	5	0.75
7	7	6	0.75
8	6	6	−0.25
9	7	6	0.75
10	7	5	1.75
Mean (X)	6.60	5.70	0.65
Std dev (S)	0.84	0.67	0.99

d = (Product 1) − (Product 2) − 0.25.

if a central location test is to be conducted, there is a very good chance that
product 1 should perform equally or even better than product 2 due to the
inclusion of the specified difference in the null hypothesis.

What is the power of the test under the null hypothesis? From formula (5),

$$\text{Power} = 1 - \Phi[(0.650/0.314) + 2.262]$$
$$= 1 - \Phi(4.332)$$
$$= 1 - [1 - (0.500 - 0.499)]$$
$$= 0.01$$

indicating that the null hypothesis is most likely true because of the low power
for the alternative hypothesis, thus demonstrating the parity claim over-
whelmingly.

CONCLUDING REMARKS

In this chapter, we have discussed the statistical problem of substantiating parity.
There is no easy answer for the problem because of the need to accept the null
hypothesis, which can be deliberately done through the use of improper sensory

and statistical methods. To circumvent this problem, three key issues should be addressed when designing an experiment to substantiate parity/equivalence. These issues are power of the test, sample size, and experimental design.

The power of the test for rejecting the null hypothesis when it is false should be very high to guarantee parity. We have illustrated that the power of the test can be increased with larger sample size. It also can be increased by lowering the variability in the data through the use of appropriate sensory experimental design. Thus, the role of sensory professional in the design of experiments is critical to the success of parity substantiation. The literature abounds with sensory methods that one can choose from, and one must be certain that the appropriate method is selected for the intended purpose. In addition, a good working relationship between a sensory analyst and a statistician further assures the success of a claim substantiation case. An effective statistician must cross the boundary between statistics and sensory, and vice versa, to make the relationship successful.

The application of quality control chart on mean difference ($\bar{d}$ control chart) to substantiate parity was discussed. This application is new and is best suited for sensory data because of its high variability. As sensory professionals will attest, it is difficult to completely repeat results from a consumer test due to several sources of variation present in consumer data. Combining the consumer test data in the form of a control chart provides better estimate of variability and also facilitates an easier interpretation of results, as opposed to looking at the results of a repetition singly.

REFERENCES

Barr, D. R. 1969. Using confidence intervals to test hypotheses. *J. Quality Technol.* 1: 256–258.

Blackwelder, W. C. 1982. Proving the null hypothesis in clinical trials. *Controlled Clinical Trials* 3: 345–353.

Blackwelder, W. C. and Chang, M. A. 1984. Sample size graphs for "proving the null hypothesis." *Controlled Clinical Trials* 5: 97–105.

Cochran, W. G. 1983. *Planning and Analysis of Observational Studies*. John Wiley, New York.

Dixon, W. J. and Massey, F. J. 1957. *Introduction to Statistical Analysis*. McGraw-Hill, New York.

Dunnett, C. W. and Gent, M. 1977. Significance testing to establish equivalence between treatments, with special reference to data in the form of 2 × 2 tables. *Biometrics* 33: 593–602.

Fisher, R. A. 1960. *The Design of Experiments*. 7th ed. Hafner Publ. Co., New York.

Freiman, J. A., Chalmers, T. C., Smith, H., and Kuebler, R. R. 1978. The importance of beta, the type II error, and sample size in the design and interpretation of the randomized controlled trial: Survey of 71 "negative trials." *N. Engl. J. Med.* 299: 690–694.

Gacula, M. C. Jr. 1987. Some issues in the design and analysis of sensory data: Revisited. *J. Sensory Studies* 2: 169–185.

Gacula, M. C. Jr. 1990. Statistical procedures for claim substantiation in sensory testing (submitted for publication).

Gacula, M. C. Jr. and Singh, J. 1984. *Statistical Methods in Food and Consumer Research*. Academic Press, Orlando, FL.

Hogg, R. V. and Craig, A. T. 1970. *Introduction to Mathematical Statistics*. The Macmillan Co., New York.

Jones, D. A. and Karson, M. J. 1972. On the use of confidence regions to test hypotheses. *J. Quality Technol.* 4: 156–158.

Lehmann, E. L. 1986. *Testing Statistical Hypotheses*. John Wiley, New York.

Metzler, C. M. 1974. Bioavailability—a problem in equivalence. *Biometrics* 30: 309–317.

National Advertising Division of CBBB. 1990. *NAD Case Report* 20(6): 18–22.

Natrella, M. G. 1960. The relation between confidence intervals and tests of significance. *The Am. Statistician* 14: 20–22, 38.

Quenouille, M. H. 1965. The use of confidence intervals in biopharmaceutics. *J. Pharm. Pharmacol.* 28: 312–313.

Ryan, T. P. 1989. *Statistical Methods for Quality Improvement*. John Wiley, New York.

Schilling, E. G. 1982. *Acceptance Sampling in Quality Control*. Marcel Dekker, New York.

Spriet, A. and Beiler, D. 1979. When can "nonsignificantly different" treatments be considered as "equivalent"? *Brit. J. Clin. Pharmacol.* 7: 623–624.

Stone, H. and Sidel, J. L. 1985. *Sensory Evaluation Practices*. Academic Press, Orlando, FL.

Westlake, W. J. 1972. Use of confidence intervals in analysis of comparative bioavailability trials. *J. Pharmaceutical Sci.* 61: 1340–1341.

Westlake, W. J. 1976. Symmetrical confidence intervals for bioequivalence trials. *Biometrics* 32: 741–744.

Wyckham, R. G. 1987. Implied superiority claims. *J. Advertising Res.* 27: 54–63.

Index